VISION AND NAVIGATION

THE KLUWER INTERNATIONAL SERIES IN ENGINEERING AND COMPUTER SCIENCE

ROBOTICS: VISION, MANIPULATION AND SENSORS

Consulting Editor

Takeo Kanade
Carnegie Mellon University

Other books in the series:

Robotic Grasping and Fine Manipulation, M. Cutkosky
ISBN 0-89838-200-9

Shadows and Silhouettes in Computer Vision, S. Shafer
ISBN 0-89838-167-3

Perceptual Organization and Visual Recognition, D. Lowe
ISBN 0-89838-172-X

Robot Dynamics Algorithms, R. Featherstone
ISBN 0-89838-230-0

Three Dimensional Machine Vision, T. Kanade (editor)
ISBN 0-89838-188-6

Kinematic Modeling, Identification and Control of Robot Manipulators, H.W. Stone
ISBN 0-89838-237-8

Object Recognition Using Vision and Touch, P.K. Allen
ISBN 0-89838-245-9

Integration, Coordination and Control of Multi-Sensor Robot Systems, H.F. Durrant-Whyte
ISBN 0-89838-247-5

Motion Understanding: Robot and Human Vision, W.N. Martin and J.K. Aggrawal (editors)
ISBN 0-89838-258-0

Bayesian Modeling of Uncertainty in Low-Level Vision, R. Szeliski
ISBN 0-7923-9039-3

VISION AND NAVIGATION

The Carnegie Mellon Navlab

edited by

Charles E. Thorpe
Carnegie Mellon University

with a foreword by

Takeo Kanade

KLUWER ACADEMIC PUBLISHERS
Boston/Dordrecht/London

Distributors for North America:
Kluwer Academic Publishers
101 Philip Drive
Assinippi Park
Norwell, Massachusetts 02061 USA

Distributors for all other countries:
Kluwer Academic Publishers Group
Distribution Centre
Post Office Box 322
3300 AH Dordrecht, THE NETHERLANDS

Library of Congress Cataloging-in-Publication Data

Vision and navigation : the Carnegie Mellon Navlab / edited by Charles E. Thorpe.
p. cm. — (Kluwer international series in engineering and computer science)
Includes bibliographical references.
ISBN 0-7923-9068-7
1. Robot vision. 2. Robots—Control systems. 3. Mobile robots.
I. Thorpe, Charles E. II. Series.
TJ211.3.V58 1990
629.8'92—dc20 89-78043
CIP

Table of Contents

4. An Approach to Knowledge-Based Interpretation of Outdoor Natural Color Road Scenes

Takahiro Fujimori and Takeo Kanade

5. Neural Network Based Autonomous Navigation

Dean A. Pomerleau

6. Car Recognition for the CMU Navlab

Karl Kluge, Takeo Kanade, and Hideki Kuga

7. Building and Navigating Maps of Road Scenes Using Active Range and Reflectance Data

Martial Hebert

8. 3-D Vision Techniques for Autonomous Vehicles

Martial Hebert, InSo Kweon, and Takeo Kanade

Contributing Authors
Carnegie Mellon University

J. Crisman
K. Dowling
M. Hebert
T. Kanade
K. Kluge
H. Kuga
I. Kweon
H. Pangels
D. Pomerleau
S. Shafer
D. Shin
S. Singh
A. Stentz
C. Thorpe
W. Whittaker

Carnegie Mellon University
5000 Forbes Avenue
Pittsburgh, PA 15213

Other Authors

Taka Fujimori
Sony Corporation
Worksation Division
6-7-35 Kiashinagawa
Shinagawaku, Tokyo 141
JAPAN

Yoshimasa Goto
Tokyo Research Laboratory
Matsushita Electric Industrial Co. Ltd.
3-10-1, Higashimita, Tama-Ku
Kawasaki 214 JAPAN

Robert Guzikowski
Ready Systems
470 Potrero Avenue
Sunnyvale, CA 94086

Jimmy Lee Ladd
55 Elkins Lane
Keller, TX 76248

Foreword

Mobile robots are playing an increasingly important role in our world. Remotely operated vehicles are in everyday use for hazardous tasks such as charting and cleaning up hazardous waste spills, construction work of tunnels and high rise buildings, and underwater inspection of oil drilling platforms in the ocean. A whole host of further applications, however, beckons robots capable of autonomous operation without or with very little intervention of human operators. Such robots of the future will explore distant planets, map the ocean floor, study the flow of pollutants and carbon dioxide through our atmosphere and oceans, work in underground mines, and perform other jobs we cannot even imagine; perhaps even drive our cars and walk our dogs.

The biggest technical obstacles to building mobile robots are vision and navigation—enabling a robot to see the world around it, to plan and follow a safe path through its environment, and to execute its tasks. At the Carnegie Mellon Robotics Institute, we are studying those problems both in isolation and by building complete systems. Since 1980, we have developed a series of small indoor mobile robots, some experimental, and others for practical applications. Our outdoor autonomous mobile robot research started in 1984, navigating through the campus sidewalk network using a small outdoor vehicle called the Terregator. In 1985, with the advent of DARPA's Autonomous Land Vehicle Project, we constructed a computer controlled van with onboard sensors and researchers. In the fall of 1987, we began the development of a six-legged Planetary Rover.

The Navlab (Navigation Laboratory) Project is the core of this long-term autonomous mobile robot research at Carnegie Mellon. It is developing the technologies for mobile robots which are capable of autonomously navigating in natural outdoor environments. The project manifests unusual breadth and depth. For sensing, many new algorithms have been developed for a large suite of sensors: sonars, color TV cameras, a 3-D range finder, an inertial navigation sensor, and a satellite navigation system. The vehicle itself has gone through several generations of computing, control, and support hardware. Nonconventional computing engines have been studied and integrated, including highly parallel systolic computers and neural net computing. We have developed a comprehensive planning and integration architecture together with its software tools. Finally, integrated working systems have been demonstrated for multiple tasks, such as road following, open terrain navigation, and fast navigation by position control. Very importantly, in addition to its own accomplishments, the Navlab Project has played a catalytic role in studying, advancing and integrating many facets of intelligent robotics.

This book consists of a collection of papers from the Navlab Project, edited by Dr. Charles E. Thorpe who has been playing a leading role in the project since its inception in 1984. The papers included here are the manifestation of this highly successful and influential project. This book represents a major milestone of autonomous mobile robot research in 80's. It provides a starting point for the research of the 90's, as we answer the important questions faced by mobile robots and build successful systems for hazardous environments, whether under water, on land, in orbit, or beyond.

Chapter 1: Introduction

Charles E. Thorpe

1.1 Mobile Robots

Vision and navigation are the essential elements of mobile robots. In more general terms, perceiving the world and finding one's way around in it are the hallmarks of all intelligent mobile creatures, both artificial and biological. The Navlab robot (figure 1.1) is our testbed for vision and navigation research. This book describes our experiments using the Navlab to understand vision, navigation, and intelligent mobile systems.

Figure 1.1: The Navlab

Mobile Robots are where Artificial Intelligence meets the real physical world. Many individual components can be developed and tested in the laboratory: planning, image understanding, control theory, neural networks, even system architectures. But the outdoor world holds countless surprises, and robots by their very mobility encounter an enormous range of situations, and are sure to find the flaws in academic theories. Only by building real robots, running tests in realistic environments, then going back to the lab to start anew, are we able to claim confidently that our theories are appropriate and that our systems work.

We have been building mobile robots at the Carnegie Mellon Robotics Institute since 1980. Our first robots were indoor creatures, used for research in sensing, planning, and mechanisms, and designed to operate in labs, classrooms, and buildings. In 1984 we first ventured outdoors, following the evolution of our interests and those of our sponsors, using vehicles such as the Terregator (figure 1.2). We were, of course, surprised and disappointed by our robots' performance. Our sensing was blinded by the sun, our image understanding inadequate, our planning unrealistic, our computing woeful; even the logistics of doing experiments proved daunting.

The chapters of this book describe our efforts to overcome those first harsh experiences, and the Navlab robot as it now exists and works. We have better sensors, including 3-D ranging; many new approaches for image interpretation; planners for following roads or traversing rough terrain; experimental supercomputers; and the Navlab vehicle, which carries sensors, computers, and researchers in air conditioned comfort as it autonomously roams its test routes and sites.

The Navlab now drives on narrow trails at speeds up to 8 mph; in more open areas over 20 mph; down rural roads; and across the rough terrain of a slag pile. It builds and uses maps of roads and landmarks, and can recognize certain objects, albeit slowly. There is an increasing sense that the Navlab really works, at least in a few realistic settings, and that it will continue to improve.

There are two general motivations for mobile robots, broadly, engineering and scientific. The engineering motivation is to build useful systems, capable of performing dangerous or repetitive tasks and freeing mankind for more important or interesting jobs. The scientific goal is broader: to understand the nature of intelligence by discovering the methods and structure of intelligent agents, which, in the realm of flora and fauna, are almost all mobile creatures. We embrace both motivations. Our continuing engineering work is leading to smart vehicles for underwater exploration, hazardous waste sites, dangerous military situations, construction, mining, our streets and highways, and even for planetary exploration with the AMBLER walking robot (figure 1.3).

Figure 1.2: The Terregator, a predecessor of the Navlab

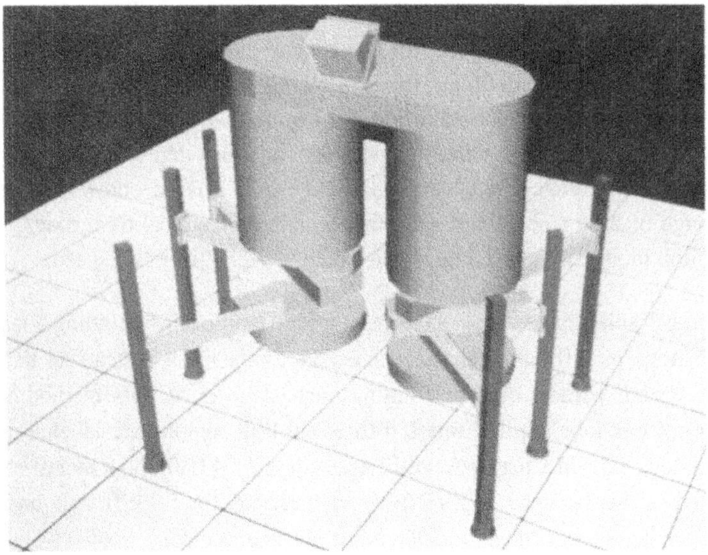

Figure 1.3: AMBLER, a walking robot for planetary exploration

At the same time, we continue to host scientific experiments (and debates) about the nature of perception, the organization of intelligent systems, and intelligent planning. The Navlab is used each year for a wider and wider variety of experiments, and is increasingly busy for experimental runs as our systems move out of the lab. We look forward to continued progress as research projects mature into building blocks, and to both practical mobile robots and a deeper understanding of intelligent mobile systems.

1.2 Overview

The chapters in this book are roughly grouped according to the areas of research required to build a mobile robot. Chapters 2 through 6 describe the variety of approaches we have taken to interpreting color images. Chapters 7 and 8 are concerned with interpreting range and reflectance data from a scanning laser rangefinder. We have also built path planners and systems software, which are described in Chapters 9, 10, and 11. Chapters 12 to 14 describe hardware, including both the Warp[1] computer and the Navlab itself, and control systems. Finally, Chapter 15 summarizes our view of vision and navigation, and previews our systems of the future.

Interpreting Color Images: Our Image Understanding has used a wide variety of approaches, from pattern recognition to explicit models to rule-based systems to neural nets. In Chapter 2, Crisman and Thorpe describe the color vision program that is the main road following perception software. Its main hallmark is supervised color classification, together with a voting scheme to find the most likely road even with the noisy segmentations that come from shadows and varying lighting. In Chapter 3, Kluge and Thorpe show how to take advantage of the stripes, lines, and other structure of typical rural roads. Using an explicit model of those features helps increase reliability and speed.

Fujimori and Kanade took an expert system approach to finding roads and intersections in difficult scenes, outlined in Chapter 4. Instead of the usual expert system paradigm of performing each stage once, this method iterates between scene labeling and model fitting until all regions are labeled and the road is detected. In Chapter 5, Pomerleau presents ALVINN, a neural network that drives the Navlab. ALVINN is remarkable for its efficient on-the-fly training scheme, as well as for its high-speed performance.

---

[1]Warp is a service mark of Carnegie Mellon.

Not all of our color vision was focused on road following. LASSIE, described by Kluge et al in Chapter 6, is a landmark recognition program. It relies on bottom-up grouping (edges to lines, lines to parallel sets, parallel lines to trapezoids) to find cars in color images, without using a strong geometric model.

Direct Sensing and Interpretation: Hebert, in Chapter 7, describes a method for finding and following the road using active scanning, with our ERIM laser rangefinder. Since the ERIM provides its own light source, it is insensitive to shadows, and drives the Navlab equally as well by day or by night. The Navlab ventures cross country as well as along roads. Chapter 8, also by Hebert, shows how the range information from the laser scanner is used for a variety of purposes, including building 3-D terrain maps at several levels of resolution.

Systems and Planning: Robot navigation needs more than just perception. The various modules for perception, mapping, planning, and control need to communicate with each other. In Chapter 9, Stentz describes CODGER, the COmmunication Database for GEometric Reasoning. CODGER provides architectural tools, including synchronization, geometric transforms, and a vehicle position history. The Driving Pipeline of Chapter 10 (Goto et al) uses the CODGER tools to sequence data from prediction to perception to planning to path execution, controlling vehicle speed to guarantee safety. And in Chapter 11, Stentz uses terrain maps, along with explicit models of vehicle shape, kinematics, and motion accuracy, to plan and execute cross-country paths.

Hardware and Control: Navlab hardware includes both the vehicle itself and its onboard computing. The Navlab hardware, described by Dowling et al in Chapter 12, provides an unparalleled mobile robot testbed. Modifications to the chassis and drive train, and computer interfaces to the control system, make the Navlab interesting in its own right as well as an excellent research platform.

At high speeds, kinematics no longer suffice to describe and control the Navlab. The control formulation of Chapter 13 models and corrects for the dynamics of the Navlab at speeds over 20 mph.

The last part of the Navlab's hardware is the onboard Warp supercomputer. Webb and Crisman describe the Warp in Chapter 14, and outline its uses during the Navlab's history. The Warp has been used for a large variety of programs, from stereo vision to 3-D range data interpretation to road following and neural network back-propagation training.

Conclusions and the Future: Finally, Thorpe provides a perspective on outdoor autonomous navigation. Chapter 15 describes our lessons on how to build simple, reliable systems, and our plans for the future.

1.3 Acknowledgments

The Navlab project has been singularly fortunate in its people: in the vision from the top, though the dedication of those in the trenches. Thanks are due to the "big picture" supporters of mobile robots: at CMU, Richard Cyert, Angel Jordan, Raj Reddy, Duane Adams, and Takeo Kanade; and at DARPA, our principle sponsor, to our program mangers Ron Ohlander, Bob Simpson, Bill Isler, and Eric Mettala, along with their colleagues including Saul Amarel, Barry Boehm, Rodger Cliff, Ray Colladay, Steve Cross, Clint Kelly, Jack Schwartz, Jim Tegnalia, and Rand Waltzman. All these people have not only believed in us enough to support us, but have also had enough faith in our research to ride along on autonomous Navlab runs, without even flinching as we passed trees.

During its career, the Navlab has had five Principal Investigators: Steve Shafer, Hans Moravec, William "Red" Whittaker, Takeo Kanade, and Chuck Thorpe. Other faculty who have greatly contributed include Martial Hebert and Jon Webb. Perhaps more importantly, we have attracted a small army of graduate students, undergrads, engineers, programmers, and visiting researchers: Arun Aggarwal, Paul Allen, Omead Amidi, Didier Aubert, Mike Blackwell, Tom Chen, Jill Crisman, Kevin Dowling, Thad Druffel, Alberto Elfes, Jim Frazier, Taka Fujimori, Yoshi Goto, Jay Gowdy, Keith Gremban, Rob Guzikowski, Eric Hoffman, Ralph Hyre, Jennie Kay, Karl Kluge, Hide Kuga, Bala Kumar, InSo Kweon, Jim Ladd, Jim Martin, Kichie Matsuzaki, Jim Moody, Taka Obatake, Masatoshi Okutomi, Tom Palmeri, Henning Pangels, Dean Pomerleau, Richard Redpath, Doug Reece, Jean-Christophe Robert, Ken Rosenblatt, Bruno Serey, David Simon, Jeff Singh, Tony Stentz, Hans Thomas, Rich Wallace, Chuck Whittaker, and Eddie Wyatt. Each made a mark on the Navlab; many were or are part of the backbone of its success.

Funding for the Navlab has come from several contracts. The principle support has come from DARPA, under contracts DACA76-86-C-0019, DACA76-85-C-0003, and DACA76-85-C-0002, all monitored by the Engineer Topographic Laboratories. Portions of this research were also partially sponsored by the Digital Equipment Corporation External Research Program and by the National Science Foundation contract DCR-8604199. The Planetary Exploration work is funded by NASA under contract NAGW-1175. The ALVINN work was supported by the Office of Naval Research under Contracts N00014-87-K-0385 and N00014-87-K-0533, by National Science Foundation Grant EET-8716324, and by the Defense Advanced Research Projects Agency (DOD) monitored by the Space and Naval Warfare Systems Command under Contract N00039-87-C-0251. The Warp project was supported in part by the DARPA under Contract N00039-85-C-0134, monitored by the Space and Naval Warfare Systems Command, and in part by the Office of Naval Research under Contracts N00014-87-K-0385 and N00014-87-K-0533. Taka Fujimori was supported by a grant from Sony Corporation, Japan. The views and conclusions in this document are those of the authors and should not be interpreted as representing the official policies, either expressed or implied, of the funding agencies or the US government.

Thanks to the following organizations for permission to republish articles, in whole or in part, as chapters of this book: The Society of Photo-Optical Instrumentation Engineers, Chapter 2; The IEEE, Chapter 3; the IEEE Computer Society, for Chapter 7; The International Journal of Robotics and Automation, for Chapter 10; The International Autonomous Systems Conference, Chapter 15; The Robotics Institute, Carnegie Mellon University, Chapters 4, 6, 8, and 12; and the Carnegie Mellon School of Computer Science, for Chapter 5.

Chapter 2: Color Vision for Road Following

Jill D. Crisman and Charles E. Thorpe

Abstract

At Carnegie Mellon University, we have two new vision systems for outdoor road following. The first system, called SCARF (Supervised Classification Applied to Road Following), is designed to be fast and robust when the vehicle is running in both sunshine and shadows under constant illumination. The second system, UNSCARF (UNSupervised Classification Applied to Road Following), is slower, but provides good results even if the sun is alternately covered by clouds or uncovered. SCARF incorporates our results from our previous experience with road tracking by supervised classification. It is an adaptive supervised classification scheme using color data from two cameras to form a new six dimensional color space. The road is localized by a Hough space technique. SCARF is specifically designed for fast implementation on the WARP supercomputer, an experimental parallel architecture developed at Carnegie Mellon.

UNSCARF uses an unsupervised classification algorithm to group the pixels in the image into regions. The road is detected by finding the set of regions which, grouped together, best match the road shape. UNSCARF can be expanded easily to perform unsupervised classification on any number of features, and to use any combination of constraints to select the best combination of regions. The basic unsupervised classification segmentation will also have applications outside the realm of road following. [1]

2.1 Introduction

At Carnegie Mellon University, we have been building successful, color vision based, road following systems for several years [6,7,9,10]. The main emphasis of our road following research is to find unstructured roads in images that are complicated by shadows, leaves or dirt lying on the road, lighting changes, and the like. We initially used edge based techniques, that searched for edges in the image to correspond with road edges in the scene. This proved inadequate for our Schenley Park test site, since often image edges caused by shadows were more distinctive than edges formed from road boundaries. Currently we have been using a color classification system, SCARF (Supervised Classification Applied

[1] Reprinted courtesy of Society of Photo-Optical Instrumentation Engineers (SPIE) from Proceedings of Mobile Robots III, November 1988.

to Road Following), where each pixel in the image is labeled as road or non-road based on the match of its color to previously learned colors. The road is found by looking for the road shape that contains the most 'road' labeled pixels. We also use an unsupervised classification algorithm, UNSCARF (UNSupervised Classification Applied to Road Following), that groups pixels that have similar color and location, and then searches for the combination of groups that best matches the road shape. This paper discusses these two systems.

Other groups have also been working on road-following. In Germany, Dickmanns and Grafe [3,4] view road following as a control problem. They have developed an elegant control formulation that incorporates a simple road edge detector with the vehicle model to drive their vehicle at speeds up to 100 kph. They also use constraints of the autobahn shape and markings. The autobahns are of constant width and are either straight, constant curvature, or clothoid in shape. The rapid processing and structured road model help to limit a search window in the image, and discard the extraneous edges normally found by edge detectors. However, it seems that their trackers could be distracted by the shadows, puddles and road imperfections that plague our test site.

The University of Maryland [2] has also been working on road following. Their system drove an autonomous vehicle based on edge detection. Image edges where tracked from the bottom of the image to the top using an edge detector in a window of the image. Once an edge is located, it is used to constrain the position and orientation of the next window. Then the edges were grouped using a Hough transform to determine which image edges form the best road edge. This system worked well when the dominant edges in the image are road edges, but similar systems at CMU have failed when tracking edges in strong shadows or when leaves or dirt lie on the roads.

At Martin Marietta, the VITS system [8] has achieved impressive speeds on fairly unstructured roads. It projects the three-dimensional color space (red, green, blue) into one dimension, or in later systems two dimensions. It then differentiates the road from non-road by a linear discriminant function. The road/non-road threshold is selected by sampling a part of the image that is guaranteed to be road. This system is similar to CMU road following, but emphasizes speed rather than general capability. Their system works fast, up to 20 kph, on their test site, but it is doubtful that it will work on other test sites, since the color projection is tuned for the features that are best to discriminate their road from their non-road.

Our goal is to build general color vision algorithms that work in a wide variety of situations. In particular, we are working on recognizing unstructured roads in various types of illumination and weather conditions. To give our system general capabilities, we must address the following problems:

- *The objects in the scene undergo spatial changes in appearance.* For example, under sunlight, roads appear to be a different color than they appear in the shade.
- *Objects in the scene undergo temporal changes in appearance.* This may occur when clouds pass over the sun for instance. The change in illumination will cause identical road segments to have different colors from frame to frame.
- *The dynamic range of our cameras is limited.* We cannot digitize meaningful data in dark shadows of a brightly sunlit image, nor can we capture data in the brightly sunlit regions of a dark image.
- *The roads in Schenley Park are very unstructured.* There are no center or bordering lines painted on our roads, as on highways. Many of the road edges are obscured or uneven. The pavement of our roads is deteriorating in places, and the pavement may be covered with the same leaves, dirt, or snow, that appear off road.

Our two new systems, SCARF and UNSCARF, were built to address these problems. Both systems deal with the limited dynamic range of the cameras by using two cameras with different iris settings to capture both dark and bright areas of the scene. SCARF is designed to be a fast, adaptive system. Even though algorithm speed is not a goal of our research, faster algorithms have the advantage of more overlap between frames, if the vehicle speed is constant. When the images are processed closer in time and distance, the lighting conditions are less likely to change dramatically and the road position in the image will not move far between frames. UNSCARF tackles the temporal and spatial changes by processing each image independently of the others. No color models are tracked from frame to frame, making this algorithm insensitive to spatial or temporal changes in color.

In the next section of this paper, we describe the SCARF algorithms and discuss results. UNSCARF is detailed and discussed in the following section, and finally, general conclusions are drawn in the final section of this paper.

2.2 SCARF

SCARF has evolved by adding more and more capabilities to a simple road following system. A block diagram of this system is shown in Figure 2.1. SCARF uses two cameras to digitize one frame of data. The two color images

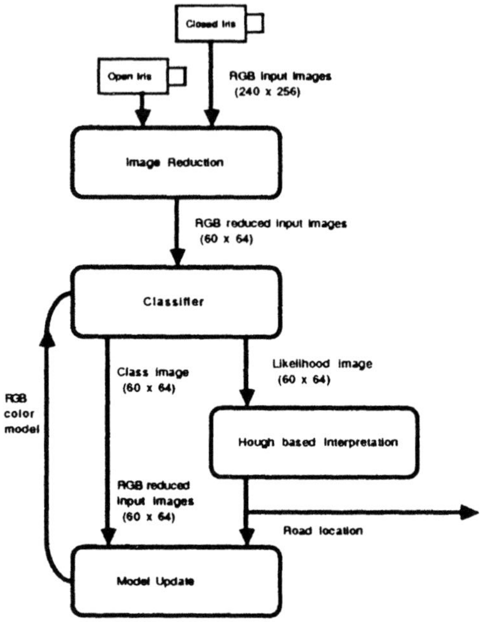

Figure 2.1: Block diagram of SCARF

are reduced by an averaging filter and sent to the classifier. For each pixel in the reduced images, the classifier calculates a likelihood value that describes how well the pixel matches remembered road and non-road colors. The interpreter uses the likelihood values to find the road shape that contains the most likely road pixels. The road location is then used to update the remembered colors for the next frame. The road location is also used to steer the vehicle. This system has been implemented on the WARP supercomputer.

2.2.1 Two Camera System

To extend the dynamic range of a single camera, we are using two images of the same scene digitized from the two cameras mounted on our test vehicle, the Navlab. The cameras were positioned as closely together as possible, and bore-sighted, minimizing the difference between the camera images. To avoid calibrating the two cameras, we treat the images as if they were taken from the same camera. This approximation is adequate for our purposes since the baseline of our cameras is much smaller than the distance to the road.

The improvement in dynamic range results from the different iris settings

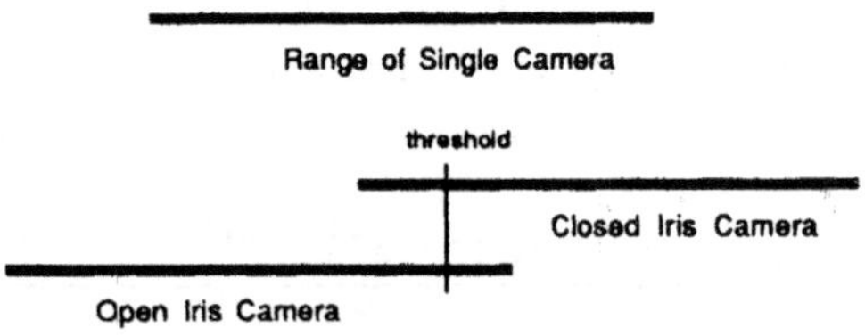

Figure 2.2: Extending the dynamic range using 2 cameras

of the two camera as shown in Figure 2.2. One of the cameras is set to capture the shadowed area of the scene by opening its iris. The second camera captures the sunny areas by using a closed iris.

When the two color images are digitized, they are first reduced by using an averaging filter. This not only reduces the data size, but will also reduce the noise content of the image. The reduced input images are used throughout the program, to increase the speed of the processing.

Two different methods for using the two input images were tested. The first approach is to combine the two reduced images into one. We used a simple thresholding technique to extend the dynamic range as shown in Figure 2.2. If the closed iris image pixel was too dark, then the pixel was selected from the open iris image, otherwise it was copied from the closed iris image. The second approach was to use both reduced input images to form a six-dimensional color space. Then all six features, the red, green, and blue bands from the two images, are used in SCARF.

2.2.2 Classifier

In standard pattern recognition theory, a classifier takes a d-dimensional measurement vector, $\mathbf{x}$, and chooses the best class label, ω_j, from a set of K classes, using a previously computed, *class conditional* probability, $P(\mathbf{x}|\omega_j)$, for each class [5]. The best class is the class that maximizes the *a posteriori* probability, $P(\omega_j|\mathbf{x})$. The expression for the *a posteriori* probability is normally derived from Bayes rule:

$$P(\omega_j|\mathbf{x}) = \frac{P(\mathbf{x}|\omega_j)P(\omega_j)}{P(\mathbf{x})}.$$

In our case, each pixel provides a 6 dimensional measurement vector (d = 6), $\mathbf{x} = [R_1 G_1 B_1 R_2 G_2 B_2]^T$, formed by concatenating the red, green, and blue bands of the two reduced input images. We use several classes to model road and non-road colors, typically 12 road models and 12 non-road models, giving 24 total color models (K = 24). We assume that the *class conditional* probability models for each class are Gaussian distributions, therefore, $P(\mathbf{x}|\omega_j)$ can be completely characterized by the mean vector and the covariance matrix of the sampled points for class ω_j. We also assume that the $P(\omega_j)$ is the ratio of the number of samples in ω_j, N_j, over the total number of samples, N. Therefore,

$$P(\omega_j|\mathbf{x}) = \frac{2\pi^{-d/2}||\mathbf{C}||^{-1/2}e^{-\frac{1}{2}(\mathbf{x}-\mathbf{m}_j)^T\mathbf{C}_j^{-1}(\mathbf{x}-\mathbf{m}_j)}}{P(\mathbf{x})}\frac{N_j}{N}.$$

Rather than calculating $P(\omega_j|\mathbf{x})$ at each pixel, we simplify the calculations by chosing the class, ω_j, that has the maximum $\ln P(\omega_j|\mathbf{x})$. This can be further simplified by noticing that $P(\mathbf{x})$ is identical for all of the classes, so that it can be eliminated from all of the terms. Therefore our classifier selects the class that maximizes the following likelihood measure:

$$\lambda_j = \ln(\frac{N_j}{N}) - \frac{1}{2}\ln((2\pi)^d||\mathbf{C}_j||) - \frac{1}{2}(\mathbf{x} - \mathbf{m_j})^T\mathbf{C}_j^{-1}(\mathbf{x} - \mathbf{m_j}).$$

2.2.3 Interpretation

The interpretation receives a likelihood image, containing λ_j, and a classification image, containing ω_j, from the classifier. By looking at the classification image, we can label each pixel in the image as either 'road' or 'non-road'. The interpreter searches for the road having the highest combined likelihood using a voting scheme similar to the Hough technique. The standard Hough algorithm searches for a line by voting for all of the lines passing through an edge point. However, we find a road by voting for all the possible roads containing 'road' pixels and by voting against all possible roads containing 'non-road' pixels. The main difference is that all of our pixels vote, not just pixels lying on the edge of the road. We also use the likelihood measure to determine the weight of each vote.

We assume the road is locally nearly straight, and can be parameterized using v, the column where the road center line crosses the horizon row, (or the vanishing point) and θ, the angle between the road center line and a vertical image line. These two parameters are the dimensions of an accumulator used for collecting votes. Each pixel in the likelihood image votes for all the roads that contain that pixel by adding its likelihood to the proper positions in the accumulator.

For each angle θ_i, a given pixel location (r, c) will vote for a set of vanishing points lying between v_s and v_e as shown in Figure 2.3. The starting column position, v_s, corresponds the interpretation that pixel (r, c) lies on on the right hand edge of the road in the image, and the ending column position, v_e

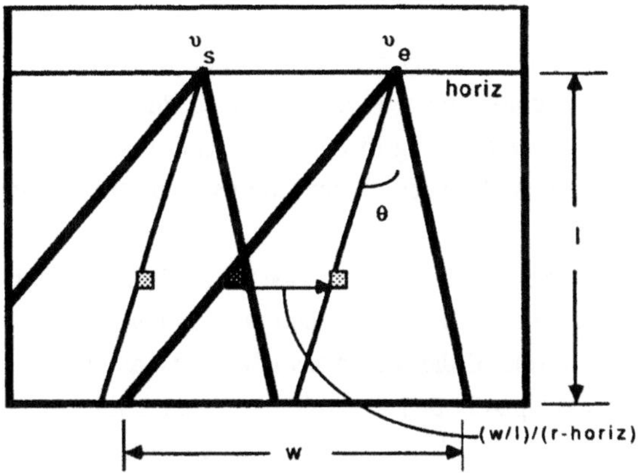

Figure 2.3: Hough voting scheme

corresponds to the (r, c) pixel lying on the left hand edge of the road. The v positions are calculated by:

$$v_s = c + (r - horiz)\tan\theta - (w/l)(r - horiz)$$

$$v_e = c + (r - horiz)\tan\theta + (w/l)(r - horiz).$$

where *horiz* is the horizon row in the image, w is the road width at the bottom of the image, and l is the length from the horizon row to the bottom of the image. The maximum value of the accumulator is chosen to be the road.

2.2.4 Model Formation

The new model can be calculated using standard statistical equations for mean and covariance:

$$\mathbf{m}_j = \frac{1}{N_j}\sum_{i=1}^{N_j}\mathbf{x}_i \tag{2.1}$$

$$\mathbf{C}_j = \frac{1}{N_j}\sum_{i=1}^{N_j}\mathbf{x}_i^T\mathbf{x}_i - \mathbf{m}_j^T\mathbf{m}_j. \tag{2.2}$$

First we have to decide which pixels belong to each class. Once we have a 'road'/'non-road' labeling, we calculate statistics for the road and non-road classes. Then we reclassify each 'road' or 'non-road' pixel using only the road classes for 'road' pixels and only the non-road classes for 'non-road' pixels. We iterate the calculate statistics and reclassify steps until the classes converge.

The road location is given from the user in initialization or from the interpretation on subsequent steps. Using this location, each pixel can be identified as road or non-road. The road region and the non-road region of the image are shrunk, forming a safety margin at the edge of the road. This is important so that the new color model is not corrupted due to the discretization of road locations or inaccurate fitting of a straight road model to a gently curving road. The reduced road and non-road areas are used to sample the colors of road and non-road.

An iterative clustering technique is applied to the road region and an identical procedure is followed for the non-road region. First, the road pixels are arbitrarily given one of the road class labels. We assign the classes by indexing through the road pixels and assigning the next road class. The color model, consisting of $\{N_j, \mathbf{m}_j, \mathbf{C}_j\}$ is calculated for each of the classes using equations (1) and (2). Then all of the road pixels are relabeled by the class whose mean color is closest to the pixel value, and a new color model is calculated using the new labeling. This 'label/sample' loop is repeated until most of the pixels remain in the same class.

2.2.5 WARP Implementation

We have implemented one of our supervised classification systems on the wire-wrapped, prototype WARP supercomputer [1]. The increase in processing speed, although significant, was limited by the small memory on each cell. Much time was spent down-loading code and data, at each function call, typically 14 calls per step. Our new PC WARP has more memory on each processing unit, allowing larger programs and larger global data structures. Therefore, we have one large WARP function rather than multiple WARP functions, taking advantage of the larger program memory. This results in greater speeds since data is only downloaded once and the WARP start-up sequence is executed once per image frame.

The inputs to the WARP program are the six reduced images and the statistical model for each class. The WARP function segments, interprets, and produces the new color model. It outputs the road location and the new color model.

2.2.6 Discussion

This program has been tested on several sequences of images. The SCARF system has driven over all of the roads of our test site successfully. We have driven the system in a variety of weather conditions, on different road surfaces, and under different lighting conditions. It adapts very well to different road

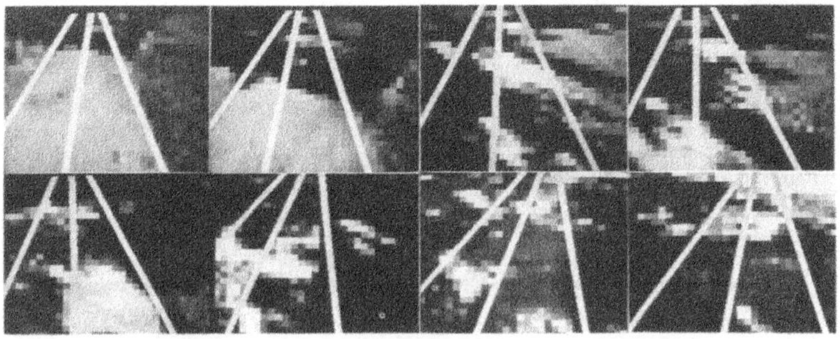

Figure 2.4: SCARF examples in dark shadows: The lines show the resulting road location.

surface types and differing off-road objects. Figure 2.4 shows SCARF running through severe shadow conditions from our test site.

Using two input images does increase the dynamic range beyond that of a single camera. We found that combining the data into a single image provided a fast means of extending the dynamic range, however, using both input images was more reliable. Not only does the use of two images increase the effective dynamic range, it also increased the data available to classify each pixel, thereby increasing classification confidence and accuracy. Moreover, using all the data from both cameras avoids the potential problems of picking a threshold for selecting data to form a single image.

The classification works well as long as the lighting conditions or the road type does not change drastically between frames. As the time and distance between frames decrease, the results from the classification improve.

The Hough interpretation provides the robustness of the SCARF system. Since it is an area based technique, there is more data used in the interpretation than an edge based technique. This makes the system less sensitive to noise. Using this interpretation, we have been able to drive our Navlab vehicle in a variety of weather conditions. The results are good even when the road may be partially covered with the same leaves, dirt, or snow that is on the non-road parts of the image.

2.3 UNSCARF

UNSCARF was designed to attack the problem with temporal and spatial color appearance changes. In SCARF, models of road and non-road colors, taken from

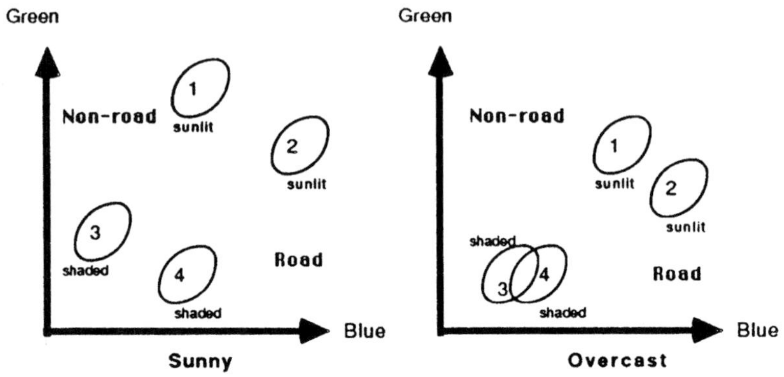

Figure 2.5: Effects of rapid illumination change

the previous image, were used to label pixels in the current image. However, if the color appearance of these objects changed dramatically, for any reason, then the color models no longer represented the colors of the objects in the new image. SCARF performed well as long as the illumination did not change too quickly. An example of a failure situation due to rapid illumination changes is shown in Figure 2.5. A classifier is calculated for sunny and shaded road and non-road classes in a sunlit scene as shown on the left. In the next image, that classifier will fail, since the sun is hidden by clouds and the colors of the road and non-road classes have shifted.

UNSCARF does not use pre-computed color models, instead it groups pixels that are similar in color and location in the image using an unsupervised classification algorithm as shown in Figure 2.6. Then the pixels with the same labels, or classes, are collected into regions using a connected components algorithm, and polygon approximations are fit to the pixel regions. Finally we search for the combination of polygons that, taken together, forms the best road shape.

2.3.1 Unsupervised Classification Segmentation

The unsupervised classification algorithm groups pixels having similar colors and locations by an iterative clustering technique similar to the model update of the SCARF system. The main difference is that none of the pixels have a 'road' or 'non-road' label. The pixels are given an initial classification. Then a statistical model is calculated for each class, and the pixels are reclassified using the new model. This is repeated until the classes converge.

Each pixel of the input image has five features (d = 5) that are used in

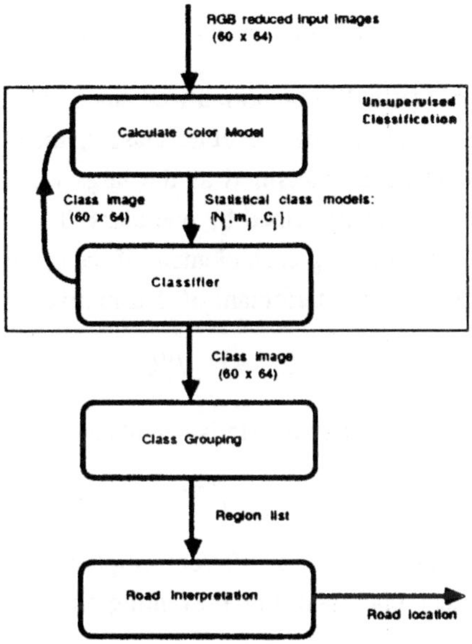

Figure 2.6: UNSCARF block diagram

the clustering:

$$\mathbf{x} = [RED, GREEN, BLUE, row, column]^T.$$

This can easily be expanded to a eight dimensional space by using the RGB bands of a second image. The system has a fixed number of classes in each image, in ours typically 24. First it labels each pixel of the image with a class, insuring that the classes are evenly scattered throughout the image. Next, a statistical color model, $\{N_j, \mathbf{m}_j, \mathbf{C}_j\}$, is formulated for each class, ω_j, for this class assignment using equations (1) and (2). Then the pixels are labeled using a classification scheme similar to that of SCARF and a new statistical model is calculated. The 'classify/sample' loop is repeated until most of the pixels in the image remain in the same class. This usually converges quite quickly, taking between 8 and 15 iterations until 95% of the pixels remain in the same class.

The classification scheme can have several different flavors. The first scheme used was a nearest mean classifier. In other words, the pixels were labeled with the class whose mean was closest to the pixel value. This has a tendency to form spherical clusters in the feature space. Since we were using the spatial parameters of (*row*, *column*) all of the regions formed from the final class labeling have approximately circular shapes in the image. To allow elliptical shapes to represent elongated or linear objects we used the nearest class as given by the Mahalanobis distance:

$$d_j = (\mathbf{x} - \mathbf{m}_j)^T \mathbf{C}^{-1} (\mathbf{x} - \mathbf{m}_j).$$

This distance metric needed to be normalized since once one of the classes had a larger covariance than any of the other classes, then in the reclassification, even more pixels would be classified as the large class. This would balloon until all of the pixels were described by one class. To avoid this, we normalize the distance metric, by dividing each element of the covariance matrices of each class by the dth root of the determinant of that matrix :

$$\mathbf{C}'_{i,j} = \mathbf{C}_{i,j}/||\mathbf{C}||.$$

Then the Mahalanobis distance metric is calculated using C':

$$d_j = (\mathbf{x} - \mathbf{m}_j)^T \mathbf{C}'^{-1} (\mathbf{x} - \mathbf{m}_j).$$

This allows each class to have a different elliptical shape, while maintaining the same size for each class and thus preventing one class from dominating the others.

Selecting the initial classes scattered throughout the image, causes the $(row, column)$ parameter statistics to be almost identical for all classes. Therefore the initial classification is based solely on color. In later steps, however the $(row, column)$ parameters are valuable. By clustering with color, we assume that an object will have a similar color throughout. By adding the $(row, column)$ parameters, we are exploiting the constraint that objects are localized in the image. The positional constraint made segmentations cleaner than strictly color constraints, by eliminating small noisy regions.

2.3.2 Interpretation

The interpreter is based on evaluating all possible roads that could appear in the image. The evaluation function looks at the difference between the road shape and the region edges in the image. The interpreter uses the same two road parameters as in SCARF: v, the column at which the road crosses the horizon, and θ, the angle between the center line and a vertical line in the image. However, instead of building an accumulator, we step through all of the interpretations and evaluate how well that interpretation fits the regions of the image.

To evaluate a candidate road, we first decide which regions would be part of the candidate road. This is done by testing if the center of mass of the region lies on the road. All of the regions lying in the candidate are then grouped

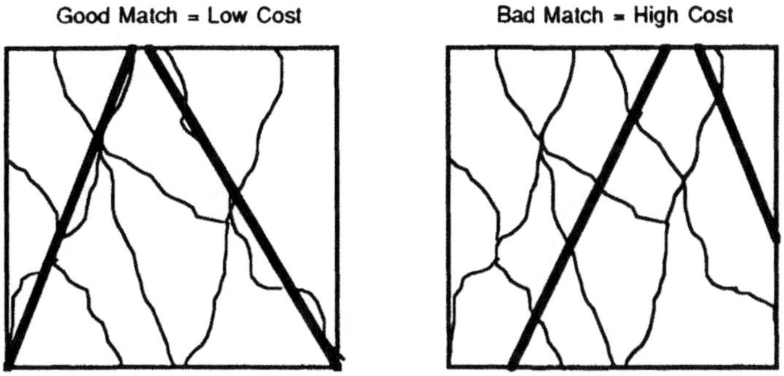

Figure 2.7: Road costs

together and approximated with a conglomerate polygon. The area between the road model and the edge of the conglomerate polygon is used as a cost metric of the interpretation. The candidate road whose conglomerate polygon has the lowest cost is selected as the result. Figure 2.7 shows the cost metric of a good fit and a bad fit for the road.

2.3.3 Discussion and Future Work

In this system, the low-level segmentation uses mainly color constraints to segment the image, while the high-level interpretation uses only geometric constraints to localize the road. Therefore, the different levels of the system are using completely different constraints. Figure 2.8 shows an example of the unsupervised classification segmentation running on a road scene. The images to the right are the class image, where each class label is represented by a different intensity value. The left image shows the pixels colored by the mean values of their class labeling. The top pair of images is the initial scattered classification used to build the initial models. The middle pair of images shows the classification and mean class colors after two 'classify/sample' loops, and the bottom pair shows the results after five 'classify/sample' steps.

The advantage of the cost evaluation scheme of our interpretation is that new constraints can easily be added to the total cost. For example, we could add costs so that all of the road regions should have a similar color, different that those of the non-road regions. We could also add a cost insuring that the road region is similar in color to the road seen in the previous image. A cost can be added so that conglomerate polygons with straight edges are preferred over those with jagged edges. Although these additional cost have not been

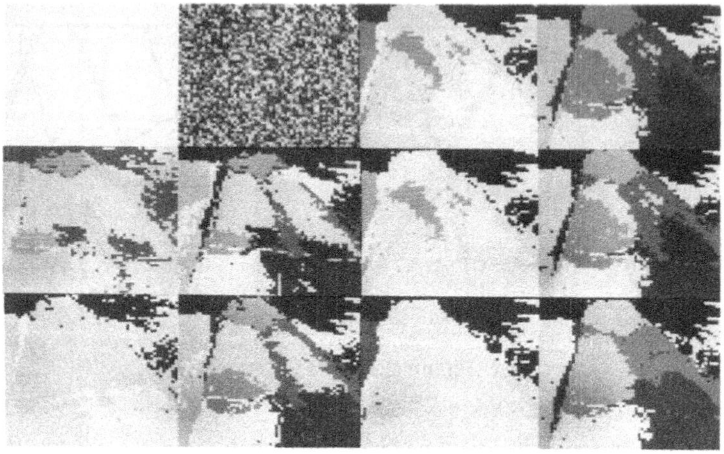

Figure 2.8: Example Clustering: Top left image is the original image. Each pair of class images corresponds to an iteration of the clustering algorithm. The right images have each class assigned a different intensity, and the left images have each pixel colored by the mean RGB class value. The bottom right pair is the final class images.

necessary on the images tested, they may become more important as we become more experienced with this interpretation system.

The system takes about 3-20 minutes to process one frame of the sequence. To speed up the processing, we have implemented the unsupervised clustering algorithm distributed on a multiple number of Sun workstations. Using this method, we have achieved a speed up proportional to the number of Suns used.

We will expand the unsupervised classification algorithm in several ways.

- First, if the system could decide the number of classes needed to characterize the data, rather than having a fixed number of classes specified, then the regions would be more representative of the data. As an initial attempt, we will split and merge regions after each reclassification step. Large regions will be split, and regions with very close mean values will be merged. This way, the system will decide how many regions it needs to adequately represent the data.
- We will expand the road interpretation to detect intersections. We will

apply the road searching that we have currently implemented. Then we will enumerate all of the possible branches from the road, and search for intersection branches with the same cost evaluation method used for the main road. We may need to add a color constraint to the algorithm, since in our test site sometimes the shadows of the trees can form intersection shapes.

- We believe that the basic unsupervised segmentation algorithm described here can be used for many different vision applications. To show this we will use this system to do terrain typing for our cross-country navigation experiments.

2.4 Results and Conclusions

SCARF and UNSCARF have been prototyped and tested individually. Our current efforts include speed and algorithmic improvements to each system. We are also considering cooperation between SCARF and UNSCARF. One idea is to use UNSCARF as a bootstrapping program and use SCARF as the general road-follower. If SCARF should realize that its results are not good, then control can be returned to UNSCARF.

The second possibility is to use lessons learned from one system to improve the other. We intend, for instance, to track the colors of the road regions detected by UNSCARF. Analyzing the changes in colors over time may provide cues which can improve the supervised classification in SCARF. This combination of different methods for scene segmentation will continue to expand the Navlab's capabilities for paved roads, dirt roads, and for terrain typing for cross-country navigation.

2.5 References

[1] M. Annaratone, E. Arnould, T. Gross, H. T. Kung, M. Lam, O. Menzilcioglu, and J. Webb. The warp computer: architecture, implementation and performance. In *IEEE Transactions on Computers*, pages 1523–1538, December 1987.

[2] D. DeMenthon. *Inverse Perspective of a Road from a Single Image*. Technical Report CAR-TR-210, University of Maryland, 1986.

[3] E. Dickmanns and A. Zapp. Autonomous high-speed road vehicle guidance by computer vision. In *Proc. 10th IFAC*, Munich, 1987.

[4] E. Dickmanns and A. Zapp. A curvature-based scheme for improving road vehicle guidance by computer vision. In *Proc. SPIE Conference 727 on Mobile Robots*, Cambridge, 1986.

[5] R. Duda and P. Hart. *Pattern Classification and Scene Analysis*. John Wiley and Sons, Inc., 1973.

[6] T. Kanade, C. Thorpe, and W. Whittaker. Autonomous land vehicle at CMU. In *Proc. ACM Computer Conference*, Cincinnati, February 1986.

[7] C. Thorpe, M. Hebert, T. Kanade, and S. Shafer. Vision and navigation for the Carnegie-Mellon Navlab. *PAMI*, 10(3), 1988.

[8] M. Turk, D. Morgenthaler, K. Gremban, and M. Marra. VITS–a vision system for autonomous land vehicle navigation. *IEEE Transactions on Pattern Analysis and Machine Intelligence*, May 1988.

[9] R. Wallace, K. Matsuzaki, Y. Goto, J. Webb, J. Crisman, and T. Kanade. Progress on robot road following. In *IEEE Conf. on Robotics and Automation*, San Francisco, 1986.

[10] R. Wallace, A. Stentz, C. Thorpe, H. Moravec, W. Whittaker, and T. Kanade. First results in robot road following. In *Proceedings of IJCAI 85*, August 1985.

Chapter 3: Explicit Models for Robot Road Following

Karl Kluge and Charles E. Thorpe

Abstract

Robots need strong explicit models of their environment in order to reliably perceive and navigate. An *explicit* model is information directly available to the program itself, used for reasoning about what to look for, how to look for it, and how to interpret what it has seen. We discuss the need for explicit models in the context of road following, showing how road followers built by our own and other groups have suffered by not having explicit models. Our new road tracking system, FERMI, is being built to study explicit models and their use. FERMI includes explicit geometric models and multiple trackers, and will use explicit models to select features to track and methods to track them.[1]

3.1 Implicit Models Considered Harmful

We claim that vision systems need to have strong explicit models in order to do reliable recognition. This is especially true in difficult situations, such as perception for an outdoor robot operating in an environment with no control over objects or illumination. Our particular domain is color vision for road following.

During the last four years there has been intense research on robot vision for following roads. Several different systems have been developed, many of them under sponsorship of DARPA as part of the Autonomous Land Vehicle program. Although there have been many solid contributions to road following, there is still no reliable general purpose road tracking vision system. Most existing road trackers work well for only a particular road, or only under benign illumination. They have impoverished models that do not allow them to reason about failures in their low level feature trackers. Weak models and weak or nonexistent high levels make them brittle in the presence of disturbances such as disappearing features or illumination changes.

Each system has a model of the road, including expectations about road shape and appearance, and the changes in shape, location, and appearance from one location to the next. The models are used to guide recognition, predicting how and where a road should appear and what methods should be used to find it. The models are also used for vehicle guidance, providing continuity while digitizing and processing each image.

A complete model of the road encompasses assumptions made by the programmer, and procedural knowledge for road recognition, as well as the data structure used by the program for road description. The assumptions made in road modeling fall into three loose categories: *subconscious* models which are implicit to the programmer; *implicit* models, representing decisions made by the programmer but not available to the program; and *explicit* models which the program itself can access and modify.

Each kind of assumption is appropriate in some circumstances. However, the more information is made explicitly available to the program, the wider the range of circumstances the program can handle autonomously. This is especially true for models of highly structured roads, such as well-marked streets and highways.

Typical subconscious assumptions, for instance, are that the road doesn't move, doesn't change color at any one location, is continuously connected, doesn't bend so sharply that it goes entirely out of the camera field of view, doesn't fold violently in 3-D. Many of these assumptions derive from the functionality of a road: if a narrow road makes a sudden right-angle bend, it is impossible for a vehicle to follow, and therefore is no longer a "road". Assumptions at that level are safe, and are applicable to a wide variety of roads. Other subconscious assumptions are much more insidious. One road following program begins with the (correct) implicit assumption that road edges are locally parallel, then (incorrectly) makes the subconscious assumption that feature-extraction routines will find the correct edges. This leads to drastic errors in inferred geometry when the subconscious assumptions are violated. Such an assumption may be not only wrong, but also hard to pinpoint and eliminate, since it was never consciously made or documented.

Implicit models show up in papers and in documentation, but not in code or data structures in any form that the program itself can access or modify. Typical implicit models are that the road is locally nearly straight, that the road is always brighter than its surroundings, or that the dominant edges in the scene are the road borders. Such implicit assumptions are often used by the programmer to select a single algorithm for recognizing that particular road type, or for calculating road geometry under that assumption. Well-constructed programs that rely on those road models are understood by their authors to only work in those cases where the underlying assumptions hold. In particular, for unstructured roads that do not have lane or edge markings and that do not follow rules of curvature or shape, the road model is very limited. With such a limited road model, it may not be possible or practical for the program itself to use an explicit model. If there is only one feature that can be tracked reliably, and only one algorithm for feature tracking, then there is no need for explicit program reasoning.

Explicit models are most useful in the opposite case, in which the road follows strong rules of shape and appearance, and there are many possible features and a variety of recognition algorithms. Then the program itself can select the correct features and algorithms, and can watch for changes in the road and change its strategy accordingly. Moreover, an explicit model that includes road semantics can help tie together separate phenomena. By "semantics" we mean labels such as "intersection" or "right turn lane", and the associated rules and descriptions that prescribe road appearance and shape in those situations. For instance, a program with only implicit models may notice that a feature it had been tracking has now disappeared. Only with an explicit model will it be possible for the program to understand that the feature was a double yellow line, that its disappearance might mean an approaching intersection, and that it is now past time to start looking for crossing traffic.

Road following programs to date use only subconscious and implicit models. This is due partly to the kinds of roads being tracked, which often do not have enough structure to make strong models necessary or possible. In other cases, however, the road has strong structure, but the designer has made all the decisions implicitly. Many road following systems have only a single road-tracking algorithm, and have a fixed road model. The designer uses an implicit model to pick the "best" method for following the road. The resulting system appears relatively simple and efficient, since it has only one algorithm to code and needs no higher-level reasoning.

Simple appearances are deceptive. Such implicit road models detract from system performance, and contribute to brittleness, and to difficulty in debugging and making enhancements. Furthermore, systems based on such a preprogrammed model of the world tend *not* to be as simple as they would at first appear. Since the world is rarely as static as an implicit, preprogrammed, model, those programs need many special cases, exceptions, recovery mechanisms, and other complications.

We contend that it is possible and advantageous to make the road model explicit, and to not only model appearance and shape information but also to include semantics in the model. Moreover, using such a model will make it easier to program and debug a road follower, and will lead to efficient programs. The bulk of the processing can be done by simple operators that needn't be concerned with special cases, while the costlier recovery procedures and switching between operators will occur infrequently.

The first half of this paper reviews other road followers, and outlines the road models and hidden assumptions used by each program. In the second half, we introduce FERMI, the Following Explicit Road Models Intelligently, and describe its construction and performance.

3.2 Systems, Models, and Assumptions

In this section, we describe several systems, describe their road models, and critique the implicit models in each.

3.2.1 SCARF: Color Classification

Implicit model: road colors mostly constant from one image to the next, known road shape (either known width, locally straight and parallel for Hough interpretation, or arbitrary but known for ground search)

Subconscious model: constant lighting and cameras so that constant road colors map to constant road images

SCARF, for Supervised Classification Applied to Road Following, has been developed over the last three years at Carnegie Mellon [6]. SCARF keeps color models for typical road and nonroad features, typically 8 road and 8 nonroad features. Each color model represents the means, and covariances of the color values for that feature, and the number of pixels in that class. An image is classified by comparing each pixel to each class, and determining the most likely classification for that pixel as well as the probability of that classification. The most likely road is found by convolving a known road shape with the classification output, looking for the road position that has the greatest sum of road probabilities inside the road shape and nonroad probabilities outside the road shape. In practice, this can be done efficiently using an adaptation of the

Hough transform that votes for areas instead of lines.

Once the most likely road has been found, SCARF builds new color models by supervised classification. The area inside the road is used to build new road models, and the area outside the road for the new nonroad classes.

SCARF was designed for use on a narrow, twisting, tree-lined bicycle path near the CMU campus. With constant illumination, it works well. Various color classes typically represent shady road, sunny road, leaves, wet patches, dirt, and so forth. As the vehicle moves onto a new type of road, classes adjust to represent the new appearances, as long as there is enough overlap between scenes that the majority of the road has been seen and modeled in the previous image.

The biggest weakness of SCARF is in changing illumination. If the sun goes behind a cloud between images, the appearance of road and nonroad features can change, rendering color models incorrect. A second weakness is the reliance on known road shape. If the road curves sharply, or if it changes width, the assumed shape model (locally nearly straight, known width) is invalid. Finally, SCARF suffers from the lack of features in its environment. It is difficult to build explicit models, since its environment has few features: the bicycle path has no lines, stripes, guard rails, or shoulders.

3.2.2 Maryland

Implicit model: small-scale road edges dominate the scene, edges are parallel, vehicle motion is accurate

Subconscious model: edge-finding is accurate, edges are clean and linked, limited curvature

Davis, LeMoigne, Kushner, and Waxman, at the University of Maryland, have a long history of research in perception for road following. Their strongest system, and the only one to actually drive an autonomous vehicle, is based on finding edges and grouping them into lines with Hough transforms [8]. During road tracking, an initial window is placed at the bottom of the image on the predicted road location. The search for the road edge in this window has two degrees of freedom, for location and orientation. Once this edge is located, other windows are placed above the initial window. In each succeeding window the road edge position is constrained by the lower window, so the Hough search need only look for orientation. This technique can work adequately for scenes in which the dominant edges are road borders. Similar techniques at CMU were defeated by strong, straight, shadow edges from trees and buildings, and by scenes in which road edges were obscured by leaves or dirt.

Besides road tracking, the Maryland research also considered 3-D shape reconstruction. The higher-level attempts at 3-D interpretation of road scenes were extremely sensitive to noise. DeMenthon [3] shows how Waxman's model can lead to perceived roads folding back over themselves, and proposes a new geometry that ameliorates some of those problems.

3.2.3 VITS

Implicit model: consistent colors within one image (road has at most 2 classes, for sunny and shaded), known vehicle motion and road model to seed process

Subconscious model: The color combination chosen is assumed to be always adequate despite changing illumination & dirt on road; this implies road appearance is constant from day to day

The Martin Marietta VITS system [7] has achieved some impressive goals. It has followed roads at speeds up to 20 kph, and detected and avoided obstacles on the road. Their system projects the 3-D color of each pixel onto a single dimension or, in later systems, onto a 2-D plane. Pixels are classified into road or nonroad based on a linear discriminant function. Once each pixel is classified, blob coloring gives the location of the road. The most interesting part of the Martin Marietta research is in selecting the road / non-road threshold. In each new scene, vehicle motion is combined with the previous road model to calculate the portion of the image guaranteed to contain road. This road area, called a power window, is sampled to determine the typical road color for this image. The Martin system is a tightly-engineered combination of perception, control, modeling, and highly tuned hardware. In many ways, their system is similar to some of the CMU road-following, but driven by speed constraints rather than generality of experiments. Where CMU's SCARF uses full color (or even 6 channels of color from two cameras) to track a variety of road appearances, they have selected the best combination of colors for their particular road. SCARF keeps many different possible appearances for both road and offroad, while VITS has at most two, again sacrificing general capability for speed.

3.2.4 Dickmanns and Grafe

Implicit models: gray-level edges of roads dominate the scene, road follows clothoid shape, physical constraints and fast processing limit feature motion, known relationship between road features to be tracked.

Subconscious models: all interesting features are oriented edges, no simultaneous distractions

Dickmanns and Grafe have demonstrated road following with a Mercedes van equipped with special-purpose computing [4]. They have achieved impressive performance, tracking a new section of the autobahn at speeds up to 100 kph. The heart of their system is an elegant control formulation, in which road geometry, vehicle turning radius and speed, and the location of visually tracked features are all fed into a single filtered state model. When running at high speeds, their system takes advantage of the geometry of the German autobahn. The road consists of straight lines, constant radius curves, and clothoids smoothly connecting curves and straights. German roads have known lane widths and well-defined markings.

The major weakness of this system is its extremely simple perception model. They use a monochrome camera and do simple edge detection. Their rapid processing and structured road model help them to detect and discard anomalous edges, but it nevertheless appears that their trackers could get distracted by shadows, puddles, road imperfections, or changing illumination.

3.3 FERMI

All of the above road followers have implicit and subconscious models of the road. But none of them has more than one means of tracking the road, or does any higher-level reasoning about the road, or has any explicit road model available to the program. Yet it is important to build and to use explicit road models. Highways, freeways, rural roads, even suburban streets have strong constraints. Modeling these explicitly makes reasoning easier and more reliable. When a line tracker fails, for instance, an explicit model of road and shoulder colors adjacent to the line will help in deciding whether the line disappeared, became occluded, turned at an intersection, or entered a shadow. This kind of geometric and photometric reasoning is vital for building reliable and general road trackers. We are now building the FERMI road tracking system to study explicit modeling, and to study the use of those models in building reliable vision.

3.3.1 Explicit Models

Our goal in constructing FERMI is to follow structured roads as reliably as possible. Our central principle is to make explicit as much as possible: road features, geometry, and other effects. We are first of all building individual knowledge sources that know how to model and track specific features:

- road edge markings (white stripes)
- road center lines (yellow stripes)
- guard rails
- shoulders
- type and color of road surface

We also have an explicit geometric model of the road. This model consists of a series of *generalized stripes*. A generalized stripe is the 2-D analog of a generaized cylinder. It consists of a spine curve (currently restricted to arcs of constant curvature), and the description of a cross-section which is translated along the spine. The model of the road in Figure 3.1, for instance, looks something like

- Spine: Curvature = 0.0.
- Feature 1: starts -304 cm ± 15 cm, height 0 cm, type shoulder, description asphalt.
- Feature 2: starts -273 cm ± 0 cm, height 0 cm, type line, description white.
- Feature 3: starts -262 cm ± 0 cm, height 0 cm, type road, description asphalt.
- Feature 4: starts -24 cm ± 0 cm, height 0 cm, type line, description yellow.
- Feature 5:...

The program will explicitly note transient road phenomena such as

- shadows
- local changes in road surface, e.g. patches
- global illumination changes, such as the sun going behind a cloud
- camera changes (auto-iris, auto-gain)
- 3-D effects such as going up and down hills

Explicitly modeling all these different features will be the basis for efficiency and reliability. The system will be efficient because the geometric constraints can specify subwindows of the image for each feature tracker, and tracker history from frame to frame can predict appearance and shape. Another reason for efficiency is that many simple trackers can be easily implemented on parallel hardware. Reliability will come first because of the strong geometric constraints among trackers, and the ability to detect and ignore anomalous outputs. The ability to use a strong geometric model of the road to focus on a small area of the image to look for a feature reduces the chances of being misled by extraneous image features. More importantly, the system will be reliable because one tracker, on discovering a shadow edge or road curvature change, can pass that information to other trackers and keep them from being caught by the same phenomenon.

3.3.2 Trackers

Many of the individual feature trackers have already been developed. We have done some preliminary experiments using, for example, the oriented edge operator used to drive the Terregator in 1986 and a simplified version of the color classifier developed in 1987. Customizing these feature trackers to follow lines, stripes, and edges will make them faster and more robust than the general-purpose trackers needed for our park scenes.

We currently have five trackers implemented:

- Oriented edge profile: Intensity profiles are extracted from a training window oriented perpendicular to the direction of the feature. These oriented templates are matched by correlation with intensity profiles from later images. The implicit model is that the color intensity profiles of an edge are roughly uniform along the length of the edge.
- Ideal oriented edge profile: This tracker searches for an ideal step edge in intensity profiles extracted from an oriented window. The implicit model is that there is a single strong oriented step edge separating two features.
- Extended linear feature tracker: Intended for use tracking such features as white and yellow road stripes. An unsupervised clustering algorithm is used on the RGB pixel values in a training window to split the image window into two clusters: the line and the background. The mean RGB values for the two clusters are used in later images to classify the pixels in a search window. A line is fit to the pixels which are classified as being part of the linear feature, giving an estimate of the location and orientation of the linear feature in the image. Implicit model: that the dominant color phenomenon in the training and search windows arises from the contrast between the line and the background, and remains approximately constant from image to image.
- Color boundary tracker: Used on ragged edges such as a grass/road boundary. Performs the same sort of clustering as the previous tracker, splitting the pixels in the training window into two classes. The pixels which have neighbors that have a different label are

marked, and a line fit to these boundary points to estimate the edge position and orientation. Implicit model: assumes that the dominant color phenomenon in the training and search windows is the contrast between the colors of the two features whose boundary is being tracked.

- Matched filter tracker: A small training window is selected. In later search windows the training window is correlated with the search window. The maximum correlation value in each row of the search window is selected as an edge point, and a line is fit to the edge points. Implicit model: the appearance of the feature is constant enough for correlation to be used.

Our current method of selecting a tracker looks at the size in the image of each feature. If the feature is narrow (i.e. a line or stripe), it selects a linear feature tracker. If the feature is wide (e.g. a lane of the road), it chooses to track the edge of that feature, and selects an edge operator such as the oriented edge tracker. Figure 3.1 shows the road described earlier, with boundary and oriented edge trackers tracking the white lines on the left and right side of the road and the left edge of the right lane.

Figure 3.1: Road with oriented edge & boundary trackers

We expect that various road phenomena (shadows, patches of different color, wet spots, etc.) will give rise to characteristic sorts of changes in the behavior of various feature trackers. For instance, if a color boundary tracker suddenly fails and the pixels in the current search window contain more blue than the current color models would predict, then that area may be in shadow.

This requires finding a good set of measures to characterize how appropriate a given tracker is for a particular feature (for instance, how well separated the color classes are in a color-based tracker), as well as how well the tracker worked on a particular search window (how high is the best peak and how does it compare to other local maximal correlation values in a correlation-based tracker; given a road/shoulder boundary found by a color-based tracker, how many pixels were misclassified on each side of the edge, etc.). We are currently working on formulating appropriate sets of measures for each of the trackers.

3.3.3 Tracker Fusion

It is necessary to merge the estimates of feature locations and orientations returned by the trackers placed on various features at various points in the image into a single estimate of where the vehicle is relative to the spine of the generalized stripe that is currently being transversed. The method of fusion needs to take into account the possibility of trackers failing or returning erroneous estimates.

The first approach to tracker fusion being considered is a Hough technique. Let us suppose that the spine of the current stripe is a straight line (the technique extends in a straightforward way to arcs of known constant curvature). Since the road is likely to be almost straight ahead of the vehicle, let's represent it as a line of the form $y = m * x + b$, where the x-axis points straight ahead of the vehicle and the y-axis points to the left. Let's suppose we have a feature tracker tracking a white stripe whose center is offset from the road spine by $offset_{stripe}$, and that the tracker has returned (x_i, y_i) as it's estimate of the location of the center of the stripe. For a given m value, the y-intercept of the white stripe center line is given by $b_{stripe} = y_i - m * x_i$, and the y-intercept of the spine by $b_{spine} = b_{stripe} + offset_{stripe} / \cos(\mathrm{atan}(m))$.

Each tracker votes for all possible spines that are consistent with its position estimate for its feature. The largest peak in the accumulator array is taken as the position of the road spine. Trackers whose position estimates are not consistent with that spine estimate are anomalies which need to be explained.

An alternate approach is to formulate tracker fusion as a linear least squares problem. This is the approach taken by Mysliwetz and Dickmanns [5]. They position their trackers in a way that allows them to formulate equations that are linear in the vehicle offset from the lane center and the average curvature of the road. They don't attempt to detect rogue trackers, however. We are experimenting with applying statistical methods for outlier detection in linear regression ([1], [2]) to the problem of detecting incorrect tracker responses.

3.3.4 Interpretations

The system does not currently attempt to explain anomalous tracker responses. At a higher level, we can use the semantics of the model to interpret tracker failure. Tracker failure may be noticed by the tracker itself, or the tracker may give a response that is inconsistent with the output of other trackers. In either case, the monitoring system will notice the failure and will try to explain the underlying cause, and use that explanation to update its model. Examples of such reasoning include:

- double yellow -> single dashed yellow: no change
- double yellow -> none: intersection appearing, predict all other lines disappear, start intersection-traversal behavior
- white line disappears -> <many possibilities>
    - road / shoulder: nothing
    - all road with no border -> possible side road turning off
    - dark scene: check for shadow
    - uninterpretable: check for occlusion

3.3.5 Current Status

The program which currently exists contains

- Code for dealing with an explicit road model described as generalized stripes with spines which are arcs with constant curvature.
- The five trackers described above.
- A simple tracker selection mechanism to decide which tracker should track which feature.
- Prediction code that positions each tracker correctly based on the perceived position of the road in the previous image and the vehicle's motion.
- Two methods of tracker fusion. One uses a Hough technique to determine the vehicle position relative to the spine of the current road stripe. The other assumes that the road is locally almost straight and uses statistical techniques to detect outliers in a least squares formulation.

- A simple facility for producing synthetic road images in order to test the effects of errors in calibration and the road model separate from the image processing problems.

We have run this program on the CMU Navlab on a section of path in the park near campus. Each digitize-track-fusion-steer cycle took about 10 seconds, running on a single Sun 3/280. The main goals of our initial work have been to develop a family of trackers that work well in many situations, and to check auxiliary functions such as path planning and camera calibration.

We are just beginning the second phase of our programming, which will exploit our explicit feature models. The first item on our agenda is fully automatic selection of features to track and tracker type and placement. The user currently decides which features of the road model should be tracked. Possible heuristics for automatic feature selection include both a priori reasoning (what is the expected contrast between these two adjacent features?) and run time reasoning (what is the actual contrast in the initialization image?).

A second step is implementation of mechanisms for explaining tracker failures. Once tracker fusion has detected that a tracker has failed the system needs to determine why it failed, both to prevent other trackers from falling into the same trap, and to allow detection of interesting road phenomena like intersections.

We also need to model the semantics of road markings. Cues such as a double yellow line turning into a dashed yellow line can predict the road becoming straight and flat.

We also will build and test additional simple feature trackers as we gain experience with failure modes. No one tracker is likely to be reliable in all circumstances, so the greater variety of trackers available the greater the chance of having one that works for a particular condition. Perhaps more important than the proliferation of operators is implementing them efficiently, using faster hardware or several cooperating computers. Most of our processing time is consumed in local image processing operations which are relatively easy to parallelize.

3.4 References

[1] Barnett, Vic, and Lewis, Toby.
Outliers in Statistical Data.
John Wiley & Sons, 1984.

[2] Beckman, R. J., and Cook, R. D.
Outlier..........s.
Technometrics 25(3):119-149, May, 1983.

[3] D. DeMenthon.
Inverse Perspective of a Road from a Single Image.
Technical Report CAR-TR-210, University of Maryland, 1986.

[4] E. Dickmanns and A. Zapp.
A curvature-based scheme for improving road vehicle guidance by computer vision.
In *Proc. SPIE Conference 727 on Mobile Robots.* Cambridge, 1986.

[5] Mysliwetz, Birger D., and Dickmanns, E. D.
Distributed Scene Analysis for Autonomous Road Vehicle Guidance.
In *Proceedings SPIE Conference on Mobile Robots.* November, 1987.

[6] C. Thorpe, M. Hebert, T. Kanade and S. Shafer.
Vision and navigation for the Carnegie-Mellon Navlab.
PAMI 10(3), 1988.

[7] M. Turk, D. Morgenthaler, K. Gremban and M. Marra.
VITS--A Vision System for Autonomous Land Vehicle Navigation.
IEEE Transactions on Pattern Analysis and Machine Intelligence , May, 1988.

[8] A. Waxman, J. LeMoigne, L. Davis, B. Srinivasan, T. Kushner, E. Liang and T. Siddalingaiah.
A visual navigation system for autonomous land vehicles.
Journal of Robotics and Automation, Vol. 3 , 1987.

Chapter 4: An Approach to Knowledge-Based Interpretation of Outdoor Natural Color Road Scenes

Takahiro Fujimori and Takeo Kanade

4.1 Abstract

This paper describes an approach to robust road scene interpretation in high-level vision. The two key ideas are adjustable explicit scene models, and an interpretation cycle of evaluation, modeling, and extrapolation. The interpreter first picks up some tractable segment regions and generates initial hypotheses, then iterates its interpretation cycle until every region is labeled. In each interpretation cycle, labeled regions play a survival game. Those which are consistent get their plausibility value increased. Those which are not get their plausibility value decreased and some of them die (become unlabeled). Surviving regions propagate the interpretation to neighboring similar regions. In the mean time, explicit scene models are also adjusted with the latest interpretation in each cycle. The models gradually extract global information from the scene, for instance hypothesizing the location of the road. When the plausibility values of the models become high, the system begins global evaluation, using the models, in addition to the local evaluation. that uses region features and adjacencies. The combination of global and local interpretation allows even deceptive regions to eventually be correctly labeled. More than 20 road scenes, some destructively shaded, are reliably interpreted by the system INSIGHT-III which implements this approach. [1]

4.2 Introduction

The goal of this study is to buid a robust road scene interpreter. Our approach is to interpret explicitly the components found in road scenes, such as roads, grass, soil, shadows, sun-lit spots, tree trunks, tree foliage, vertical objects, and the sky. Though one of its applications is naturally autonomous vehicle navigation, this study focuses on theoretical aspects of natural scene understanding, rather than practical aspects such as how quickly a road is extracted in order to navigate an actual vehicle. In general, there are three substantial difficulties in the field of

[1]An earlier version of this paper appeared as part of CMU Robotics Institute report CMU-RI-TR-89-5. Reprinted by permission.

natural scene understanding:

1. How to obtain reliable low-level region segmentation,
2. How to interpret oversegmented or undersegmented regions consistently with domain knowledge,
3. How to organize domain knowledge.

This paper concentrates on the second theme *interpretation* though the three themes are mutually dependent. Since each one of these is a strenuous AI topic, the first and third topics needed to be moderately simplified.

Low-level segmentation is done with available tools. The segmentation tool is histogram-based and semi-automatic, which means threshold values are interactively determined by a human. Automatically segmented images are bad and sometimes even a human cannot understand them. Insisting on fully-automatic segmentation could have obscured potential problems of *interpretation*. The domain chosen is road scenes from a narrow, winding, trail. In this domain the kinds of objects in the scenes are moderately limited, currently 10, which alleviates the difficulty of domain knowledge organization. On the other hand, complex road scenes were intentionally selected so as to uncover the potential problems in *interpretation*. Several scenes are severely shaded by trees. A couple of them are not understood even by a human without predictions. Another intention was to accumulate knowledge out of many scenes (more than 20 to date).

Outdoor natural color road scenes are ill-structured. Trees and objects often shade the road destructively. The edges of shadows and sun-lit spots are much stronger than those of roads. Natural objects are not usually made with straight lines. Old road pavements have many cracks and spots. The color intensities of the road pavements vary significantly depending on their surface conditions, the weather, the direction of the sun, the camera condition, and other circumstances. In the *interpretation* of natural outdoor color scene, there are two main questions:

1. How to extract global information which depicts a scene roughly,
2. How to generate reliable initial hypotheses.

If the system has global information about the scene, this reduces the search space and local region labeling becomes unambiguous. For example, if the system knows the rough location of the road in a scene, only the regions along the road edges are ambiguous. More reliable road edges will be obtained by checking intensity steps crosswise in long narrow windows along the expected locations of the edges. Of course, global information is not available in advance, but must be extracted by the interpreter. In our approach, explicit scene models represent global information and the interpretation cycle is a mechanism for adjusting those models.

Good initial hypotheses usually lead to successful interpretation while bad initial hypotheses often do not. It is hard to change the labels of the initially labeled regions. This is natural because initially labeled regions are the seeds for the following interpretation. This problem is partly beyond this paper, because it also depends on the low-level region segmentation. However, several ideas for generating reliable initial hypotheses are described in Section 4.7.2 A new region segmentation objective in low-level vision is also suggested and discussed in Section 4.8.1

4.3 Related Work

Draper [6,7] has experimented knowledge-based road scene interpretation in the "Schema System" architecture, which evolved from the VISIONS system [9]. This system comprises independent object-specific interpreters, called schemata, and a central blackboard for communication. The system interprets a road scene in a top-down manner by starting from the road-scene schema and then hierarchically calling subpart schemata such as roof, sky, road, or tree. Each schema generates initial hypotheses based on the color, texture, shape, area, and position of segmented regions, and corroborates or denies its hypotheses by checking positive and negative evidence, extending hypotheses, resolving conflicts, and calling subpart schemata. Information about the latest interpretations with confidence values is exchanged through the central blackboard.

Ohta [19] has built a knowledge-based outdoor color scene interpreter which recognizes 4 objects and 2 subobjects: sky, trees, buildings with subobjects (windows), and roads with subobjects (cars), using both bottom-up and top-down interpretation. In this system, each large region initially has 4 hypotheses: sky, tree, building, and road. The confidence of each label is modified by evaluation, and the hypothesis which gets the highest confidence value wins. It first tentatively merges small regions with one of the large regions and calculates confidences for the 4 hypotheses on each large region by bottom-up methods, checking the color, texture, shape, and position of segmented regions; and relations, contrast, lines, etc. After this, these hypotheses are evaluated. Then, the interpretation goes to rule-based top-down analysis, which selects the most appropriate label for each large region and finds a label for small regions. Whenever the top-down analyzer makes a significant decision such as the position of the scene horizon, the interpretation goes back to the bottom-up process and re-evaluates the labels for the large regions. In this way, the bottom-up process and top-down process work cooperatively.

McKeown [14,15] has demonstrated a rule-based system called SPAM with around 500 OPS5 rules for interpreting aerial airport imagery containing terminals, parking aprons, hangars, runways, taxiways, parking lots, roads, tarmac, grassy areas, and functional areas. SPAM is a region-based bottom-up system with top-down predictions. It allows multiple initial hypotheses on each region, based on local properties such as 2-D shape, location, orientation, texture, area, and perimeter, and then checks consistency locally, groups local interpretations into functional areas, and generates one or more airport models. Height information derived from stereo images is also utilized when necessary. Recently he also developed tools for knowledge acquisition, machine rule generation, and performance analysis [14] in order to ease and expedite the knowledge acquisition processes [13]. Suburban house scene knowledge was collected for the SPAM architecture with these tools.

Brooks [2] has developed a viewpoint-insensitive 3-D model-based system called ACRONYM, which was applied to jet plane recognition in aerial photographs. ACRONYM has an intermediate level represented by ribbons (elongated trapezoids) and ellipses, to which both 3-D models and low-level edges are transformed. Interpretation takes place in this level in a top-down manner. Recently, ACRONYM was transferred to industry [25], and uncertainty management [20] has been incorporated into it.

Binford[1] surveyed model-based image analysis systems, including work by Nagao and Matsuyama [17], Ohta [19], Brooks [2], and Shirai [22].

4.4 Adjustable Explicit Scene Models and the Interpretation Cycle

The two key ideas of our system for extracting global information from an interpreted scene are adjustable explicit scene models and the interpretation cycle. The global information makes global evaluation possible and thus leads to reliable consistent interpretation. In scene interpretation, a chicken and egg problem arises: a scene can hardly be interpreted locally and correctly without global information of the scene; yet the global information can not be obtained without local interpretation fragments. Adjustable models and the interpretation cycle are an approach to this problem.

4.4.1 Adjustable Explicit Scene Models

Adjustable explicit scene models explicitly approximate the appearance of the objects in a road image, using simple geometric figures. These models have four or five degrees of freedom, and are readjusted in each interpretation cycle to fit the latest interpretations. Figure 4.1 (a) depicts a triangular explicit scene model for a road or a long narrow strip on the ground level. It has 4 degrees of freedom: the vanishing point moves horizontally and vertically, and the lower corners move horizontally. Figure 4.1 (b) is a parallelogram explicit scene model for a tree trunk or a vertical object protruding from ground. Since its upper and lower bases remain horizontal, it has 5 degrees of freedom: location (2 degrees of freedom), height, width, and lean.

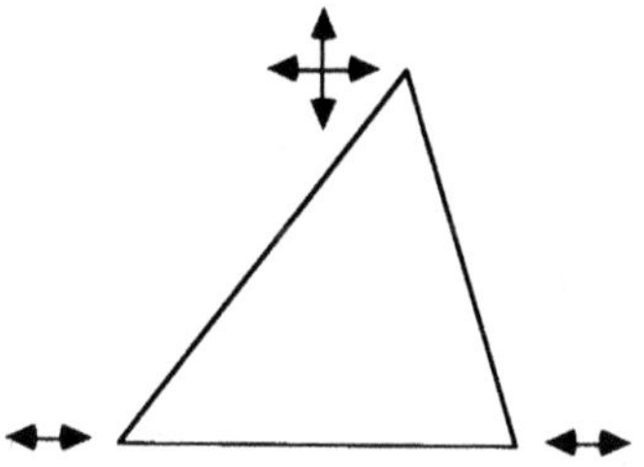

(a) Road Model

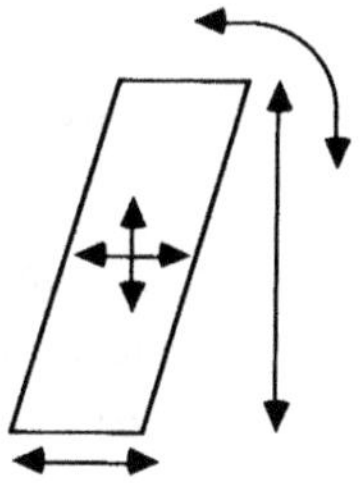

(b) Tree Trunk Model

Figure 4.1: Adjustable explicit scene models

Figure 4.2 illustrates how the road model is adjusted to an interpreted scene. To measure the appropriateness of the road model, a discrepancy value is defined as the sum of the road area outside the model + the grass area and the soil area inside the model. The road model is adjusted such that the discrepancy value is locally minimized. Figure 4.3 shows the tree trunk model adjustment. The discrepancy value for the tree trunk model is the sum of the tree trunk area outside the model + the non-tree-trunk area inside the model. Because the tree trunk model is also adjusted to the local minimum of the discrepancy value, it is not affected by the trunk area of other trees. Finding the local minimum of the discrepancy value is relatively inexpensive. It only requires checking the pixels along the model boundaries. Section 4.7.5 explains the details.

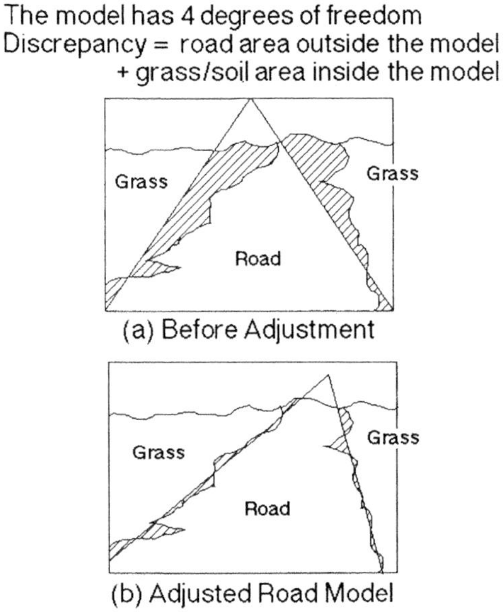

Figure 4.2: Road model adjustment

The explicit scene models can be applied to a scene whose segmented regions are partly labeled and partly unlabeled, in the early stages of interpretation. Though the adjusted models may be less reliable, they can represent some useful scene information. For instance, unlabeled regions in the middle of the adjusted road model are likely to be fragments of the road, while those away from the adjusted road model are more likely to be fragments of the grass or the soil. If the plausibility value of the model becomes higher for some reason, the model can label the unlabeled regions and even discard the contradicting labels, which may, in turn, require more model adjustments.

These explicit scene models are simple and cannot express detailed shapes or details of roads and tree trunks, such as curves or tree branches. Complex or flexible models with more degrees of freedom may be applied afterwards if

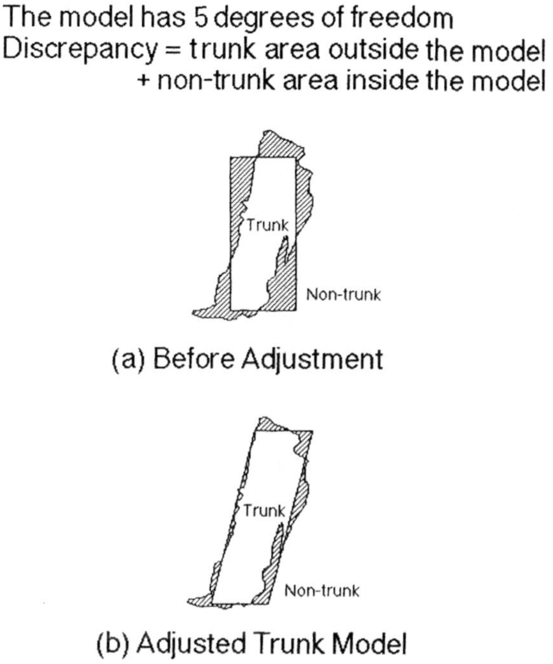

Figure 4.3: Tree trunk model adjustment

necessary. With the global information extracted by these simple models, those complex or flexible models will be adjusted quickly and reliably to the scene.

An interesting aspect of the explicit scene models is that they help detect misinterpretation. If the width of the road is too wide or too narrow, if its vanishing point becames too high or low, or if a tree trunk is found in the middle of the road, it implies that something bad has taken place. Then the system will be able to retry the interpretation of the scene with different low level parameters, or activate a contingency planner which could decide to decelerate the autonomous land vehicle and take an image with another camera iris and angle. The acceptable range of the adjusted models will be determined heuristically or by the interpretation result of the previous scene.

4.4.2 Interpretation Cycle

Figure 4.4 shows the structure of the high-level scene interpreter, which consists of an initial hypothesis generator, an evaluator, a modeler, and an extrapolator. First, roughly speaking, the initial hypothesis generator picks up some tractable regions by checking their area, location, intensities, adjacent regions' intensities, shape, border straightness, parallel lines, trapezoidal lines, etc. It uses those features to make initial hypotheses, as described in Section 4.7.2. Then, the interpretation cycle iterates until every region is labeled. The evaluator checks the local consistencies of the current interpretation by looking at the labels of adjacent regions. If they are consistent, their plausibility values are increased. If they are not, their plausibility values are decreased. When the plausibility values of the models become higher, it checks the global consistencies of the current interpretation by looking at both labels and models. If the labels are contradictory, their plausibility values are decreased drastically. If the plausibility value of a labeled region becomes negative, the region becomes unlabeled. Next, the modeler adjusts the models according to the current interpretation. If the area of a model becomes zero, it becomes eliminated. Then, the extrapolator runs. It picks up some moderately tractable unlabeled regions that are next to labeled regions and hypothesizes their labels by basically propagating known regions.

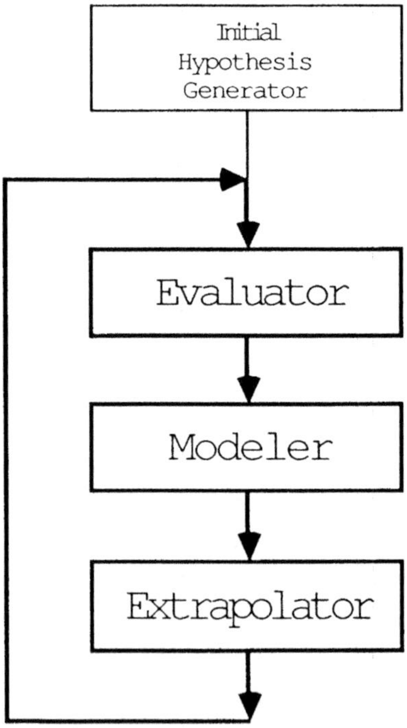

Figure 4.4: Interpretation Cycle

Figure 4.5 illustrates the interpretation cycle with a simple example, which only has a region map, color data, and 5 interpretation cycles. Figure 4.5 (a) is the region map which has 10 segmented regions. Figure 4.5 (b) shows the colors of the regions. It includes one ambiguous region and two deceptive regions. R9 is ambiguous since its color is in-between. R6 and R10 are deceptive because their colors differ from those of their neighbor regions. The initial hypothesis generator picks up regions R1, R4, R5, and R7 as tractable regions, and makes hypotheses: sky, grass, road, and grass, respectively. Then, the interpretation cycle starts.

In the first cycle, the evaluator slightly raises the plausibility values of R1, R4, and R5 by checking the consistency between adjacent labeled regions. The modeler adjusts the triangular road model. The adjusted road model is far from the ideal one since its vanishing point is too high. But it has collected very rough global information of the road. Then, the extrapolator runs and the labeled regions propagate. Both R4 and R5 try to propagate their labels to R9 since its color is between them. In the second cycle, many of the regions get labels, but R6, R9, and R10 are mislabeled. The evaluator raises the plausibility values of the labeled regions except for those mislabeled. The plausibility value of R9 becomes negative and R9 is unlabeled. The modeler adjusts the model and gets much better global information. The vanishing point is almost correct. The plausibility of the model is now usable. The extrapolator partly uses the model and R5 propagates to R3 and R9. In the third cycle, all regions except deceptive ones are labeled. R9 survives through the evaluation. The model extracts the proper road location and the plausibility of the model becomes reliable. Nothing happens in the extrapolation. In the fourth cycle, the evaluator decreases the plausibility values of R6 and R10 by using the reliable road model. The road model does not change in the modeling. The extrapolator finally labels R6 and R10 correctly and the interpretation ends in the following cycle.

Unlike some other region based interpretation systems [19,15], a segmented region can take at most one label at a time. With this restriction, the explicit scene models can be adjusted unambiguously. Since this restriction temporarily excludes alternative labels, the evaluator has a local and global negative evaluation mechanism which refuses inconsistent labels and gives chances to alternative labels.

The robustness of the interpreter results from a balance among the four modules depicted in Figure 4.4 Some strong rules in a module may work very well on a particular road scene, but not on other scenes. Rules should be simple and straight-forward so that the obvious regions are labeled quickly. Occasional mistakes should be taken care of by the mechanism of the interpretation cycle. For example, the initial hypothesis generator should label a few easiest regions but can leave slightly ambiguous regions unlabeled, since the following interpretation cycle is capable of disambiguating them. The extrapolator can make

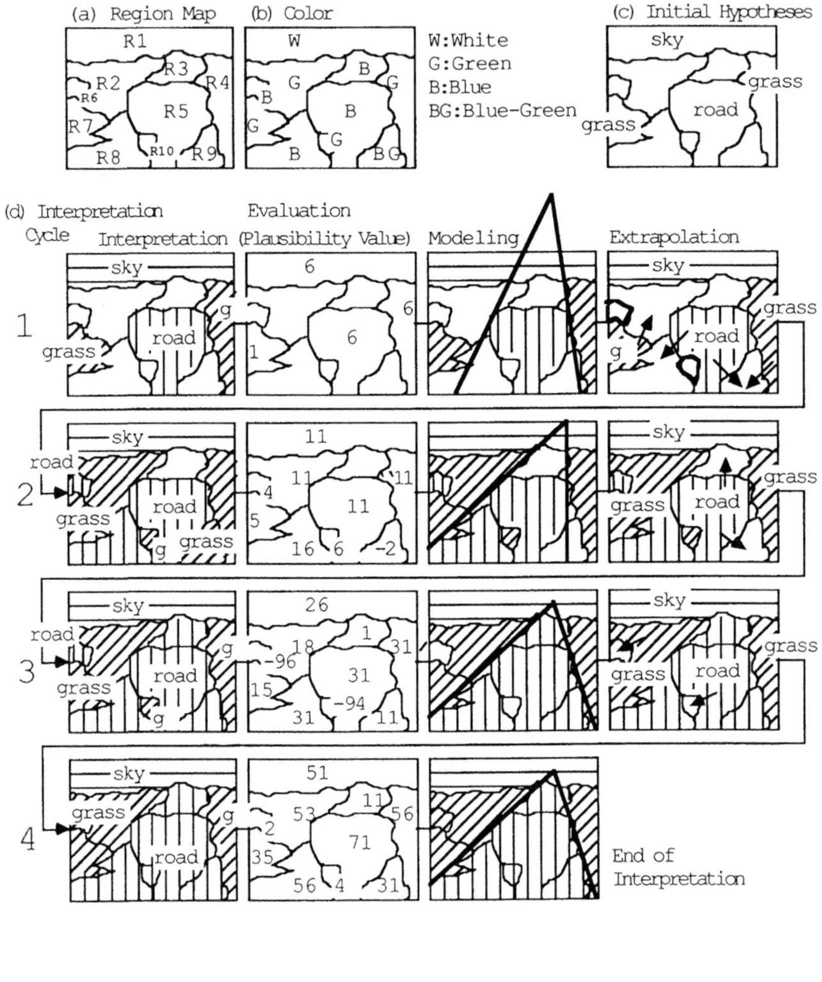

Figure 4.5: A simplified example of interpretation

easy hypotheses since the negative evaluation is able to reject bad hypotheses. The evaluator does not have to be nervous because the modeler is grabbing global information and will settle the interpretation at the end. If the scene is very difficult and the interpretation does not converge, it can be detected by monitoring the number of interpretation cycles iterated and the plausibility level of the models. Then a contingency planner may be activated.

4.5 System Overview

The knowledge-based road scene interpretation system INSIGHT-III is capable of extracting roads, grass, soil, shadows, sunny spots, tree trunks, tree foliage, vertical objects, sky, and unknown areas without predictions. Its only assumptions are that a road is in the scene, and that the scene horizon is approximately horizontal in the image. It also detects pairs of a tree trunk and its shadow, the direction of sun (right or left), and the weather (sunny or not), and uses this information in interpretation. By inverse projection (back projection), it generates a road map of the scene, though its depth information is relative. Its high-level interpretor has around 190 rules written in OPS83 [8] (a descendant of OPS5 [3]), and C++ [23] functions for model adjustment. The system runs on SUN workstations. One road scene takes about 40 to 70 minutes to interpret, depending on the complexity of the scene. An experimental fully-automatic low-resolution version of the system interprets it a scene around 8 minutes. 10 road scenes, some badly shaded, are interpreted by the system. In addition, more than 15 road scenes are interpreted by the fully-automatic low-resolution version of the system, using exactly the same high-level knowledge-based interpreter.

Figure 4.6 shows the system organization. A color median filter, which considers color pixels as vectors and keeps color edges sharper than the scalar median filter, is first applied to the raw color road scene image of 512×480 spatial resolution with 8 bit pixels. Second, the noise reduced image is semi-automatically and recursively segmented by the PHOENIX histogram-based region segmentation tool [21,11]. At each step, PHOENIX calculates red, green, and blue histograms of the scene portion, and a human selects one histogram and a threshold value by looking for the deepest valley in the histograms. Then the scene portion is divided into several smaller regions by using the threshold value. This process recurs until all the regions become homogeneous or small. Third, the color Canny edge detector [18] extended from Canny edge detector [4] extracts edge information from the raw image. Fourth, the feature extraction program calculates the intermediate representation of each region from the region map, the edge data, and the raw image. This intermediate representation includes the region map, mean intensities, standard deviations, texture, area, perimeter, mass center, minimum bounding rectangle (MBR), scatter matrix, boundaries, vertices, and lines of each region, as well as relations among regions, boundaries, vertices, and lines. Fifth, the extracted intermediate representation data are fed to the high-level knowledge-based interpreter and the road scene is interpreted. Last, the adjusted scene models proceed to the map generation and a road map is drawn by inverse projection.

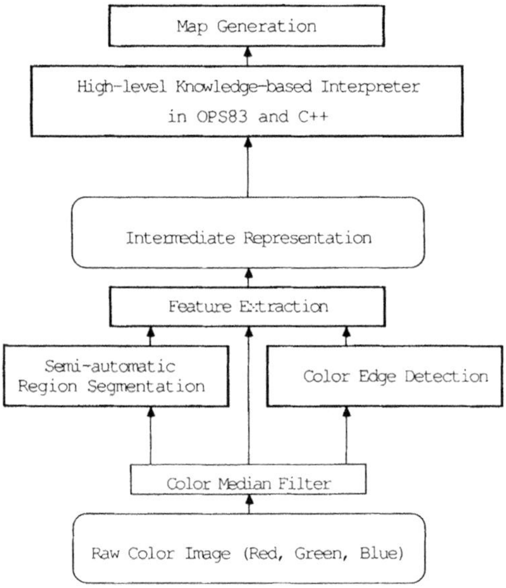

Figure 4.6: System organization

In the case of the fully-automatic low-resolution version of the system, only the region segmentation tool is different. Instead of PHOENIX, a region segmentation tool [24] is used that is based on the distances among pixels in color space. It generates a region map of 64×60 spatial resolution by grouping similar pixels. This version is designed to be integrated with the Navlab.

Figure 4.7 illustrates the high-level knowledge-based interpreter configuration. The intermediate representation data enter the working memory of the rule-based system. Then the rules of the initial hypothesis generator are activated and label some regions. Then, the rules of the evaluator, the modeler, and the extrapolator examine the scene interpretation data cyclically until every region is labeled. All the intermediate representation data, the temporal data, and the explicit scene models reside in the working memory.

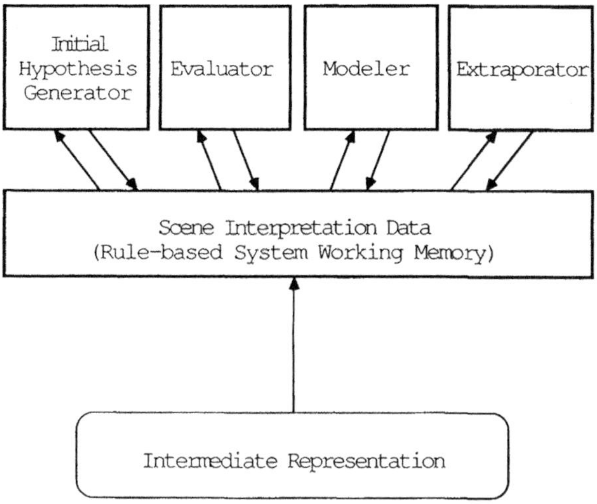

Figure 4.7: High-level knowledge-based interpreter configuration

4.6 Results of the Road Scene Interpretation

Figures 4.8 to 4.11 show typical results of scene interpretation by INSIGHT-III. Figure 4.8 is an evenly-lit scene that is easy to interpret. Figure 4.8(a) shows the original scene, (b) the segmented image, (c) the solution classes, (d) the interpretation, (e) the adjusted models, and (f) the generated map. In the solution class images, horizontal lines indicate regions in the *ground level* class, while vertical lines indicate *above ground* regions. In this figure, the sky is the only component labeled as it above ground. Figure 4.9 shows the processing and results for a scene with trees. Note the *above ground* solution class of the tree, and its location on the map.

Figure 4.10 is a much more difficult scene, which requires several iterations of the interpretation cycle to converge. Figure 4.10 (a) shows the original scene, and (b) shows the solution classes. The next images, (c) through (h), show the evolving interpretation and road model as the scene becomes more completely interpreted. As more and more regions become labeled, the road model becomes more and more accurate. Lastly, images (i) and (j) show the resulting interpretation and map.

Finally, Figure 4.11 shows intersection interpretation. Images (a), (b), (c), and (d) show segmentation, solution class, interpretation, and first road fitting. Since there is a significant uninterpreted road region, another road model is hypothesized and fit in image (e). The resulting map, complete with trees, is shown in (f).

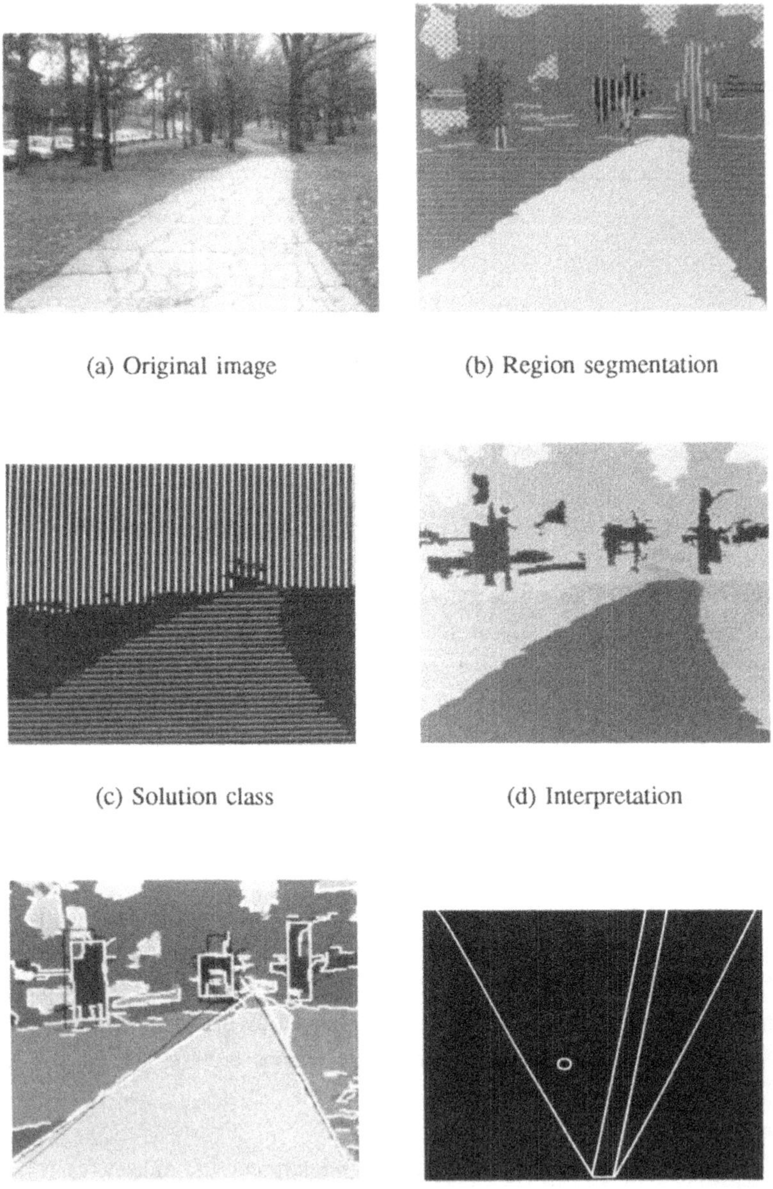

(a) Original image

(b) Region segmentation

(c) Solution class

(d) Interpretation

(e) Adjusted model

(f) Generated map

Figure 4.8: Evenly-lit road

(a) Original image

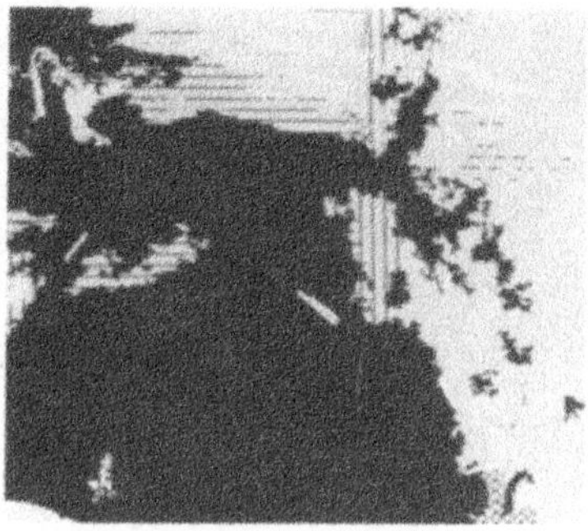

(b) Region segmentation

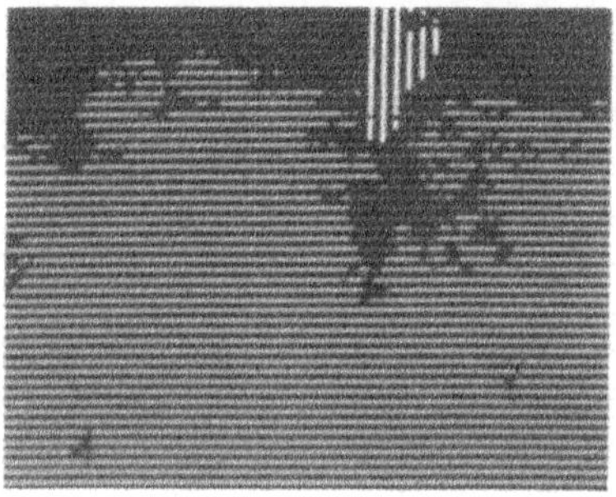

(c) Solution class

(d) Interpretation

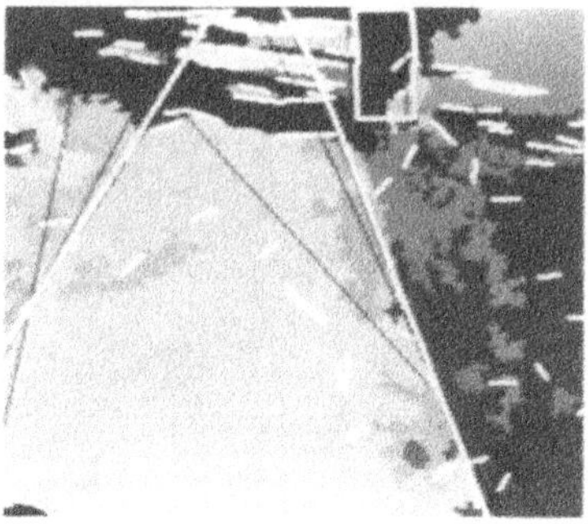

(e) Adjusted model

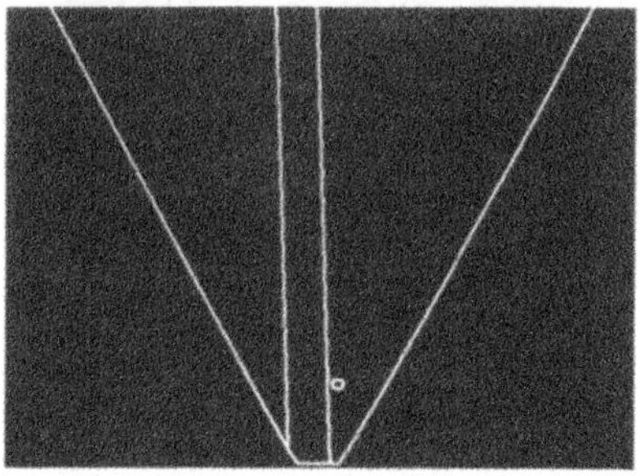

(f) Generated map

Figure 4.9: Road scene with trees

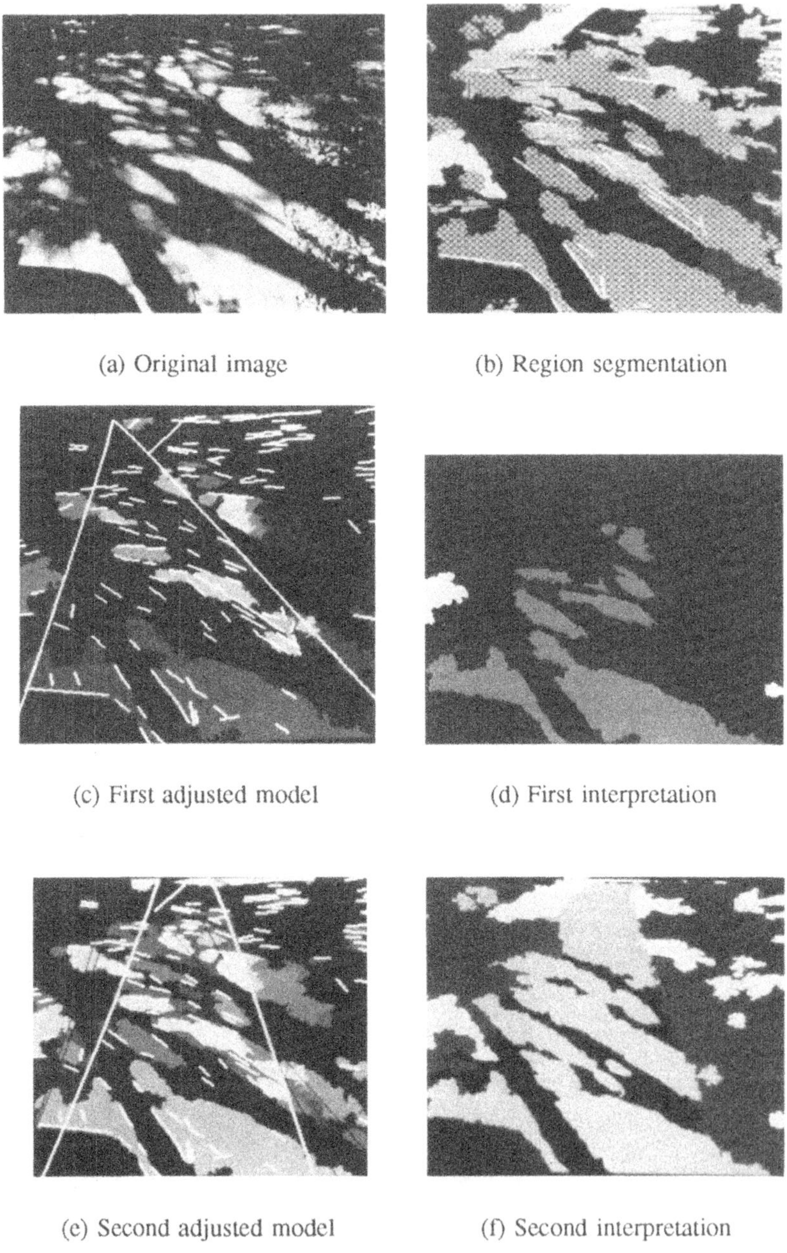

(a) Original image (b) Region segmentation

(c) First adjusted model (d) First interpretation

(e) Second adjusted model (f) Second interpretation

Figure 4.10: Beginning of difficult scene

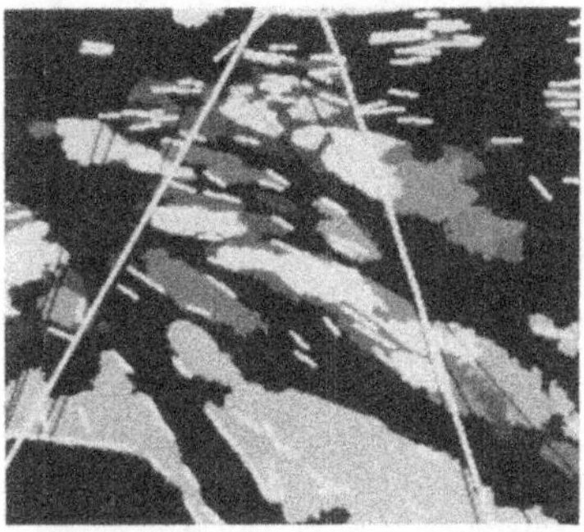

(g) Third adjusted model

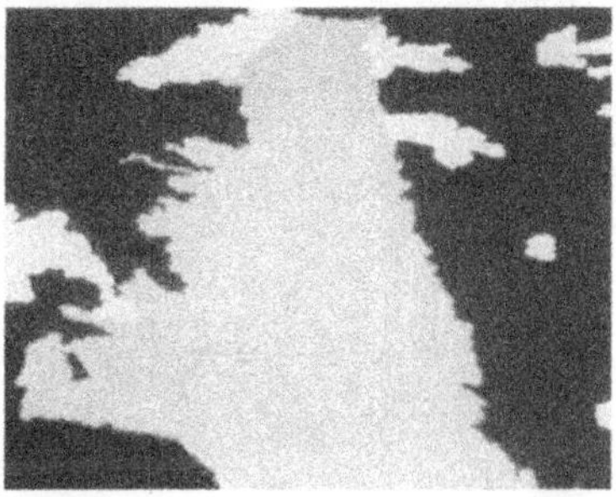

(h) Third interpretation

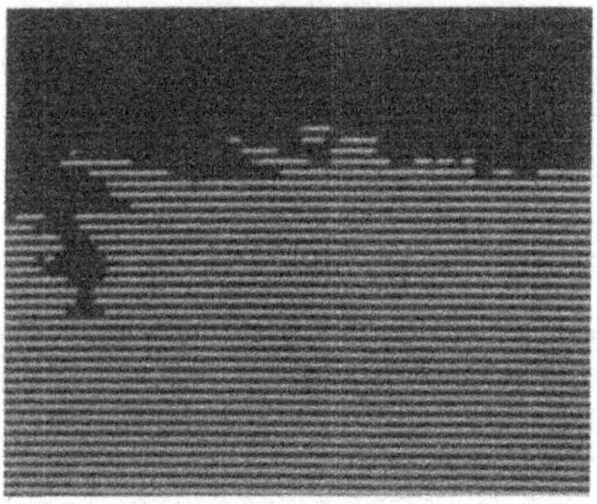

(i) Interpretation class

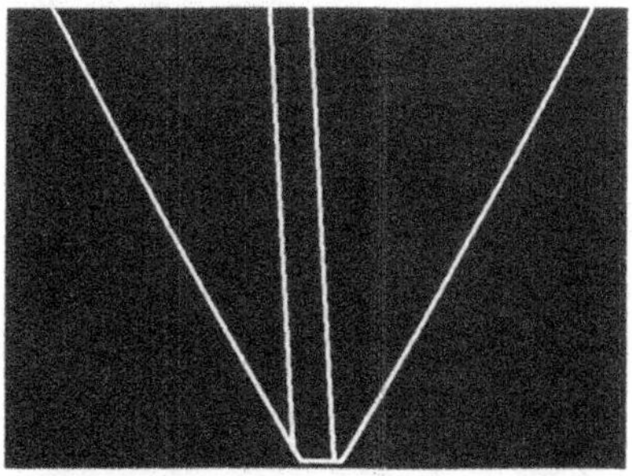

(j) Map

Figure 4.10: End of difficult scene

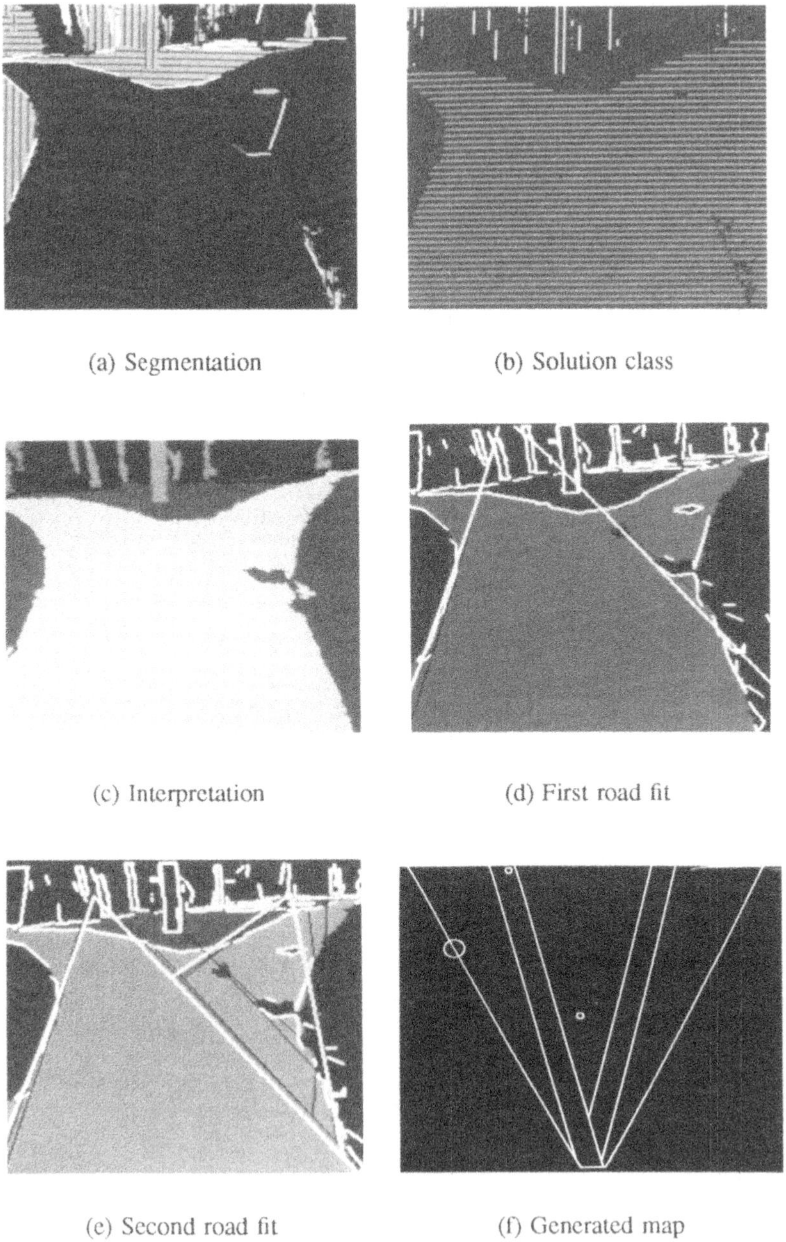

(a) Segmentation

(b) Solution class

(c) Interpretation

(d) First road fit

(e) Second road fit

(f) Generated map

Figure 4.11: Intersection fitting

4.7 The Road Scene Interpretation System in Detail

4.7.1 Feature Extraction and Intermediate Representation

The quality of the semi-automatic region segmentation by PHOENIX is better than fully-automatic segmentation, but still far from ideal. Especially when the image is shaded and complicated, it often generates hair-like long and narrow portions extending from the region bodies along their edges. Some regions are not separated. Some are fragmented. The boundaries are jagged. In the feature extraction, first, the region trimmer is run on the region map to alleviate these problems. It fragments hair-like portions, cuts bridges connecting two big bodies, and then merges small noisy fragments into the adjacent regions which share the longest boundaries. After the region trimmer runs, the intermediate representation data is calculated.

```
struct region{
    region identification number
    number of neighbors
    neighbor identification numbers
    number of holes
    hole identification numbers
    number of boundaries
    boundary identification numbers
    mean intensity (Black&White, R, G, B, R-B)
    standard deviation (Black&White, R, G, B, R-B)
    texture value
    area
    perimeter
    mass center row, column
    minimum bounding rectangle (MBR)
    scatter matrix
}
```

Figure 4.12: Region data structure

In addition to the region map, regions, boundaries, vertices, and lines are the primary elements of the intermediate representation, whose structures are shown in Figure 4.12, Figure 4.13, Figure 4.14, and Figure 4.15. The relations among regions, boundaries, and vertices are depicted in Figure 4.16. Several boundaries, each terminating at two vertices, enclose a region. A line is linked to the nearest boundary.

```
struct boundary{
   boundary identification number

   vertex identification numbers
   region identification numbers
   number of lines
   line identification numbers

   length
   contrast
   chain code length
   chain code
}
```

Figure 4.13: Boundary data structure

```
struct vertex{
   vertex identification number

   number of boundaries
   boundary identification numbers

   position row, column
}
```

Figure 4.14: Vertex data structure

```
struct line{
   line identification number

   boundary identification number

   start row, column
   end row, column
   distance from origin
   orientation
   length
}
```

Figure 4.15: Line data structure

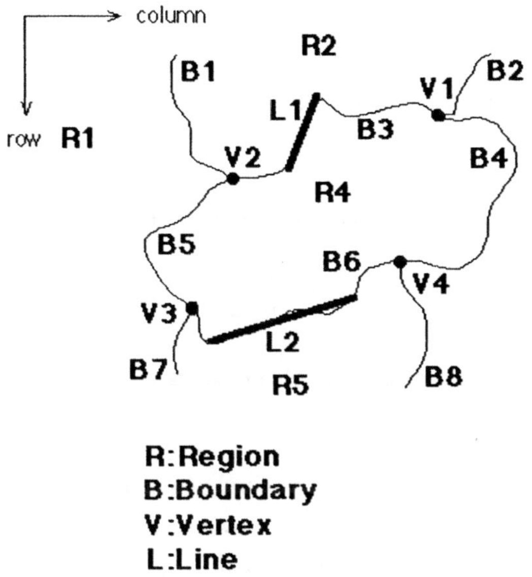

Figure 4.16: The relation description

Region Description

Figure 4.17 describes a region, including both geometrical and appearance properties.

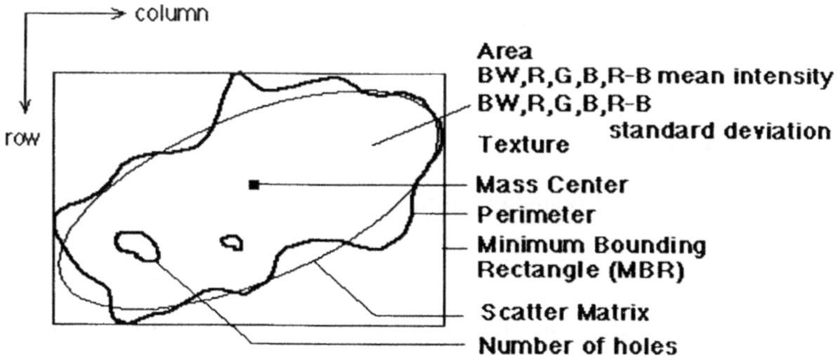

Figure 4.17: The region description

Texture The texture value extraction algorithm is as follows:

1. derive a binary edge image from the black & white image using a 3 by 3 Laplacian operator and a threshold T ($T = (standarddeviation) \times 1.4$),
2. calculate the number of 1's in each 9 by 9 window of the binary image

Perimeter The perimeter is the sum of the arc lengths of all boundaries, counting horizontal and vertical moves as 1 and diagonal moves, which need 2 steps, as $\sqrt{2}$.

Mass Center The mass center of a region is computed as follows:

$$M = \frac{1}{N} \sum_{i=1}^{N} P_i,$$

where N is the number of pixels in the region, P_i is the position vector of i-th pixel.

Minimum Bounding Rectangle The Minimum Bounding Rectangle (MBR) is defined as the set of four numbers: rowmin, rowmax, colmin, and colmax. They define the spatial extent of the region.

Scatter Matrix The scatter matrix represents the elliptical area which approximates the shape of the region. The scatter matrix C of a region can be calculated as follows:

$$C = \frac{1}{N} \sum_{i=1}^{N} (P_i - M)(P_i - M)^t,$$

where N is the number of pixels in the region, P_i denotes the position vector of i-th pixel, M denotes the mass center of the region.

Boundary Description

Figure 4.18 describes a boundary. Boundaries are the borders of the regions in a region map. They are not related to the color edge detector.

Length Length is the sum of the arc lengths of the boundary, counting horizontal and vertical moves as 1 and diagonal moves, which need 2 steps, as $\sqrt{2}$.

Contrast Contrast is the difference of `mean_intensity_BW` between the regions forming the boundary.

Vertex Description

Only 3 or 4 boundaries can meet at a vertex because of the square tesselation.

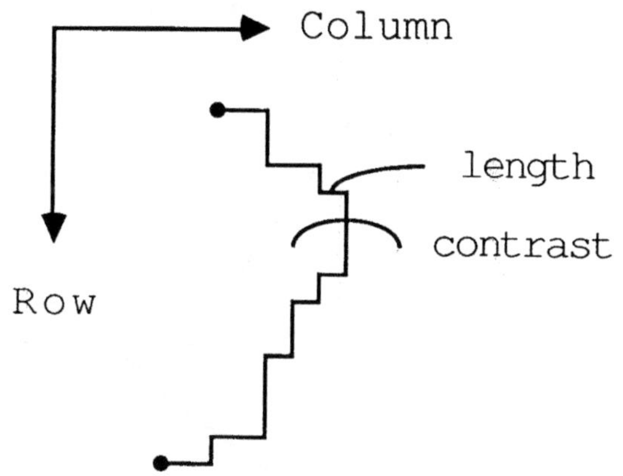

Figure 4.18: The boundary segment description

Line Description

Figure 4.19 describes a line. Lines are obtained by applying the Miwa line extractor [16] to the edges derived with the color edge detector [18]. The Miwa line extractor has a pair of connected arms tracking along the edges and finds break points in a line by checking the angle formed by the arms.

4.7.2 Initial Hypothesis Generation

The initial hypothesis generator picks up some tractable regions and makes initial hypotheses. It is skeptical and tries to avoid erroneous hypotheses.

Clancey's classification problem solving model [5] is applied to initial hypothesis generation. By analyzing more than 10 expert systems such as MYCIN, SACON, The Drilling Advisor, GRUNDY, and SOPHIE-III, he discovered a basic problem solving strategy for classification problem solving, which these systems have in common. The classification problem is the problem which classifies input data into known solutions. If the number of solutions is fairly limited, the model works very well. It has three steps. The first step is data abstraction: the input data are abstracted and converted to more tractable forms. The second step is heuristic match: data abstractions are projected heuristically into solution classes. Each solution class groups similar solutions. The last step is refinement: one of the known solutions is selected by refining the solution abstraction.

Figure 4.20 shows how the classification problem solving model is ap-

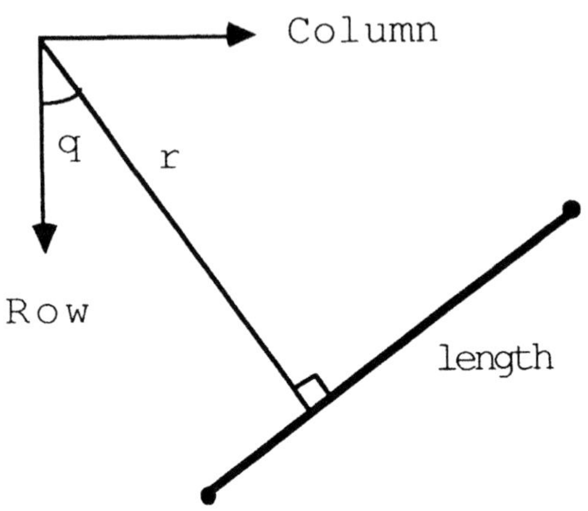

Figure 4.19: The line segment description

plied to the initial hypothesis generation for color road scene interpretation. The intermediate representation data are first abstracted by simple rules which generate data abstractions such as large area, green area, lengthwise area, parallel lines, etc. Each region gets as many data abstractions as possible. Second, some regions with data abstractions are projected onto solution classes: *ground level* or *above ground. Ground level* groups roads, grass, soil, shadows, and sunny spots. *Above ground* groups tree trunks, tree foliage, object, and sky. Last, each region which has fallen into a solution class is refined and labeled with a solution selected from the members of its solution class.

Solution classes are determined since the appearance of the objects protruding from ground is very different from objects on the ground level. When tracing objects vertically, the appearance of the *above ground* objects such as tree trunks does not change much, while *ground level* objects such as roads change appearance drastically because of the perspective projection. When generating initial hypotheses, it is less dangerous to pick up one of the solution classes first and then refine it than to pick up one of the solutions directly.

Data Abstraction

All the data abstractions are listed in Figure 4.21 There are 6 categories: area, location (in an image), color, texture, shape, and line. The color category has 3 subcategories: normal, bright, and dark. Data abstraction rules are very simple. They just check a couple of features of regions and generate data abstractions. Color data abstractions are not very exclusive. For example, one region can be

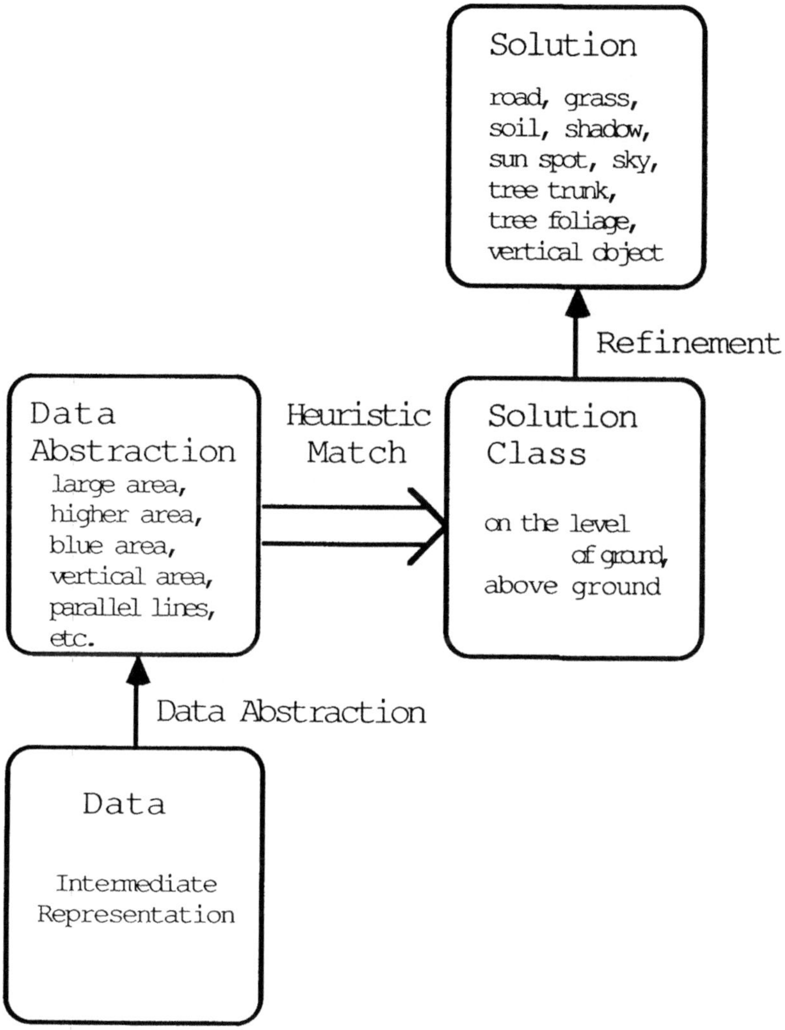

Figure 4.20: Classification problem solving model applied to initial hypothesis generator

blue area and greener area at the same time.

Category		Data Abstracion
Area		large area, moderately large area
Location		higher area, lower area
Color	Bright	bright area, bright red area, bright green area
	Normal	blue area, bluest area, green area, greener area, brown area
	Dark	dark area, darker area, darkest area
Texture		textural area
Shape		lengthwise area, crosswise area, vertical parallel area, trapezoidal area
Line		long line, linear lines, horizontal line, vertical line, vertical parallel lines, trapezoidal lines

Figure 4.21: Data abstraction rules

Data abstraction is the most important component of the initial hypothesis generation. Thus, it is one of the most important components in the road scene interpreter. Though it partly depends on the low-level region segmentation as mentioned in Section 4.2, some robust ideas are embedded in the data abstraction rules:

Look at Neighbor Regions Compare a region's features with those of its neighbors, then more reliable data abstractions can be generated. Though the global information of the scene is not available, rules can check the neighbors locally. Two actual examples are shown in Figure 4.22. If a region is darkest among its neighbor regions, the region is given the additional abstraction "really dark". If the color of a region is closest to pure green (Red=0, Green=255, Blue=0) among neighbor regions, it is guaranteed that the region is "really green".

Use Color Ratio Use the color ratio such as Red/Blue and Green/Blue. Do not use R, G, and B intensities directly except for detecting dark or bright regions. R, G, and B intensities vary significantly, depending on weather, camera iris,

```
Darker Area:
   IF (black&white < 64)
      and (20 <= black&white or 20 <= red)
      and (green <= blue + 32)
      and (darkest among neighbor regions)
   THEN The region is "darker area"

Greener area:
   IF (48 < green)
      and (blue + 10 < green)
      and (closest to pure green of neighbors)
   THEN the region is "greener area"
```

Figure 4.22: Data abstraction rules that look at neighbor regions

camera white balance, video amplifier, A/D converter, etc. R, G, and B intensities were used directly in the predecessor system INSIGHT-II, but as the number of images increased, contradicting intensities were observed. Consequently, in INSIGHT-III, the color usage was totally rebuilt by discarding direct color intensity references and introducing color ratios. An example is in Figure 4.23.

```
Bluest Area:
   IF (48 < blue and g/b <= 0.90)
   THEN the region is "bluest area"

Blue Area:
   IF (36 < blue)
      and (( g / b <= 0.97)
         or (g / b <= 1.05 and r / b <= 1.25)
         or (g / b <= 1.10 and r / b <= 1.25))
   THEN the region is "blue area"
```

Figure 4.23: Data abstraction rules that use color ratio

Detect Very Dark and Very Bright Regions Detect the very dark and very bright regions beforehand and do not use their color information, because their colors are severely distorted owing to the limited dynamic range of the camera. The two solutions, shadows and sunny spots, are in a sense for these kinds of regions, though the regions do not necessarily have to be labeled with shadows or sunny spots. Two kinds of data abstractions: *very dark* and *very bright*, are required for outdoor natural color scene interpretation.

Heuristic Match

There are three kinds of heuristic match rules: ground level, above ground, and both. Most of them are for inferring *ground level* objects. The reason is that two *above ground* objects, tree foliage and sky, are more ambiguous than those on the ground level. They could be grass and road, respectively. So, they usually remain unlabeled in the initial hypothesis generation, since they can be easily disambiguated by the position of the vanishing point of the road which will be available later. One rule, which infers an *above ground* object and a *ground level* object at the same time, is intended to detect pairs of a tree trunk and its shadow. It turned out to be a powerful rule. Some examples are in Figure 4.24.

Above Ground:
  IF no solution class is generated
    and higher area
    and (darker area or darkest area)
    and lengthwise area
  THEN it is "above ground"

Ground Level:
  IF no solution class is generated
    and large area
    and lower area
  THEN it is "ground level"

Figure 4.24: Heuristic match rule examples

Refinement

Each solution, grass, soil, shadow, sunny spot, tree foliage, or sky, has one region-based refinement rule which checks the solution class and other data abstractions. Road, tree trunk, or vertical object has a region-based refinement rule and an edge-based refinement rule, since the shapes of these objects are much more constrained than the other objects. Figure 4.25 explains some rules.

4.7.3 Context Control

Figure 4.26 illustrates the context control of the road scene interpreter. It looks slightly different from that in Figure 4.4, though they are basically the same. After the extrapolation, the interpretation cycle goes back to the heuristic match of the initial hypothesis generator. The reason is that a part of the task of the

Soil Found:
  IF ground level
    and brown area
    and not darkest
    and 32 < MBRrowmax
    and road model plausibility <= usable
  THEN the region is "soil"

Road Trapezoid Found:
  IF ground level
    and blue area
    and trapezoidal area
    and not textured area
    and road model plausibility <= usable
  THEN the region is "road"

Figure 4.25: Refinement rule examples

extrapolator is very similar to that of the heuristic match and refinement. Simply, the extrapolator shares some rules with the initial hypothesis generator.

4.7.4 Evaluation

Each labeled region has a plausibility value. In each interpretation cycle, labeled regions are evaluated and get their plausibility values increased or decreased. If a plausibility value becomes negative, its label is removed and gives chances to alternative labels. Each region has two slots for memorizing the two previous labels. A removed label cannot be immediately retried, but must wait until at least two alternative labels have been tried. Figure 4.27 shows the composition of the evaluation rules. Evaluation rules are categorized into local and global evaluation rules. Local and global evaluation rules each have both positive and negative evaluation rules. A local evaluation rule picks up a pair of labeled regions at a time and checks their consistency. A global evaluation rule picks up one labeled region at a time and checks its consistency with one of the explicit scene models. Four of the negative global evaluation rules are dominant. When the plausibility values of explicit scene models become reliable, these four rules enforce the model by rejecting labeled regions that violate the models.

Some local evaluation rules are in Figure 4.28. If a pair of labeled regions are consistent, the rule gives 5 points to each labeled region. If a pair of labeled regions are inconsistent, the rule takes 8 points from the less plausible

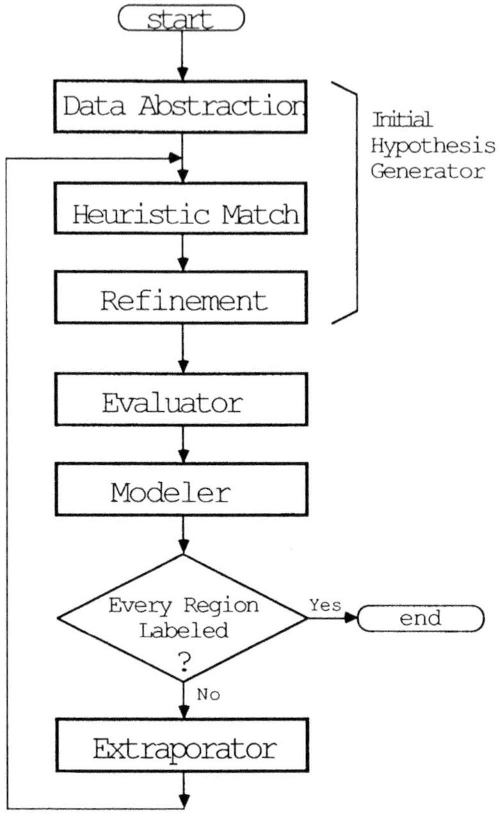

Figure 4.26: Context control

Category		Positive	Negative	Total
Local		7	11	18
Global (by Models)	Ordinary	4	7	11
	Dominant	-	4	4
Total		11	22	33

Figure 4.27: Evaluation rules

labeled region and 2 points from the more plausible region or, if they are equally plausible, it takes 5 points from each labeled region. When a pair of labeled regions is inconsistent, one labeled region could be good and the other one could be bad. The negative evaluation rule can not figure out locally which one is wrong. But the properly-labeled region must survive. Each labeled region has a plausibility value. If the plausibility value is high, this means the region has survived some interpretation cycles and got its plausibility value increased by positive evaluation rules. Thus, the more plausible labeled region should have an advantage over the less plausible labeled region.

```
Local Positive Evaluation:
  IF a foliage and a tree trunk are adjacent
    and foliage mass center is higher than that of tree trunk
  THEN give each region 5 points

Local Negative Evaluation:
  IF a road and a tree trunk are adjacent
    and the trunk is highest among adjacent trunks
    and top of the road is higher than top of the trunk
    and usable <= road model plausibility
  THEN take 8 points from less plausible region
    and take 2 points from more plausible region
    if equally plausible, take 5 points from each
```

Figure 4.28: Local evaluation rule examples

Some global evaluation rules are in Figure 4.29. After a couple of interpretation cycles, the plausibility of the models becomes moderately high, and then the global evaluation starts in addition to the local evaluation. If a labeled region is consistent with a related model, the rule gives 10 points to the labeled region. If a labeled region is inconsistent with a related model, the rule takes 10 points from the labeled region. When the plausibility of the models becomes reliable, the four dominant negative global evaluation rules can be fired. If a labeled region is inconsistent with a related model, the dominant negative global evaluation rule takes 10000 points from the labeled region, so that the label is removed.

4.7.5 Modeling

Initially, one road model is already generated and ready to be adjusted as depicted in Figure 4.1 (a). The model is adjusted to the latest interpretation in each cycle

Global Positive Evaluation:
  IF a tree trunk is inside a tree model
    and carefully-usable <= model plausibility
    and it is dark area, darker area, or darkest area
    and it is not ground level
  THEN give the region 10 points

Global Dominant Negative Evaluation:
  IF a tree trunk is not inside any tree trunk model
    and reliable <= model plausibility
  THEN take 10000 points from the region

Figure 4.29: Global evaluation rule examples

by the functions written in C++. In each model adjustment, first, the vanishing point is moved horizontally until the discrepancy reaches a local minimum; then, the lower end of the left road edge is slid horizontally until the discrepancy reaches a local minimum; then, the lower end of the right road edge is slid horizontally; last, the vanishing point is moved vertically. This road model adjustment cycle is iterated one to three times depending on the plausibility of the road model.

The direction of the movement can be determined by checking only the labeled pixels along the boundary.

In the illustration, if the bottom of the left edge moves left, the discrepancy value is improved.

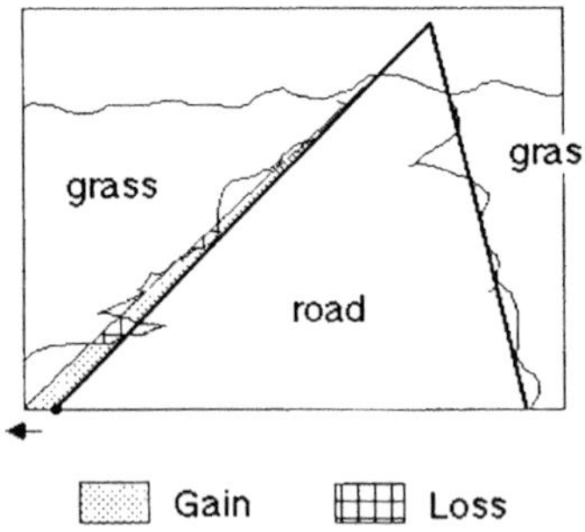

Figure 4.30: Model adjustment algorithm

Modeling is one of the most important components in the road scene interpreter. The explicit road scene models extract global information of the road scene, on which the evaluator and the extrapolator rely. If the global information of the models is wrong, the whole interpretation is confused and corrupted. Especially for the first one or two interpretation cycles, the models are unstable simply because there are only a few labeled regions. So, for the first few interpretation cycles, the models must be less sensitive until fairly many regions are labeled. We observed that the models were once oversensitive and adjusted very strangely. A couple of ideas are adopted in order to make the models less sensitive for the first few interpretation cycles:

Do Not Adjust a Road Model Edge Unless Grass or Soil is Found on That Side. Grass or soil regions prevent the road model from expanding. If there is no grass or soil region found on the right or left, the lower end of the right or left road model edge can expand very much. So, if no grass or soil is found, the lower end of the road model edge on that side should not be adjusted.

Restrict the Vanishing Point and Width of the Road Model Loosely. If the vanishing point moves far off the top edge of the image or very close to the bottom edge of the image, that is very strange. If the road width becomes twice as wide as usual, this is also unusual. These situations should be avoided for the first few interpretation cycles.

Though it belongs to domain knowledge organization, experimentally, a Y-intersection detection mechanism has been implemented to confirm the extendibility of the explicit scene models and the interpretation cycle. Figure 4.11 shows the basic idea. Since the road adjustment looks for a local minimum of the discrepancy value, a road model is adjusted to one of the intersecting roads. The other road makes the discrepancy value big. So, if there is a big discrepancy on either the right or the left, the modeler generates an additional road model in that direction. The modeler is also capable of eliminating erroneous additional road models. Elimination of an erroneous additional road model is easy since the vanishing point of the road model becomes very close to the main road model.

Adjustment of the tree trunk model is one degree of freedom after another: mass center row, mass center column, width, height, and lean. Each degree of freedom finds a local minimum of the discrepancy value, and this adjustment cycle takes place once in each interpretation cycle. Generation and elimination of the tree trunk models differ from those of the road models. Every time a fragment of a tree trunk is found, a tree trunk model is generated and adjusted independently. Thus, each tree trunk often has several tree trunk models at first. By checking the overlaps of the tree trunk models, some models are eliminated and one model remains for each tree trunk. If a tree trunk model becomes very small, it is also eliminated.

The explicit scene model adjustment is not expensive because it only tries to find a local minimum of the discrepancy value, as shown in Figure 4.30. Only the pixels along the model boundaries need to be checked to determine which direction the model should be moved. The required computing power for model adjustment is proportional to the number of degrees of freedom, since only one degree of freedom is adjusted at a time.

As the interpretation proceeds, the plausibility of the models becomes higher. Figure 4.31 shows the plausibility levels of the road model. It is determined by the area of the reliable regions whose plausibility value is greater than or equal to 10. At first, the plausibility level of the models is not usable since the models are not reliable. Then, it becomes carefully usable; the system uses the models but still checks the features of regions carefully. Then, it becomes usable; the system uses the models and checks just a few features of regions. Then, it becomes reliable; the system uses the models without checking the features of the models. Last, it becomes dominant; the system totally relies on the model and rejects all of the inconsistently labeled regions. Currently, only the plausibility of the main road model is calculated and represents all the plausibilities of the models.

The road model has 5 levels of plausibility.

	Retio of reliable regions in a scene (10<region plausibility value)
1. Unreliable	-
2. Carefully-usable	1/3
3. Usable	1/2
usually 1 interpretation cycle	
4. Reliable	1/2
2 interpretation cycles	
5. Dominant	-
(quit interpretation as soon as possible)	

Figure 4.31: Plausibility levels of the model

4.7.6 Extrapolation

Labeled regions, which have survived some interpretation cycles, propagate to their adjacent unlabeled regions by checking the location, shape, color, area, and so on. Extrapolation rules are simple and make easy guesses. Most of the guesses are good, but some of them might be wrong. They are judged by the evaluator. Figure 4.32 lists the categories of the extrapolation rules.

Category		Number of rules
Solution Class	higher/lower than vanishing level	6
	higher than a above-ground region lower than a on-the -level-of-ground region	2
	similar to	3
	surrounded by	1
	with sun direction	2
Solution	object A expects object B	9
	similar to	6
	surrounded by	5
	with sun direction	2
	by scene model	13
Total		49

Figure 4.32: Extrapolation rules

1/3 of the extrapolation rules infer solution classes. Two of them remove *ground level* solution class or *above ground* solution class. Removal of the solution classes is caused by the transition of the vanishing point of the road model. The rest generate solution classes. Figure 4.33 shows some examples.

2/3 of the extrapolation rules infer solutions. Figure 4.32 lists the categories of the extrapolation rules and some examples. Adjacency is a basic technique for extrapolation and is used by many of the rules. Two rules use the direction of the sun and try to propagate tree trunk regions to that direction, since the sunny side of a tree trunk is often bright and ambiguous. Half of the rules use scene models in their inferences.

One weather rule determines whether it is sunny or not by simply checking the existence of sunny spots or shadows. Then another rule detects the direction

Ground is lower than vanishing level:
  IF a region has no solution class
    and vanishing point level + 80 < MBRrowmax
    and usable <= road model plausibility
  THEN the region could be ground level

Road expects grass:
  IF an unlabeled region is adjacent to a road region
    and it is (green area or bright green area)
    and it is not lengthwise area
    and road model plausibility <= usable
  THEN the region could be grass

Figure 4.33: Extrapolation rule examples

of sun by finding an appropriate pair of a shadow and a tree trunk or vertical object. Unless shadows are vertical in the image, it turned out to be fairly easy to detect the direction of the sun and utilize it in interpretation.

4.7.7 Road Map Generation

Because explicit road scene models extract the vanishing point of a road and the roots of trees, a road map can be generated under the assumption that the roots of trees are on the ground level. If its view frame angle θ is known, then inverse projection becomes possible.

Figure 4.34 illustrates the inverse projection. Though the actual image plane is tilted, let us assume that the image plane (projection plane) is perpendicular to the ground. This assumption does not distort the map, since all the inversely projected objects are two-dimensional and on the ground level. Then all we have to know is where the center of projection is. In the plane view, the center of projection needs to be on the line of symmetry of the isosceles trapezoid view frame. The height of the center of projection must be the same as the level of the vanishing point of the road. The distance d between the projection plane and the center of projection is calculated as

$$d = \frac{b_{lower}}{2} \cot \frac{\theta}{2},$$

where b_{lower} is the length of the lower (shorter) base of the trapezoid view frame as depicted in Figure 4.34 (a). Note that if we draw a perpendicular line segment from the center of projection to the projection plane, the foot of this segment lies at the center of the image. Now the remaining unknowns are the height of the center of projection and the tilt angle of the projection plane. The plane determined by the center of projection and the level of the vanishing point in the image plane must be parallel to the ground plane. By tilting the projection plane until this plane becomes parallel to the ground plane, the height of the center of projection and the tilt angle are determined at the same time.

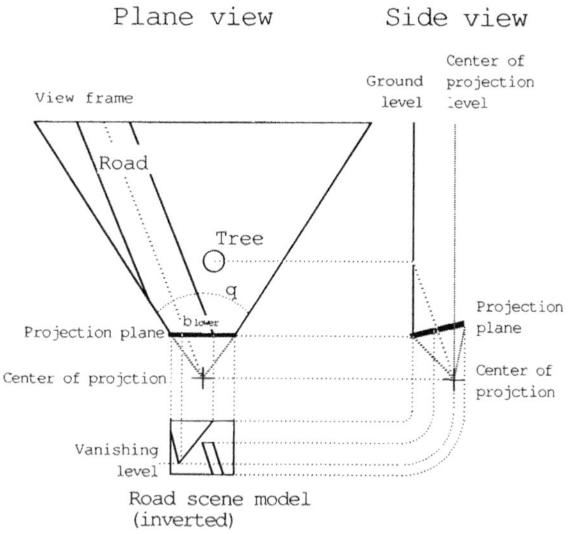

Figure 4.34: Inverse projection

4.7.8 System Analysis

Figure 4.35 shows the composition of rules. The initial hypothesis generator consists of 60 rules: 27 data abstraction, 11 heuristic match, and 22 refinement. This is the largest rule module. The extrapolator has 52 rules and is the second largest rule module. The modeler has only 16 rules, because most of it is written in C++ functions. If C++ functions are taken into account, the modeler is the biggest module. In total, the interpreter comprises 187 rules to date. The source code of the interpreter only has about 8300 lines, 2/3 in OPS83 and 1/3 in C++. If the source code of the feature extraction program is added, it exceeds 10000 lines. Since the main body of the interpreter is written in the rule-based programming language OPS83, it is still very maintainable.

There are two important components: Data abstraction and modeling. The data abstraction of initial hypothesis generation is the first knowledge base applied to the intermediate data. The very first stage always plays an influential

		Number of rules	
Context control		22	
Data Abstraction	Initial Hypothesis Generation	27	60
Heuristic match		11	
Refinement		22	
Evaluation		33	
Modeling		16	
Extrapolation		52	
Map generation		4	
Total		187	

Figure 4.35: The composition of rules

role in a total system. It connects intermediate-level and high-level, and still preserves parallelism like low-level image processing because rules in the data abstraction are independent. Modeling is significant since adjusted scene models become dominant through evaluation and extrapolation in the latter half of the interpretation. Once the models start converging, everything goes well. Every effort in knowledge acquisition is made such that the scene models converge. The first task of the knowledge engineer is, by looking at a new road image, to adjust a triangular road model on it in his mind. No matter how much the road is shaded, if the road model can be adjusted properly, it is worth working on the image.

Negative evaluation is not troublesome, as opposed to intuition. Actually, evaluation is the smallest of the four modules, as shown in Figure 4.35 (the modeler is the largest since its functions are in C++). There are three main reasons:

1. Global negative evaluation is reliable as long as models are properly adjusted.
2. Since local negative evaluation cannot avoid mistakes, the simple strategy of Section 4.7.4 is nearly optimal.
3. The global strategy of the system, based on making the easy guesses and iterating through the interpretation cycle, alleviates the evaluator's task.

The four modules, initial hypothesis generation, evaluation, modeling, and extrapolation, are in a certain balance. None of them is perfect. None of them is stupid. They help one another and perform reasonable interpretation. So, when modifying the system, the balance needs to be taken into account. Another aspect of the system is that one rule could be fired more than 100 times. For example, there can be more than 20 pairs of road and grass regions in a scene. A road-grass positive evaluation rule is fired on each pair of these regions in every interpretation cycle in order to keep them alive. This characteristic is very different from a typical production system.

Since the system adopted a cyclic mechanism, various phenomena such as oscillation, overshoot, divergence, etc. studied in control theory can potentially take place. Oscillation and divergence were observed during development.

4.8 Future Work

To enhance the initial hypothesis generator, a couple of potentially effective improvements are discussed here.

4.8.1 Inexhaustive Region Segmentation for a High-level Interpreter

Intermediate representation data need to be appropriately abstracted in order to provide good initial hypotheses for the following interpretation stages. Conversely, it is also true that perfect initial hypotheses cannot be generated locally. However, at least a few significant regions, in terms of perceptual organization [12], must be properly abstracted bottom-up as initial clues, otherwise there is no way to interpret a natural color scene without predictions. Even for a human, if a scene is very dark and does not give good initial clues, interpretation can be difficult. Therefore, pursuing reliable initial hypothesis generation bottom-up is a correct way to more robust natural color scene interpretation.

From the view point of a high-level scene interpreter, little consideration has been given to low-level region segmentation. The objective of low-level region segmentation has been roughly to segment an image as well as a human. This objective automatically includes semantics, which is not very successful [10]. But an ideal region segmentation for a high-level interpreter can be very different. The objective of low-level region segmentation should be how pure are the low-level cues it can give to a high-level interpreter. In the case of the system INSIGHT-III, the performance of low-level segmentation must be measured by the quality of the data abstraction it provides to the initial hypothesis generation.

We observed that the average colors of segmented regions differ from those of the corresponding portions of the original image. This is caused by the heterogeneous pixels along the boundaries and in the portions of a region that are undersegmented. If heterogeneous pixels are mixed, the averaged color tends to be neutral. This color contamination deforms the color cues and makes data abstraction erroneous. Since region boundaries are not important to a high-level interpreter with adjustable explicit scene models, exhaustive region segmentation loses its meaning. The objective of this region segmentation is to grab very homogeneous regions for reliable data abstraction. So, the pixels along boundaries are not segmented.

Inexhaustive region segmentation gives rise to some questions. *First, what is the image interpretation result, labeled regions or adjusted models?* It depends on the application. For autonomous vehicle navigation, adjusted explicit models or a road map is apparently better. If the global information extracted from inexhaustively segmented regions is sufficient, exhaustive region segmentation is no longer necessary. Conversely, some applications may need labeled regions. In this case, the image can be exhaustively segmented afterwards with the global information extracted. *Second, can explicit scene models be adjusted appropriately?* Basically, there is a local minimum of the discrepancy value near the best match. Although the gaps between regions may create some weak local minima, they can be overcome by modifying the adjustment algorithm slightly. This will be much easier than unlabeling misinterpreted regions caused by color contamination. *Third, how can relations among regions be expressed?* Adjacency can be defined among unconnected regions. For example, if two regions are within a certain distance at the closest points and no other region interferes with them, then those regions are adjacent. *Last, won't it lose edge information?* Edges are separately extracted with the color edge detector.

4.8.2 Adaptive Data Abstraction

Since the interpreter is capable of recognizing interpretation failure by checking the consistency of adjusted scene models or detecting infinite iteration of interpretation cycles, the parameters for data abstraction can be changed several times until the interpreter gets reasonably adjusted scene models. This meta-interpretation cycle works as a mechanism for recovering from fatal interpretation failure. If model parameters are limited, it becomes feasible to build such a mechanism. While developing INSIGHT-III, many efforts were made on differentiating road and soil regions, which was important for road extraction. Often these two regions are similar. Both are *wound level* and have little texture. The only cue is the slight difference of the colors: soil is a little redder and darker than road, though it is not always so because of camera condition, color contamination, noise, etc. This difficulty will be alleviated by adaptive data abstraction.

4.9 Conclusion

An outdoor natural color road scene interpreter with adjustable explicit scene models and an interpretation cycle has been developed, and more than 20 natural road scenes, some destructively shaded, have been reliably interpreted. By moderately simplifying the low-level region segmentation and domain knowledge organization, and intentionally selecting ill-structured scenes, some of the issues of robust natural color scene interpretation have been clarified:

1. Global information of a natural scene, which makes the interpretation substantially reliable, can be extracted by adjustable scene models and a interpretation cycle during interpretation.
2. Negative evaluation is not troublesome. It performs well in concert with global information extraction.
3. The classification problem solving model is appropriate as an initial hypothesis generator in natural scene interpretation.
4. Color contamination caused by the heterogeneous pixels along boundaries and undersegmented portions of a region weakens initial hypothesis generation.

4.10 Acknowledgement

The writers wish to thank Thomas Chen for region feature extraction, Chuck Thorpe and Jill Crisman for fully-automatic region segmentation, Steven Shafer and Karl Kluge for PHOENIX, Carol Novak for the color edge detector, Hirobumi Miwa for his line extractor, and the members of the Vision and Autonomous Systems Center at Carnegie-Mellon University for their valuable comments and discussions.

4.11 References

[1] T.O. Binford. Survey of model-based image analysis systems. In *The International Journal of Robotics Research*, pages 18–64, 1982.

[2] R.A. Brooks. Symbolic reasonig among 3-D models and 2-D images. In *Artificial Intelligence 17*, pages 285–348, 1981.

[3] L. Brownston, R. Farrell, E. Kant, and N. Martin. *Programming Expert Systems in OPS5*. Addison-Wesley, Reading, Massachusetts, 1985.

[4] J. Canny. A computational approach to edge detection. In *IEEE Transactions on Pattern Analysis and Machine Intelligence*, pages 679–698, November 1986.

[5] W.J. Clancey. Classification problem solving. In *Proceedings of the National Conference on Artificial Intelligence*, pages 49–55, Austin, Texas, 1984.

[6] B.A. Draper, R.T. Collins, J. Brolio, J. Griffith, A.R. Hanson, and E.M. Riseman. Tools and experiments in the knowledge-directed interpretation of road scenes. In *Proceedings of the DARPA Image Understanding Workshop*, pages 178–193, February 1987.

[7] B.A. Draper, A.R. Hanson, and E.M. Riseman. *A Software Environment for High Level Vision*. Technical Report, University of Massachusetts, 1986.

[8] C.L. Forgy. *The OPS83 Report*. Technical Report CMU-CS-84-133, Department of Computer Science, Carnegie-Mellon University, May 1984.

[9] A.R. Hanson and E.M. Riseman, editors. *VISIONS: A Computer System for Interpreting Scenes*, pages 303–333. Academic Press, New York, 1978.

[10] T. Kanade. Region segmentation: signal vs semantics. In *Computer Graphics and Image Processing 13*, pages 279–297, 1980.

[11] K.I. Laws. *The PHOENIX Image Segmentation System: Description and Evaluation*. Technical Report 289, SRI International, December 1982.

[12] D.G. Lowe and T.O. Binford. The perceptual organization of visual images: segmentation as a basis for recognition. In *Proceedings of the DARPA Image Understanding Workshop*, pages 203–209, 1983.

[13] J. McDermott. Making expert systems explicit. In *IFIP Congress*, pages 1–10, Austin, Texas, 1986.

[14] D.M. McKeown, Jr. and W.A. Harvey. Automating knowledge acquisition for aerial image interpretation. In *Proceedings of the DARPA Image Understanding Workshop*, pages 205–226, February 1987.

[15] D.M. McKeown, Jr., W.A. Harvey, and J. McDermott. Rule-based interpretation of aerial imagery. In *IEEE Transactions on Pattern Analysis and Machine Intelligence*, pages 570–585, September 1985.

[16] H. Miwa and T. Kanade. *Straight Line extraction*. Technical Report, Robotics Institute, Carnegie-Mellon University, February 1988. [in preparation].

[17] M. Nagao and T. Matsuyama. *A Structural Analysis of Complex Aerial Photographs*. Plenum, New York, 1980.

[18] C.L. Novak and S.A. Shafer. Color edge detection. In *Proceedings of the DARPA Image Understanding Workshop*, pages 35–37, February 1987. Section 2 of the PI-report: "Image Understanding Research at CMU" by Takeo Kanade.

[19] Y. Ohta. *Knowledge-based Interpretation of Outdoor Natural Color Scenes*. Pitman Publishing, 1985.

[20] J. Pearl. *Distributed Revision of Composite Beliefs*. Technical Report CSD 860045 R-64-III, Cognitive Systems Laboratory, UCLA Computer Science Department, April 1987.

[21] S.A. Shafer and T. Kanade. Recursive region segmentation by analysis of histograms. In *Proceedings of the International Conference on Acoustics, Speech, and Signal Processing*, pages 1166–1171, Paris, May 1982.

[22] Y. Shirai. *Recognition of man-made objects using edge cues*. Academic Press, New York, 1978.

[23] B. Stroustrup. *The C++ Programming Language*. Addison-Wesley, Reading, Massachusetts, 1986.

[24] C. Thorpe and J. Chrisman. *Region segmentation based on color proximity*. Technical Report, Robitics Institute, Carnegie-Mellon University, 1988. [in preparation].

[25] D.Y. Tseng and J.F. Bogdanowicz. Image understanding technology and its transition to military applications. In *Proceedings of the DARPA Image Understanding Workshop*, pages 310–312, February 1987.

Chapter 5: Neural Network Based Autonomous Navigation

Dean A. Pomerleau

5.1 Introduction

Autonomous navigation has been a difficult problem for traditional vision and robotic techniques, primarily because of the noise and variability associated with real world scenes. Autonomous navigation systems based on traditional image processing and pattern recognition techniques often perform well under certain conditions but have problems with others. Part of the difficulty stems from the fact that the processing performed by these systems remains fixed across various driving situations.

Artificial neural networks have displayed promising performance and flexibility in other domains characterized by high degrees of noise and variability. Examples include handwritten character recognition [3] [5] and speech recognition [10]. ALVINN (Autonomous Land Vehicle In a Neural Network) is designed to bring the flexibility of connectionist techniques to the task of autonomous navigation. Specifically, ALVINN is an artificial neural network designed to drive the Navlab, the Carnegie Mellon autonomous navigation test vehicle.

5.2 Network Architecture

ALVINN's current architecture consists of a single hidden layer back-propagation network (See Figure 5.1). The input layer of the network consists of a "retina" 30x32 units onto which a video camera image is projected. The activation level of each unit in this retina is computed from the corresponding patch of the video image by using the following formula.

$$Activation = \frac{b}{255} + \frac{b}{r+g+b}$$

where r, g and b represent the red, green and blue color intensities of the corresponding patch of the image. The blue band of the color image is used as a baseline because it provides the highest contrast between the road and the non-road. The first term in the equation serves to enhance the road vs. non-road contrast in situations where there is little color but significant intensity variation between the road and the non-road. The second term serves to reduce the effect of shadows and other lighting variation on the preprocessed image by normalizing the absolute amount of blue in a patch of the image by the brightness of that patch.

Each of the 960 units in the input retina is fully connected to the hidden layer of 9 units, which is in turn fully connected to the output layer. The output

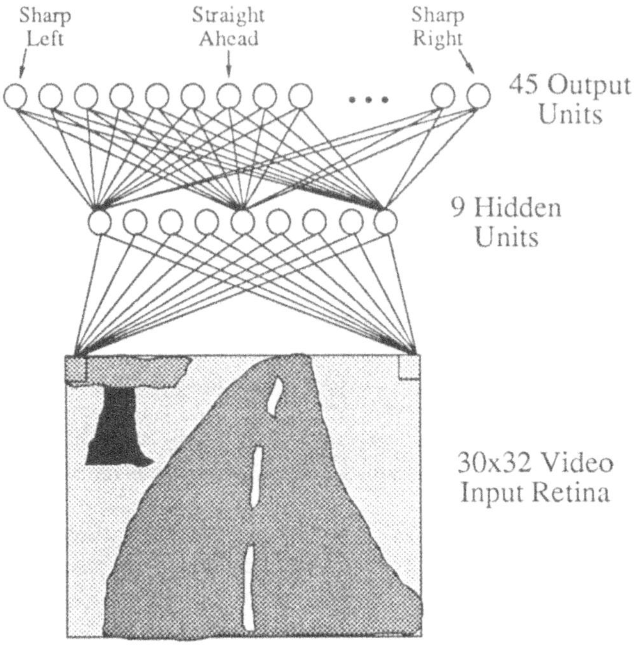

Figure 5.1: ALVINN Architecture

layer consists of 45 units and is a linear representation of the direction the vehicle should travel in order to keep the vehicle on the road. The centermost output unit represents the "travel straight ahead" condition, while units to the left and right of center represent successively sharper left and right turns.

To drive the Navlab, a video image from the onboard camera is preprocessed and injected into the input layer. After completing a forward pass through the network, a steering command is read off the output layer. The most active output unit determines the direction in which to steer the vehicle.

5.3 Training And Performance

ALVINN is trained using the back-propagation neural network learning procedure. In particular, the network is presented with a road image as input and a forward pass through the network is performed in order to determine the network's "answer" for that particular image. If the network's answer is correct (i.e. if the network dictates the correct steering direction), the weights of the connections between units in the network are left unchanged. If the network's answer is incorrect, the weights in the network as altered so that next time this or a similar road scene is presented, the network's steering direction will be closer to the correct direction. The network is trained to perform pure pursuit planning, which means the correct steering direction is defined to be that one which will bring the vehicle to the road center a fixed distance (usually 10 meters) ahead of its current position. This training procedure is applied repeatedly with a number of training images in order to teach the network to drive under a variety of conditions. For a more detailed description of the back-propagation learning procedure, see[8].

The source of training data has evolved substantially over the course of the project. Training was originally performed using simulated road images designed to portray roads under a wide variety of weather and lighting conditions. The network was repeatedly presented with 1200 synthetic road scenes, while the weights between units in the network were adjusted by the back-propagation algorithm, using the Warp systolic array supercomputer [7]. The network required between 30 and 40 presentations of these 1200 synthetic road images in order to develop a representation capable of accurately driving over the single-lane Navlab test road. Once trained, the network was able to drive the Navlab at up to 1.8 meters per second (3.5 MPH) along a 400 meter path through a wooded area of the CMU campus under a variety of weather conditions including snowy, rainy, sunny and cloudy situations.

Despite its apparent success, this training paradigm had serious drawbacks. From a purely logical standpoint, generating the synthetic road scenes was quite time consuming, requiring approximately 6 hours of Sun-4 CPU time. Once the roads scenes were generated, the 30-40 presentations of the 1200 images to train the network required an additional 45 minutes of Warp computation time. In addition, differences between the synthetic road images on which the network was trained and the real images on which the network was tested often resulted in poor performance in real driving situations. For example, when the network was trained on synthetic road images which were less curved than the test road, the network would become confused when presented with a sharp curve during testing. Finally, while relatively effective at training the network to drive under the limited conditions of a single-lane road, it quickly became apparent that extending the synthetic training paradigm to deal with more complex driving situations like multi-lane and off-road driving would require prohibitively complex synthetic training data generators.

To deal with these problems, we have developed a new training scheme which involves teaching the network to imitate a human driver under actual driving conditions. As a person drives the Navlab, back-propagation is performed to train the network using the current video camera image as input and the direction in which the person is currently steering as the desired output.

There are two problems associated with this scheme of using solely the current video image and steering direction for training "on-the-fly". First, since the human driver steers the vehicle down the road center during training, the network will never be presented with situations where it must recover from mistakes, such as being slightly off the road center. When driving for itself, the network is bound to eventually make a mistake, so it should be prepared to recover by steering the vehicle back to the road center. The second problem

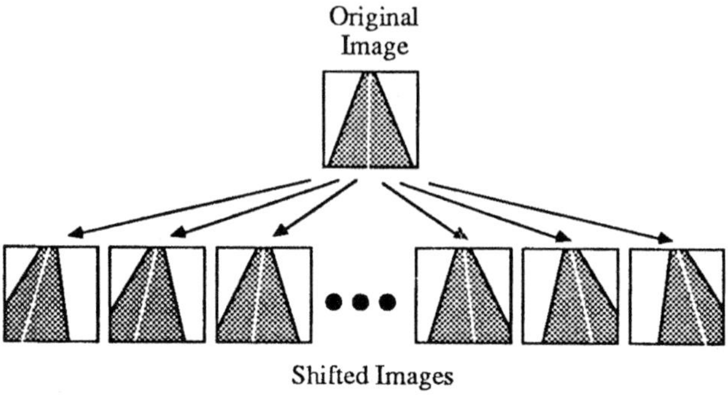

Figure 5.2: The single original video image is laterally shifted to create multiple training exemplars.

with naively training the network with only the current video image and steering direction is the danger of overlearning from repetitive inputs. If the human trainer drives the Navlab down a straight stretch of road during part of a training run, the network will be presented with a long sequence of similar images. This sustained lack of diversity in the training set will cause the network to "forget" what it had learned about driving on curved roads and instead learn to always steer straight ahead.

Both problems associated with training on-the-fly stem from the fact that back-propagation requires training data which is representative of task to be learned. To provide the necessary variety of exemplars while still training on real data, we have modified the simple training on-the-fly scheme described above. Instead of presenting the network with only the current video image and steering direction, we shift each original image laterally to create 10 additional images in which the vehicle appears to be shifted by various amounts relative to the road center (See Figure 5.2). The correct steering direction as dictated by the driver for the original image is altered for each of the shifted images in order to account for the extra lateral vehicle displacement in each. By creating multiple shifted training exemplars from the one original, we've eliminated the problem of the network never being presented with situations from which recovery is required. Also, overtraining on repetitive images will be less of a problem, since the shifted training exemplars will add variety to the training set. However as additional insurance against the effects of repetitive inputs, we further increase the training set diversity by maintaining a buffer of recently encountered road scenes.

So training on-the-fly in practice works as follows. A video image is digitized and preprocessed. This single original image is shifted 5 times to the left and 5 times to the right in 0.2 meter increments to create 11 new training exemplars. Eleven old exemplars from the current training set of 100 road scenes are randomly chosen and replaced by the 11 new exemplars. One forward and one backward sweep of the back-propagation algorithm is performed on these 100 exemplars to incrementally updates the network's weights, and then the process is repeated.

The network requires approximately 50 iterations through the above cycle to learn to drive on the test road. Running on a Sun-4, this takes approximately five minutes during which a person drives at 2 miles per hour over about 200 meters of the single lane test path. After this training phase, not only can the network imitate the person's driving along the same stretch of road, it can also generalize to drive along parts of the path it has never encountered under a wide variety of weather conditions. In addition, since determining the steering direction from the input image merely involves a forward sweep through the network, the system is able to process slightly over 2 images per second, allowing it to drive the Navlab at speeds up to 8.0 miles per hour. This is over 3 times faster than any other sensor-based autonomous system has driven the Navlab.

5.4 Network Representation

The representation developed by the network to perform the road following task depends dramatically on the characteristics of the roads it is trained on. When trained on examples of roads with a fixed width, the network develops a representation in which hidden units act as filters for roads at different positions. Figures 5.3 and 5.4 are diagrams of the weights projecting to and from single hidden units in such a network.

In Figure 5.3, the weights from the video camera retina to this hidden unit support the interpretation that this hidden unit is a filter for a single road on the left side of the visual field (See the small schematic to the right of the weights from the video retina in Figure 5.3). The single road filter interpretation is reflected in the weights from this hidden unit to the direction output units. Specifically, this hidden unit makes excitatory connections to the output units on the far left, dictating a sharp left turn to bring the vehicle back to the road center.

Figure 5.4 illustrates the weights to and from a hidden unit with a more complex representation. This hidden unit acts as a filter for two roads, one slightly left and one slightly right of center. The weights from the video camera retina along with the explanatory schematic in Figure 5.4 show the positions and orientations of the two roads. This hidden unit makes bimodal excitatory connections to the direction output units, dictating a slight left or slight right turn. Hidden units which act as filters for one to three roads are the representation structures most commonly developed when the network is trained on roads with a fixed width.

The network develops a very different representation when trained on images with widely varying road widths. A typical hidden unit from this type of representation is depicted in Figure 5.5. The first important feature to notice in Figure 5.5 is the single abrupt change in weights along the diagonal line on the left side of the weights from the video camera retina to this hidden unit. This

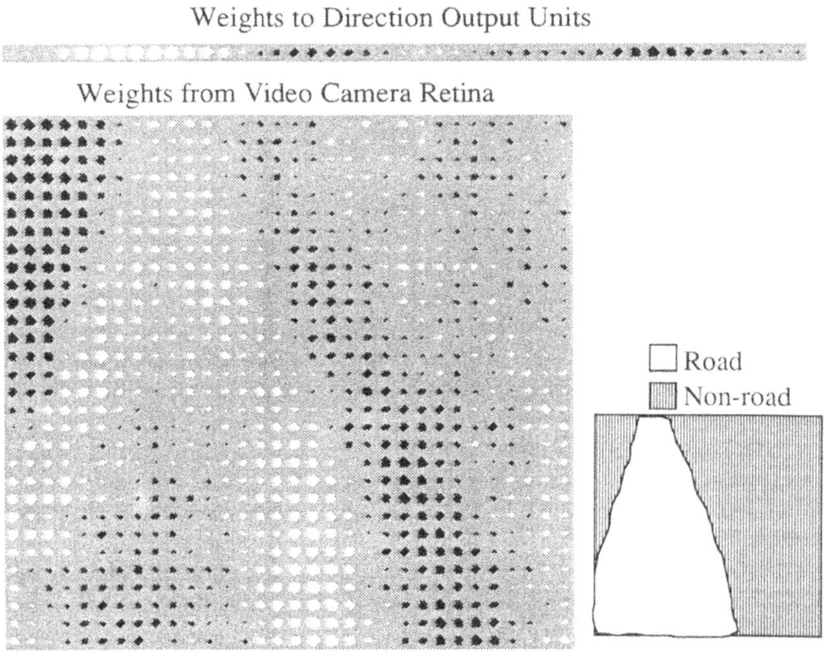

Figure 5.3: Diagram of weights projecting to and from a typical hidden unit in a network trained on roads with a fixed width. This hidden unit acts as a filter for a single road on the left side of the visual field as illustrated by the schematic.

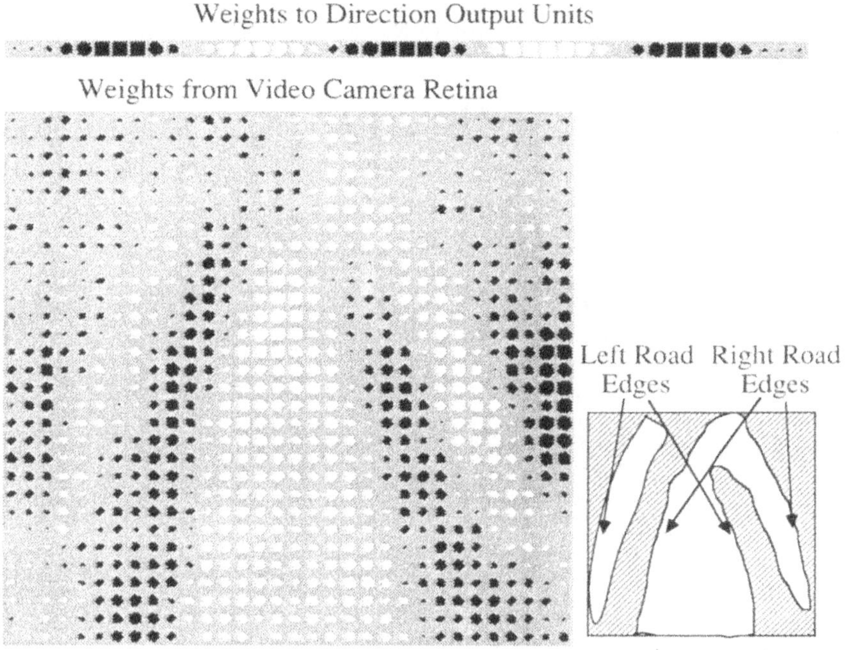

Figure 5.4: Diagram of weights projecting to and from a typical hidden unit in a network trained on roads with a fixed width. This hidden unit acts as a filter for two roads, one slightly left and one slightly right of center, as illustrated by the schematic.

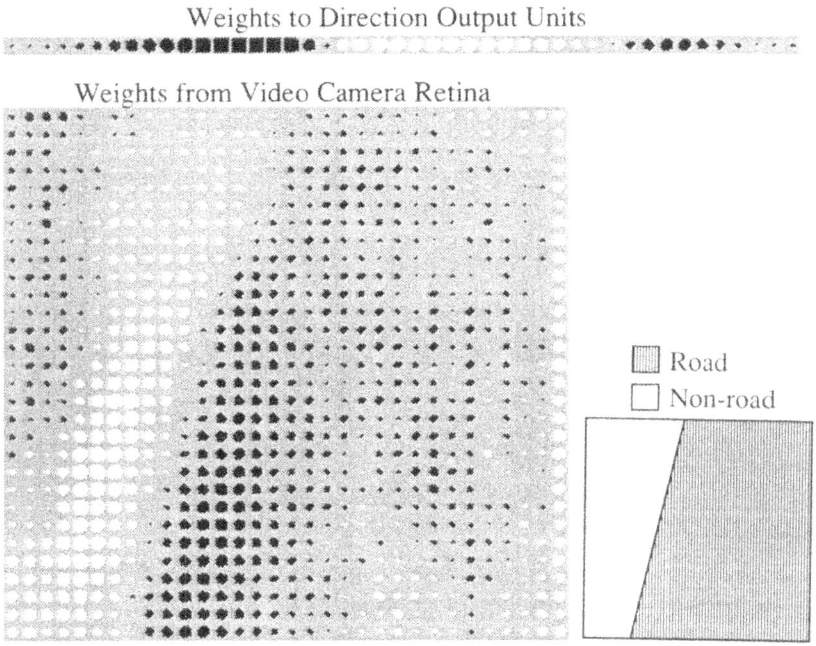

Figure 5.5: Diagram of weights projecting to and from a typical hidden unit in a network trained on roads with different widths. This hidden unit acts as a detector for the left road edge, as illustrated by the schematic.

edge clearly indicates that this hidden unit is a filter for the left edge of the road (See the schematic to the right of the video retina weights). In addition, notice that unlike in the earlier weight diagrams, the weights from the units which are supposed to represent the road are inhibitory (black) rather than excitatory (white). This pattern indicates that this hidden unit responds to roads which are darker, and hence have lower retina activation levels than the non-road. This illustrates that when the training set requires it, the network will develop units to deal with roads which can be either darker or lighter than the non-road.

The hidden unit in Figure 5.5 supports a rather wide range of travel directions, as indicated by the large number of excitatory connections it makes to the output layer. This is to be expected, since the correct travel direction for a road with an edge at a particular location varies substantially depending on the road's width. This hidden unit would cooperate with hidden units that detect the right road edge to determine the correct travel direction in any particular situation.

5.5 Discussion And Extensions

The distinct representations developed for different circumstances illustrate a key advantage provided by neural networks for autonomous navigation. Namely, in this paradigm the data, not the programmer, determines the salient image features crucial for accurate navigation. This data responsiveness allows ALVINN to quickly learn from example the driving skills which must be explicitly programmed into traditional autonomous navigation systems.

Not only are individual driving networks easier and faster to develop than image processing systems designed for the same task, they are at least theoretically better suited for driving in the circumstances under which they are trained. This "well-suitedness" results from the fact the the back-propagation algorithm keys on the features which are statistically most highly correlated with the correct steering direction when tailoring the network to perform the task. This is illustrated by the distinct internal representations developed when the network is trained on roads of fixed vs. variable width. For fixed width roads, the network found the relevant feature for road following to be road-shaped regions at different positions and orientations in the image, and therefore used these features as its internal representation. For variable width roads, the network learned to key on single road edges at particular positions and orientations, since these are a more efficient representation than road-shaped regions when road width varies substantially.

This responsiveness of processing to relevant features is in contrast with other autonomous navigation systems at CMU [9] and elsewhere [2] [1] [4]. Each of these implementations has relied on a fixed, highly structured and therefore relatively inflexible algorithm for finding and following the road, regardless of the conditions at hand.

We hope to make a major test of our flexibility claim in the near future by training ALVINN under very different circumstances including multi-lane and off-road driving. We have already achieved promising results in the multi-lane driving domain using images stored on disk. Specifically, when trained on a sequence of images of a two-lane road with a line down the middle, the network learns that the center line is the important feature to attend to for maintaining the correct vehicle orientation, and therefore develops hidden units that detect and use the position of the center line to determine the correct steering direction.

However a truly robust autonomous navigator requires not only the ability to deal with a number of different driving situations separately, it also requires the ability to combine situation specific driving capabilities into a single, general navigation system. Developing such a system is the long term goal of the ALVINN project. We are beginning to explore both rule-based and connectionist techniques for arbitrating between multiple symbolic and neural network knowledge sources.

5.6 Conclusion

Artificial neural networks appear well suited to the task of road following. ALVINN's ability to develop relevant internal representations and to learn on-the-fly from example provide a flexibility which we believe to be unique among autonomous navigators. Much work remains in determining how to exploit this flexibility to develop a more general and robust autonomous driving system.

5.7 References

[1] Dickmanns, E.D., Zapp, A. (1986) A curvature-based scheme for improving road vehicle guidance by computer vision. *"Mobile Robots", SPIE-Proc. Vol. 727*, Cambridge, MA.

[2] Dunlay, R.T., Seida, S. (1988) Parallel off-road perception processing on the ALV. *Proc. SPIE Mobile Robot Conference*, Cambridge MA.

[3] Jackel, L.D., Graf, H.P., Hubbard, W., Denker, J.S., Henderson, D., Guyon, I. (1988) An application of neural net chips: Handwritten digit recognition. *Proceedings of IEEE International Conference on Neural Networks*, San Diego, CA.

[4] Kuan, D., Phipps, G. and Hsueh, A.-C. Autonomous Robotic Vehicle Road Following. *IEEE Trans. on Pattern Analysis and Machine Intelligence, Vol. 10*, Sept. 1988.

[5] Pawlicki, T.F., Lee, D.S., Hull, J.J., Srihari, S.N. (1988) Neural network models and their application to handwritten digit recognition. *Proceedings of IEEE International Conference on Neural Networks*, San Diego, CA.

[6] Pomerleau, D.A. (1989) ALVINN: An Autonomous Land Vehicle In a Neural Network, In *Advances in Neural Information Processing Systems, Vol. 1*, D.S. Touretzky (ed.), Morgan Kaufmann.

[7] Pomerleau, D.A., Gusciora, G.L., Touretzky, D.S., and Kung, H.T. (1988) Neural network simulation at Warp speed: How we got 17 million connec-

tions per second. *Proceedings of IEEE International Conference on Neural Networks*, San Diego, CA.

[8] Rumelhart, D.E., Hinton, G.E., and Williams, R.J. (1986) Learning internal representations by error propagation. In D.E. Rumelhart and J.L. McClelland (Eds.) *Parallel Distributed Processing: Explorations in the Microstructures of Cognition. Vol. 1: Foundations.* Cambridge, MA: Bradford Books/MIT Press.

[9] Thorpe, C., Herbert, M., Kanade, T., Shafer S. and the members of the Strategic Computing Vision Lab (1987) Vision and navigation for the Carnegie Mellon Navlab. *Annual Review of Computer Science Vol. II*, Ed. Joseph Traub, Annual Reviews Inc., Palo Alto, CA.

[10] Waibel, A., Hanazawa, T., Hinton, G., Shikano, K., Lang, K. (1988) Phoneme recognition: Neural Networks vs. Hidden Markov Models. *Proceedings from Int. Conf. on Acoustics, Speech and Signal Processing*, New York, New York.

Chapter 6: Car Recognition for the CMU Navlab

Karl Kluge, Takeo Kanade, and Hideki Kuga

6.1 Introduction

LASSIE (*L*andmark *A*cquisition *S*ub*S*ystem *I*mproving *E*xploration) is a program which recognizes a car from several points of view. Trapezoids are formed from groups of lines extracted from a color image; instances of the car model are found by searching for sets of trapezoids which fit the constraints between parts of the car such as the windows, roof, and trunk. No information is used about the dimensions of the various parts, setting LASSIE apart from many other programs which use features extracted from images to recognize objects. [1] The remainder of the introduction sketches out the varied functions of object recognition in the navigation system of an autonomous vehicle and discusses the choice of domain for this work. The next section of the report discusses related work, followed by a section giving a detailed description of the LASSIE program. Results are shown for a number of images. The report closes with a discussion of directions for further work on LASSIE.

6.1.1 The function of object recognition in autonomous navigation

An autonomous vehicle navigating through the world may need to recognize a variety of natural and man-made objects for a number of reasons:

- to confirm or refine its internal estimate of its position in the world (*Am I really where I think I am?*)
- to make decisions about mission execution based on the presence or absence of objects in specified locations (*I should halt and look for conflicting traffic if there is an octagonal red sign at the intersection I'm approaching.*)
- to gather information for report at the end of a mission (*I should count the number of cars I see as I go along this stretch of road.*)
- to update the navigation system's internal model of the world (*Remember there was a stop sign at that intersection so I'll know that I should plan to stop and look for traffic the next time I approach it.*)

---

[1] An earlier version of this paper appeared as part of CMU Robotics Institute report CMU-RI-TR-89-5. Reprinted by permission.

- to make its way along the path specified by its mission. Humans tend to give directions to an objective in high level qualitative terms *(go down the road until you come to the third traffic light, turn right, keep going until you see the a big church on the left, then turn left at the next street...)* rather than giving them in terms of precise distances *(go down the road 0.75 miles, turn right, go another 2.3 miles, turn left...)*.

A navigation system may select an object to serve as a landmark in response to internal goals. As an example, the navigation system might realize that its wheels were slipping and choose a stationary object in the environment in order to keep track of its position relative to the object. Alternatively, the mission may specify objects for which the system is to watch. Examples of this would be "turn right at the third stop sign," or "when you get to the big yellow marker you will be at world position (x, y, z): update your internal position estimate accordingly."

Objects to be recognized may be artificial (such as road intersections, buildings, and vehicles), or they may be natural (such as distinctive rock formations or trees or rivers). The objects may have semantic meaning (this is a car, this is a stop sign, this is a tree), or they may simply be distinctive structures detected by the navigation system's perceptual modules and tracked through time without identifying them as any particular kind of thing.

6.1.2 Choice of domain

Object recognition may be called upon by a broad range of capabilities built into a navigation system, as discussed in the previous section. The Sidewalk Navigation System developed for the Terregator at CMU [5] used segmentations of color and range images to recognize stairs and intersections in order to navigate around a network of campus sidewalks. In order to test other types of capabilities requiring object recognition, such as deciding which way to turn at an intersection based on the presence or absence of a specified object, it was necessary to have a program capable of finding other kinds of objects. The planning part of such capabilities can be tested and demonstrated using simple objects (such as orange traffic cones). A module capable of recognizing a more challenging type of object is more desirable, both as a much more challenging recognition problem to solve, and as more likely to bring out unexpected interactions between parts of the navigation system as the recognition module competes for system resources.

Cars seemed an attractive domain for a number of reasons. The main nontechnical reasons were that cars are fairly common objects which are easily moved and weather proof, making possible the easy collection of a variety of images to use as data. The main technical reasons were

- cars differ a great deal in size and appearance (think of the differences between the appearance of a station wagon versus a four-door sedan, versus a compact car), but the variation of appearance is much less than for a domain such as tree recognition;
- there is a great deal of structure in the edges projected into an image from a car, with many of the edges forming quadrilaterals corresponding to meaningful parts of the car;
- the significant linear features that form a car can be detected reasonably well by current line extraction techniques, but not so well that the system can assume perfect or close to perfect edge data.

6.1.3 The goals and state of the research

The goal of this research is to produce an object recognition program that

- is capable of extracting the significant geometric structures (in this case, trapezoids and their subclasses such as parallelograms and rectangles) formed by the edge segments extracted from an image;
- uses the relationships between those structures and constraints from an object model to find instances of the object;
- does not use more than the crudest of information about the actual dimensions of the objects (for instance discriminating station wagons from compacts based on the length of the body relative to the length of the hood, or stating that the trunk of a car projects into the image as a parallelogram, which imposes some crude constraints on the dimensions of sides of the trunk).

This approach is in contrast to many programs that perform object recognition based on a precise or parameterized model of the shape of the object(s) to be recognized. With an exact shape model, or even a parameterized model, a previously unseen car may not be recognizable. The hope is that using only the crudest kind of metrical information about the objects will allow a more generic recognition capability.

A variety of images of a particular car were used in testing the program as it evolved to its current state. The program demonstrates a fair amount of robustness in processing those images, as will be seen from the results in section 6.4. Recognition of different makes of cars has not been attempted, and will undoubtedly require a fair amount of modification of the model of the parts of the car, but should be possible within the framework that has been developed.

6.2 Related work

In recent years much of the research in object recognition has been concerned with the recognition of objects based on data from 3-D sensors such as laser range finders. For an extensive survey of such work see Besl and Jain [1].

Brooks [2] describes ACRONYM, a system which recognized objects (represented as collections of parameterized generalized cones) using edges extracted from an intensity image. Edges were collected into "ribbons" which represented contours of generalized cones. A constraint management system (CMS) was used to verify the consistency of hypothesized matches between "ribbons" extracted from the image and object parts. A match was consistent if there was some viewpoint from which the match could have been imaged, with constraints on possible camera position accumulated based on the parameters of the generalized cones matched and the parameters of the corresponding image ribbons.

Goad [4] created a system which precompiled an exact description of the lines forming an object into a recognition program for that object. The programs would search the lines extracted from an image for matches. The object did not have to be a polyhedron, it simply had to have enough linear features for matching to take place. Matching was done via a simple backtracking search which checked sets of lines for matches based on their being consistent with some viewpoint. Rather than use a CMS as ACRONYM did, the program made assumptions which allowed it to map possible viewpoints onto a tessellation of the unit sphere. The set of cells corresponding to possible viewing directions would be restricted to reflect the new constraints on camera position as lines were added to a match.

SCERPO (Spatial Correspondence, Evidential Reasoning, and Perceptual Organization), described in Lowe [7], used estimates of the likelihood of accidental alignments to look for pairs of parallel lines and pairs of lines whose endpoints were close enough to suggest their being connected. It then looked for pairs of parallel lines connected to other pairs of lines to form trapezoids. These trapezoids were matched to parts of the object being looked for to give an initial pose estimate. The program then searched for additional lines to confirm/refute the initial match and refine the pose estimate. Like Goad's system, it required an exact model of the geometry of the lines forming the object.

Thompson and Mundy [9] describe a scheme for recognizing objects using a primitive called the "vertex pair." Approximating the projection from model to image by an affine transform, a match between a vertex pair in the model and its projection in the images completely specifies the pose of the object. Matches are found by a fairly brute-force voting scheme in the six dimensional parameter space. Once again, an exact description of the object's geometry is required.

Ohta's system [8] used color, texture, shape, and adjacency properties of regions extracted from a color image to label the regions as parts of various kinds of objects such as road, car, tree, or building. The program had no concept of the structure of the objects in a scene — a set of adjacent regions labeled "building" might be parts of one building or of several buildings.

The 3-D MOSAIC system described by Herman and Kanade [6] worked in the domain of aerial images of buildings. It extracted line segments from the scene and grouped them together into sets of segments forming faces of buildings in the scene. Monocular and binocular cues were combined with assumptions about the domain (for instance, that walls are usually vertical and roofs are usually parallel to the ground plane) to construct a wireframe representation of the scene. The system had no recognition component, but the ideas used in grouping segments into potential building faces had an influence on the grouping processes in LASSIE.

The system described by Fua and Hanson [3] works in the domain of aerial image analysis. The image is segmented into regions, and edges are extracted using the Sobel edge operator. Expectations about the geometry of "cultural regions," mainly building outlines, are used to drive the resegmentation of the image in order to extract cultural regions. These expectations are very weak and generic. The system is capable of deducing certain things about the relations between the regions it finds, for instance, that a lit region joined along an edge to a darker, non-shadow region is likely to be formed by a peaked roof.

6.3 The LASSIE object recognition program

6.3.1 Overview

The recognition process consists of the following stages:

- Extract edge points from a color image, then link them into line segments.
- Merge segments that are almost collinear and are separated by a small gap.
- Eliminate short line segments.
- Extract significant line pairs. The pairs are chosen based on their either being nearly parallel or possibly forming the two non-parallel sides of a trapezoid.
- Examine the pairs formed in the previous step to look for segments that would close the pairs off to form trapezoids.
- Examine the trapezoids found in the previous step to merge sets of overlapping trapezoids whose shapes are similar.
- Search for groups of trapezoids potentially formed by the windows, roof, and trunk of a car.
- Search, given an initial match, for other features (ellipses corresponding to wheels, lines from the back and side of the car) to confirm that match.

Apart from the final verification stage, the system is currently purely a bottom-up program, with no top-down interactions between the earlier stages.

6.3.2 Description of the segmentation and grouping stages

Segmentation

- A blue minus red (B - R) image is computed from the input color image.
- The Canny edge detector is run over the B - R image. The sigma chosen is 2.5.
- The 5% of image points with highest gradient magnitude are selected from the Canny output as edge points.
- A line tracker, written at CMU by Hiro Miwa, is run over the thresholded edge image. The line tracker places breakpoints in chains of edge pixels by looking for places where the edge curvature is changing rapidly at multiple scales. Gaps of one pixel in edges are jumped by the tracking program. Lines shorter than 10 pixels are discarded. The line tracker also finds ellipses in the edge

data which can be used to find the location of wheels forming part of a car.

Line merging

Segmentation may break up lines in the scene because of gaps in the thresholded edge image. Also, very close parallel edge segments may arise from the same linear feature in the scene. A simple merging procedure is applied to compensate for these errors in edge extraction, illustrated in figure 6.1. Given two edge segments, S1 and S2:

- A line is fit to S1 and S2 using least squares. The endpoints of S1 and S2 are projected onto that line in order to find the endpoints of the merged segment, S3.
- S1 and S2 are merged to form S3 if the following conditions are met:
    1. the distance between the endpoints of S1 and S2 and the merged segment S3 is less than some number of pixels, and
    2. S1 and S2, when projected onto S3, either overlap or are separated by a gap smaller than a threshold set by the user.

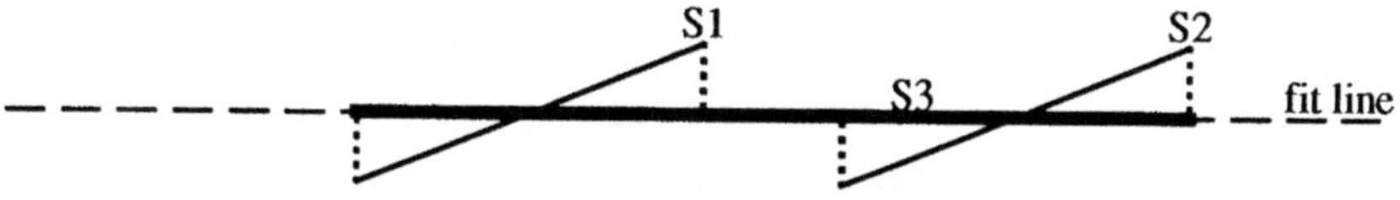

Figure 6.1: Line merging

Thresholding

The lines are thresholded on length to eliminate spurious lines created by such things as variations in the surface of the road on which the car is resting, and specular reflections of tree foliage off the surface of the car. The default threshold is 30 pixels.

Examination of line pairs

The set of lines is examined for pairs of lines which are parallel within some tolerance, or which could be the non-parallel sides of a trapezoid, i.e., the virtual lines connecting corresponding endpoints of the segments are close to parallel.

Extraction of trapezoids

Each pair of lines found in the last stage could possibly be two of the four sides of a trapezoid. To look for evidence supporting one of the sides connecting the pair to form a trapezoid:

- the list of edge segments is searched for edges which meet the following criteria:
    - edge is close to parallel to the hypothesized side;

- edge endpoints are closer than a threshold (based on the separation between the line pair) to the hypothesized side;
- probability of edge being a false match to the hypothesized side is less than some threshold (using the formula given by Lowe [7]);

- the supporting segments found are projected onto the hypothesized side and the fraction of the hypothesized side covered is computed

- a line is fit to the supporting segments if a sufficient fraction of the hypothesized side is covered (see figure 6.2). If only one of the two missing sides is found by this process the other is hypothesized as present.

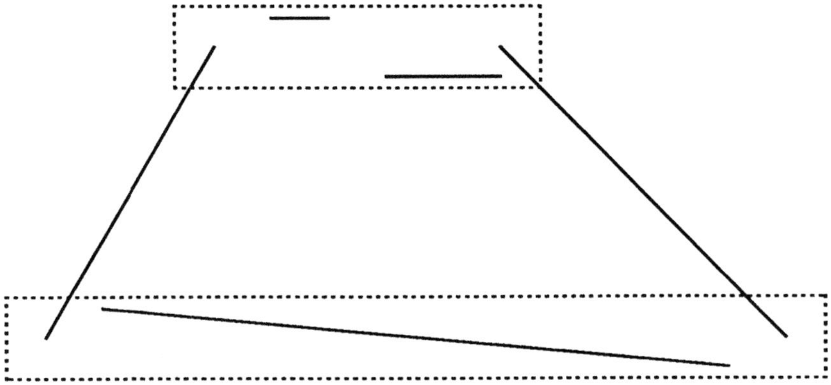

Figure 6.2: Trapezoid creation

A number of heuristics are used to eliminate some of the trapezoids found by this process. A trapezoid is discarded if any of the following conditions are met:

- one side of the trapezoid does not have sufficient support and is hypothesized, and an edge segment crosses the hypothesized side and extends more than a threshold distance outside the trapezoid;

- the shorter side of a pair of opposite sides is less than some fraction of the length of the longer side of the pair;

- the angle between a pair of adjacent sides is greater than a threshold angle;

- a side supported by one of the original pair of lines is longer than some multiple of the length of the original segment.

These heuristics are illustrated in figure 6.3.

Trapezoid merging

The line pair finding process will find several pairs of lines from the set of edges forming a trapezoid in the image. In the simplest case, the top and bottom edges will form one pair, and the left and right edges will form another pair.

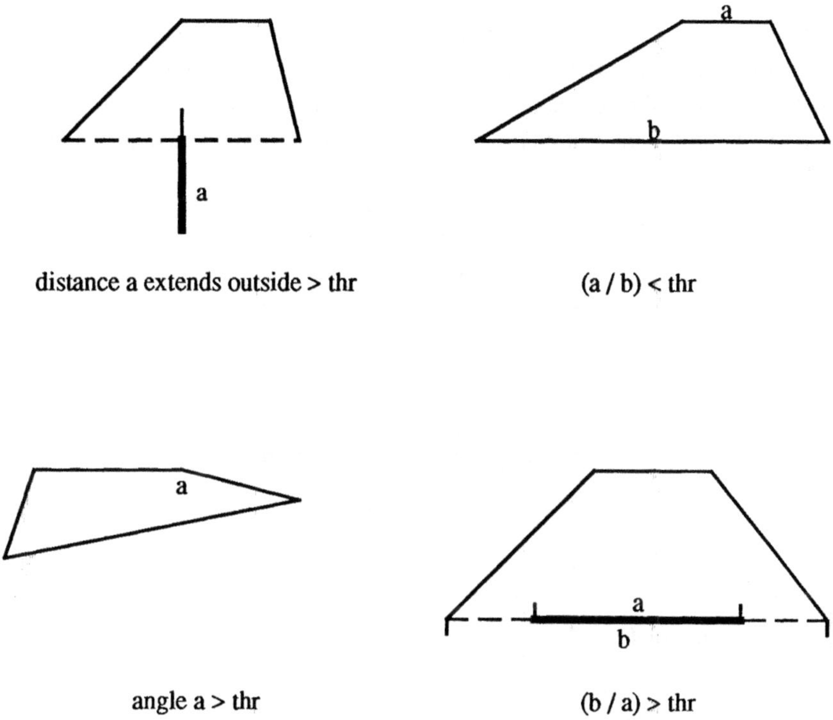

Figure 6.3: Trapezoid elimination heuristics

This will result in finding two trapezoids, each using the same set of lines. These are really the same trapezoid. The duplication should be detected, and the two trapezoids merged. In addition, things such as trim around windows may create multiple pairs of lines which then give rise to trapezoids that are very similar. This stage of processing looks through the list of trapezoids returned by the previous stage and merges pairs of trapezoids which have corresponding sides which are close to each other, nearly parallel, and less that some probability of being false matches to each other (once again using Lowe's formula [7]).

6.3.3 Feature-fetchers and the search for initial matches

The procedures used to search for initial matches of image features to instances of cars consist of a match conflict resolver and a set of procedures called feature-fetchers (see figure 6.4). The match conflict resolver keeps track of "claims" to image features by matches. Only one match using a given image feature (the match with the highest confidence) survives, with the confidence values of competing matches for an image feature getting set to zero. Feature-fetchers are procedures which return instances of particular features. The features returned may be primitive features (such as lines, ellipses, or trapezoids) or compound features (such as a pair of trapezoids which could match the side and rear windows of a car). The model of the appearance of a car is encoded procedurally in the feature-fetchers in the form of calls to functions that evaluate constraints that should hold for a match.

Figure 6.4: Hierarchy of search procedures

A feature-fetcher may be called upon multiple times to return a given feature instance. A good example of this is the trapezoid_list feature-fetcher, which is called by all the other feature-fetchers. Each instance of a feature found by a feature-fetcher is assigned a sequential index number. That allows each feature-fetcher calling trapezoid_list to keep track of which trapezoids it has seen by storing the index of the last instance returned to it by trapezoid_list without requiring trapezoid_list to know anything about what feature-fetchers are calling it. In the case of the trapezoid_list feature-fetcher all the instances of the feature are extracted before the matching process begins. In the case of the other feature-fetchers instances are found as needed and cached. An instance can be fetched from the previously cached results if another feature-fetcher calls requesting an instance with an index number below the highest index value assigned so far. This avoids duplication of work by the feature-fetchers.

The functions of the individual feature-fetcher procedures are as follows:

- trapezoid_list returns trapezoids extracted in the earlier stages of the program
- get_windows returns a partial car match containing a pair of trapezoids which could correspond to the side and rear windows of a car instance
- add_roof gets partial matches from get_windows and looks for trapezoids from trapezoid_list which could match the roof given the trapezoids in the match of the windows
- add_trunk gets partial matches from add_roof and looks for trapezoids from trapezoid_list which could match the trunk given the window and roof trapezoids.

Feature-fetchers may employ multiple strategies to do their jobs. The add_trunk feature-fetcher is a good example. The roof and windows of the car look very similar from either the left or right, appearing roughly like a truncated pyramid. Since add_roof can't disambiguate which side of the car is being looked at, add_trunk employs two strategies to look for a trunk trapezoid to extend a partial match from add_roof. In the first strategy it assumes the car is being seen from the right and looks for the trunk to the left of the windows. When no more instances are found by the first strategy, the second strategy is invoked. This strategy assumes that the car is being seen from the left side, and looks for the trunk to the right of the windows. This involves stepping through the list of instances found by add_roof a second time, taking advantage of the caching done by the feature-fetchers.

The feature-fetchers use a variety of constraints on the relationships between trapezoids and between sides of trapezoids to find matches. The routines which check if the constraints are satisfied take into account uncertainty about the objects (for instance, whether both endpoints of a segment are known with high certainty). Currently eight constraints are used in the model:

- **rightp**(seg1, seg2) -- Weak "right-of" constraint between segments. Satisfied if the maximum image column number of either endpoint of seg1 is greater than the maximum image column number of either endpoint of seg2. Otherwise not satisfied. This definition assumes that the camera is roughly vertical, and this constraint is used very sparingly to minimize the restrictions this places on the camera.

- **leftp**(seg1, seg2) -- Weak "left-of" constraint between segments. Satisfied if the minimum image column number of either endpoint of seg1 is less than the minimum image column number of either endpoint of seg2. Otherwise not satisfied. Also used sparingly for the same reason as given for rightp.

- **areparallel**(seg1, seg2, tol) -- Parallelism constraint between segments. Satisfied if seg1 and seg2 are parallel within the tolerance given by tol. May be satisfied if the segments aren't close to parallel, but both endpoints of either one of the segments are hypothesized. Otherwise it is not satisfied.

- **angbetween**(seg1, seg2, low, high) -- Included angle constraint between segments. Here seg1 and seg2 are directed. Satisfied if the angle going counterclockwise from seg1 to seg2 is between the values low and high. May be satisfied if the actual direction of either segment is not known because both its endpoints are hypothesized. Otherwise not satisfied.

- **trapform**(trapezoid, start_side, ang_type1, ang_type2, ang_type3, ang_type4) -- Rough shape constraint for trapezoids. The ang_type variables have the value of OB (for obtuse -- defined here as an angle between 70 and 180 degrees) or AC (for acute -- defined here as an angle between 0 and 110 degrees). Trapform checks the angle at each vertex (with start_side used to indicate what side to consider side zero) to see if it is of the appropriate type. Trapform is satisfied if the angles at all four vertices are of the correct type. It may be satisfied if the angle at a vertex is of the wrong type but the position of the vertex is hypothesized. Otherwise it is not satisfied.

- **phratio**(trap1, pair1, htype1, trap2, pair2, htype2, low, high) -- Relative height ratio constraint between trapezoids. Trap1 and trap2 are trapezoids, pair1 and pair2 specify which of the two pairs of opposing sides is involved from trap1 and trap2. Htype1 and htype2 specify the segment pair distance measure: either midpoint-to-midpoint for nonparallel lines or midpoint-to-opposite-segment for parallel or roughly parallel segments. The ratio between the height of the pair specified for trap1 and the height of the pair specified for trap2 is computed. If the ratio falls between the low and high values specified the constraint is satisfied. If the ratio falls outside the specified range, but one of the segments of one of the specified pairs was only hypothesized, then the constraint may be satisfied. Otherwise the constraint is not satisfied.

- **paraface**(seg1, low1, high1, seg2, low2, high2) -- Constraint on the overlap between parallel or roughly parallel segments. The endpoints of seg2 are projected onto the line specified by seg1 and the fraction of the length of seg1 overlapped by the projection of seg2 is computed. Similarly the fraction of seg2 overlapped by the projection of seg1 is computed. The constraint is satisfied if the fraction of seg1 overlapped is between low1 and high1 and the fraction of seg2 overlapped is between low2 and high2. The constraint may be satisfied if the overlapped fraction for a segment is out of range but the length of the segment is not know for sure (because one or both of its endpoints are hypothesized). Otherwise the constraint is not satisfied.
- **onside**(trap1, edge1, trap2) -- Constraint on the relative position of two trapezoids; whether trap2 is to the left of edge edge1 of trap1. Picture walking clockwise around the sides of trap1. The constraint is satisfied if the side of trap1 specified by edge1 is the only side such that all four vertices of trap2 are to your left as you walk along the side. Otherwise the constraint is not satisfied.

The above routines return a confidence factor between 0.0 and 1.0. Currently they return 1.0 if the constraint is satisfied; 0.1 if the constraint may be satisfied as described in the above descriptions; and 0.0 if the constraint was not satisfied. Any constraint on a possible match evaluating to 0.0 eliminates that match. The overall confidence of a match is just the sum of the values returned by the calls checking constraints from the model divided by the number of constraints.

6.3.4 Verification of initial matches

The verification stage goes through the list of matches which survive match conflict resolution, and checks for lines and ellipses verifying features that were not detected bottom-up. Information about expected orientation and position of lines based on a partial match is used to construct search windows to detect lines belonging to the rear and side of the car. The commitment to making no assumptions about the sizes of various parts is broken in the verification stage, which assumes that the side of the car is within some range of ratios of the height of the side window. This violation of one of the design goals can probably be eliminated by looking for the wheels and the bottom of the side of the car together. Ellipses extracted from the edge data are examined to look for wheels. The results of this stage are the final matches found by the program.

6.4 Results

Figure 6.5 shows the lines extracted from an image of the car in the upper left section; the lines after thresholding and merging in the upper right section; the extracted trapezoids in the middle left section; the initial match found bottom-up in the lower left section; and the confirmed match in the lower right section. Line merging was able to improve the quality of the line segments, for instance by linking together the lines forming the back edge of the trunk and the line forming the top edge of the side window.

The trapezoids found include the roof, the trunk, the rear window, and bits of trim along the side. There are also a number of trapezoids found which arise from accidental alignments in the image. The hypothesized fourth sides on a number of them pass through the much stronger trapezoids found for the trunk and rear window, but do not get eliminated because no edge pierces far enough through the virtual sides.

Only one match remains after match conflict resolution. Comparing the final match with the original lines, one can see that the program has done a good job of finding the various parts of the car.

Figures 6.6 and 6.7 show the result of running the program on lines extracted from an image of the car as seen from the other side. Here two matches survive the match conflict resolution, even though there is only one car in the scene. This is a result of the match conflict resolution using features rather than image areas as the primitive things claimed by a match.

Figures 6.8 and 6.9 show the results produced by LASSIE for six additional images. The original lines extracted from the images are on the left and the verified matches are on the right. As can be seen, the program does a fairly good job of locating the parts of the car despite noise in the extracted edges.

6.5 Directions for future work

Analysis of trapezoid overlap to improve match conflict resolution and heuristic discarding of trapezoids. The current match conflict resolution scheme only looks at features. The trapezoid merging may not be able to merge all of the trapezoids arising from the various parts of the car. If all of the features searched for give rise to two or more trapezoids then multiple matches will survive match conflict resolution, despite there only being one car in the image. An example of this was shown in the results section in figure 6.7.

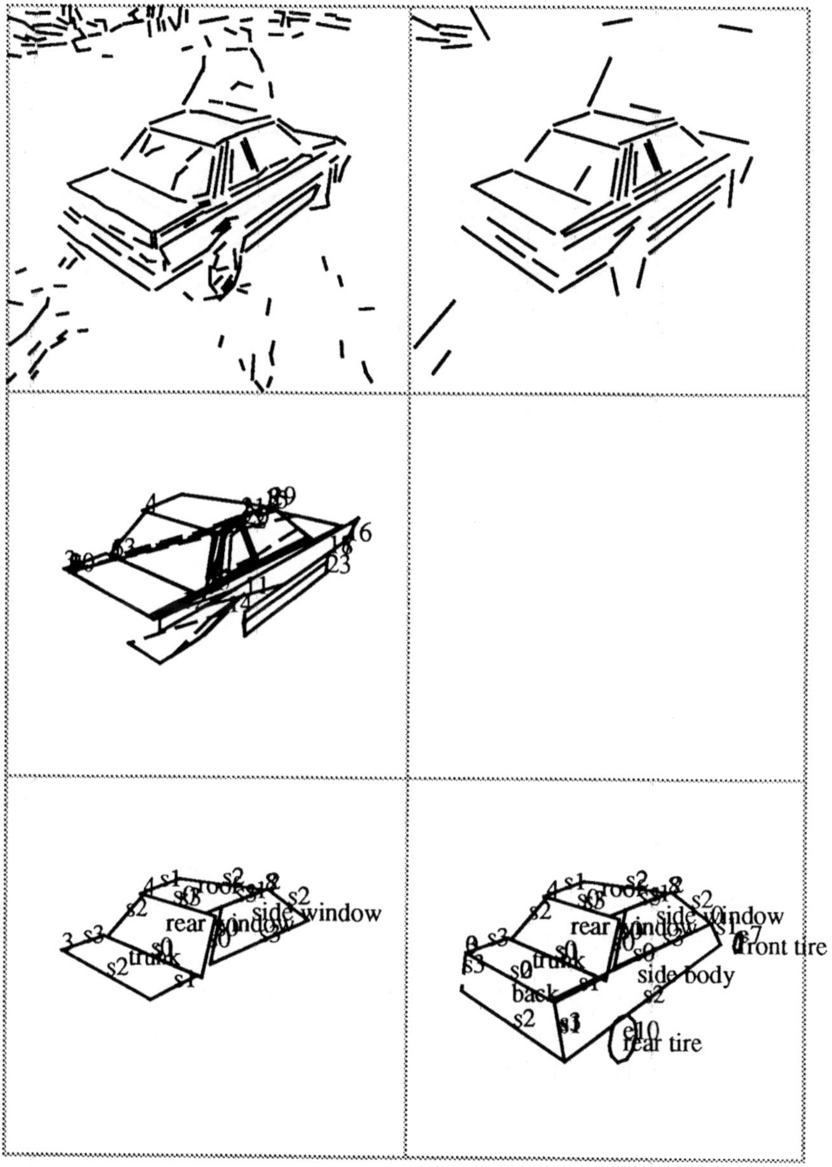

Figure 6.5: Results for the first image

Figure 6.6: Second image: original lines, merged lines, and trapezoids

Analyzing the area of overlap of the features forming the matches should allow resolution of this sort of conflict. The method will have to be designed in a way that will handle partial occlusion of one car by another.

Also, in some cases a hypothesized side which closes off a trapezoid goes through another trapezoid for which there is very strong support. There are cases where the test for lines crossing hypothesized sides fails to catch this. Looking at areas of intersection between trapezoids would help in the detection of these cases.

Improving the confidence factor scheme. The matches which survive conflict resolution are not always the most aesthetically pleasing ones. This is largely due to the overly simple confidence measure scheme currently used.

There are two major areas in which the confidence factor scheme could stand improvement. The first is in the granularity of the values returned by the

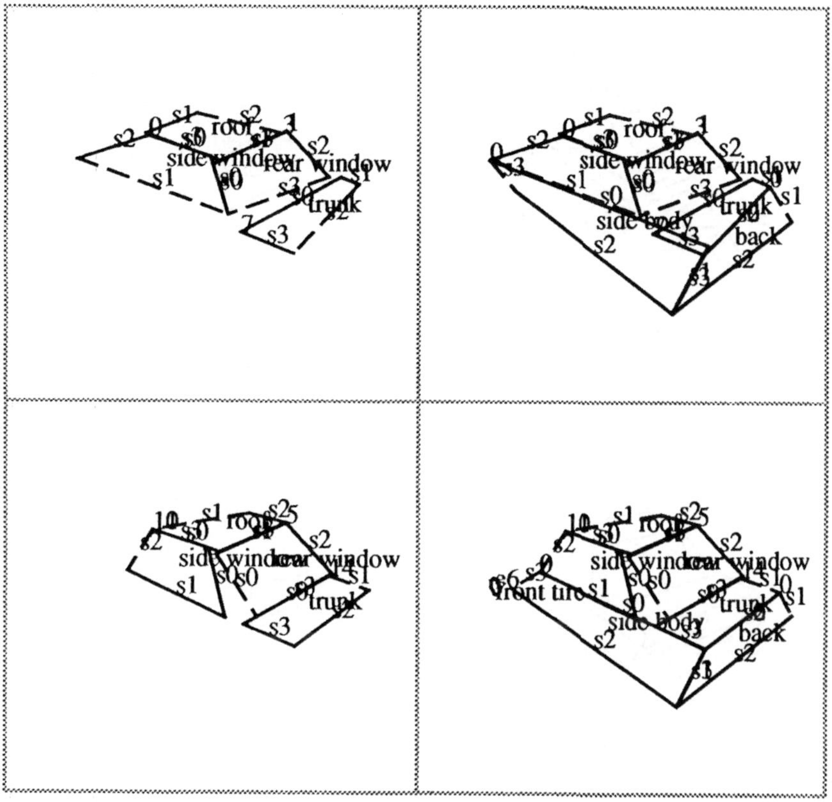

Figure 6.7: Second image: bottom-up match and match with verification

constraint tests. Currently one of three values is returned based on whether the constraint is satisfied, unsatisfied but unsure, or definitely unsatisfied. This is too coarse. For instance the parallelism constraint could return a value proportional to how close to parallel the segments were.

The second area for improvement is the evidence combination scheme, which is completely ad hoc. Taking the product of the values returned for the various constraints in the model quickly produces values which are all uniformly small. Other techniques have not been investigated at this point.

Integration of top-down search and verification into the feature-fetchers. Currently top-down verification of features not found in partial matches is done in a separate phase. Feature-fetchers should be written to try to extend a partial match by top-down search for missing features if no matching feature was found by the bottom-up feature extraction.

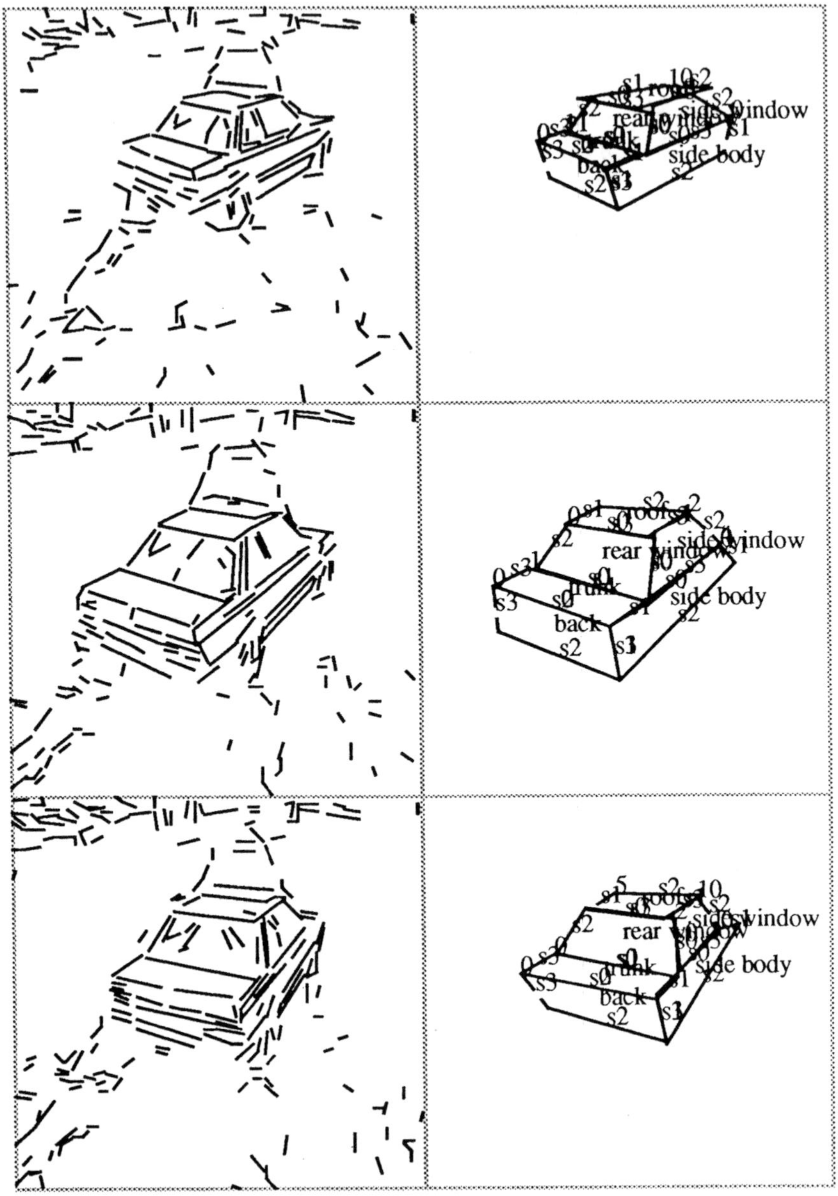

Figure 6.8: Original lines and verified matches from another three images

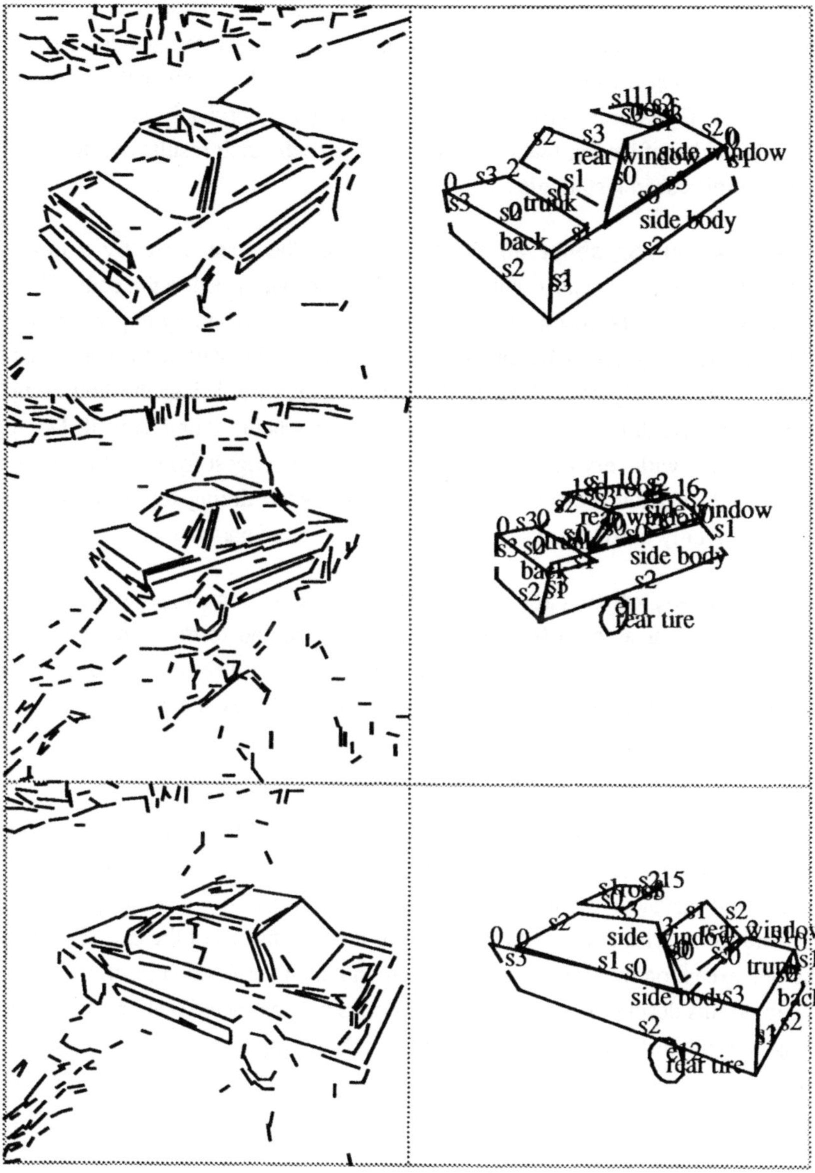

Figure 6.9: Original lines and verified matches from the final three images

Making the search more flexible. Currently there is a single fixed order in which features are searched for: first find candidates for the windows, then extend by adding a roof, then extend by adding a trunk. Even with top-down search capabilities built into the feature-fetchers this is probably not sufficient. The whole idea of having feature-fetchers which cache their results was to allow more complex search strategies.

Doing something about all the thresholds. There are currently a dozen thresholds visible to the user, with another three or four buried in the code. Their default values are the result of good initial guesses followed by experimenting with how far they could be changed before the results started to deteriorate. Setting them loosely enough to find all the features needed in the bottom-up stage of the search results in various bogus features being found as well. Doing an initial pass with very tight thresholds to find the strongest features and letting the individual feature-fetchers loosen them as needed during top-down search would hopefully make the system less sensitive to these parameters.

Other views, other cars. Currently the system is only programmed to recognize the car as seen from two views, from behind to the right and from behind to the left. Extension to the other views of the car needs to be done. Also the program needs to be extended to recognize other makes of cars.

6.6 Summary

The LASSIE program has demonstrated the ability to robustly extract significant groups of lines forming trapezoids in the test images we have used. It can find instances of the test car from two views using only the geometric constraints described, without any model of the dimensions of the car other than the one exception mentioned in the section on match verification (section 6.3.4). It can do this despite the imperfections of edge data extracted from real outdoor images of the test car. Extensions of the work to support more complex search strategies, mixed top-down and bottom-up searching in the feature-fetchers, and improved confidence measures can be expected to improve the performance of the program.

6.7 References

[1] Besl, Paul J., and Jain, Ramesh C.
Three-Dimensional Object Recognition.
Computing Surveys 17(1), March, 1985.

[2] Brooks, Rodney A.
Symbolic Reasoning Among 3-D Models and 2-D Images.
Artificial Intelligence 17:285-348, 1981.

[3] Fua, Pascal, and Hanson, Andrew J.
Locating Cultural Regions in Aerial Imagery Using Geometric Cues.
In *Proceedings of the DARPA Image Understanding Workshop*. December, 1985.

[4] Goad, Chris.
Special Purpose Automatic Programming for 3-D Model-Based Vision.
In *Proceedings of the ARPA Image Understanding Workshop*. June, 1983.

[5] Goto, Y., Matsuzaki, K., Kweon, I., and Obatake, T.
CMU Sidewalk Navigation System: A Blackboard-Based Outdoor Navigation System Using Sensor Fusion with Colored-Range Images.
In *Proc. Fall Joint Computer Conference*. November, 1986.

[6] Herman, Martin, and Kanade, Takeo.
The 3-D MOSAIC Scene Understanding System: Incremental Reconstruction of 3D Scenes from Complex Images.
Technical Report CMU-CS-84-102, Dept. of Computer Science, Carnegie Mellon Univ., February, 1984.

[7] Lowe, David G.
The Viewpoint Consistency Constraint.
International Journal of Computer Vision 1:57-72, 1987.

[8] Ohta, Y.
A Region-Oriented Image Analysis System by Computer.
PhD thesis, Kyoto University Dept. of Information Science, 1980.

[9] Thompson, D., and Mundy, J. L.
Three-Dimensional Model Matching from an Unconstrained Viewpoint.
In *Proc. IEEE International Conference on Robotics and Automation*. April, 1987.

Chapter 7: Building and Navigating Maps of Road Scenes Using Active Range and Reflectance Data

Martial Hebert

Abstract

This paper presents algorithms for building maps of road scenes using an active range and reflectance sensor and for using the maps to traverse a portion of the world already explored. Using an active sensor has some attractive advantages: It is independent of the illumination conditions, it does not require complex calibration in order to transform observed features to the vehicle's reference frame, and it provides 3-D terrain models as well as road models. Using the map built from sensor data facilitates navigation in two respects: The vehicle may navigate faster since less perception processing is necessary, and the vehicle may follow a more accurate path since the navigation system does not rely entirely upon inaccurate visual data. We present a complete system that includes road following, map building, and map-based navigation using the ERIM laser range finder. We report on experimentation of the system both on the CMU NAVLAB and the Martin Marietta ALV. [1]

7.1 Introduction

Autonomous road following using visual information is an important application of mobile robots. In addition to navigating on roads, the visual information can be used to build maps of the observed environment. An area of research that has not been previously explored is to close the loop by using the map built from previous observations to guide the navigation on a portion of the world already explored. Such a capability of map based navigation would enable us to improve the performance of the vehicle in three directions:

- Faster navigation: Perception is typically the bottleneck in autonomous mobile systems because images have to be processed as often as possible to compensate for the lack of knowledge about the world. If apriori knowledge of the environment is available from previous observations, perception is needed only to periodically check that the vehicle stays on the path prescribed by the map. The perception bottleneck is therefore reduced, thus leading to faster navigation.
- More reliable navigation: Autonomous navigation is unreliable because of the uncertainty associated with any sensor data and processing. Relying more on a map means relying less on sensor data acquired during the execution of a navigation plan. Map based navigation should therefore provide more accurate navigation.

- Simpler perception: A map can provide the expected appearance of the environment at any location. That includes the expected location of objects, and the expected position and appearance of the road. This additional knowledge allows for simpler perception processing.

Although map based navigation algorithms could be used with a man-made map (*e.g.* from surveying), using a map built from sensor information does not make any assumptions on the amount of knowledge available to the system, thus leading to a fully autonomous system. This is also important since it is difficult to obtain the resolution of a map built from sensor data by using surveying alone.

Most of the existing road following systems are based on intensity or color image processing [10,14,11]. In this paper, we investigate the use of active sensing, namely laser range finding, for both road following and map building. Using such a sensing modality has some attractive features such as its stability with respect to illumination conditions and the direct conversion to world coordinates without calibration. Our goal is therefore to build a complete system from road following to map building using active sensing, whereas previous research on active sensing for autonomous vehicles focused on 3-D map building or obstacle detection [2,3,6,4].

The images used in the experiments reported in this paper are range and reflectance images from a laser range finder, the ERIM scanner [13]. On each scan, the ERIM produces two images, one containing range and the other reflectance. Each image is 64 rows by 256 columns by 8 bits. In the range image, the 8 bits encode distance to the nearest object along that direction, to a maximum value of 64 feet. The reflectance image represents the amount of laser energy reflected back to the scanner at each pixel. The filed of view of the scanner is 30° vertical by 80° horizontal.

While all examples in this paper come from Navlab data and runs, the Martin Marietta ALV has an identical ERIM scanner, and we have ported and demonstrated our road following system on their vehicle as well.

7.2 Following roads using active reflectance images

Early work on road following from active sensing focused on the use of range data to find the edges of the road [8,1]. The drawback of this approach is that it assumes that the road is limited by edges that correspond to discontinuities of the terrain surface. This assumption severely limits applicability of the algorithms. An alternative approach is to use the active reflectance images for road following. Active reflectance images have two characteristics that make them attractive for road-following applications: First, they are insensitive to outside illumination, that is no shadows are cast by objects in reflectance images and the influence of the level of ambient light on the image is minimal (in fact, any program using reflectance images would work as well under night conditions). Second, each pixel in the reflectance image is also a range pixel whose position in space can be derived from the geometry of the scanner. This allows us to compute the position of the edges of the road found in a reflectance image in the vehicle's 3-D world without any of the calibration procedures that are typical of the video-based road following algorithms [5].

Edge detection would be the natural way of finding road edges in grey level images. The nature of the reflectance data, however, suggests the use of a region-based technique for two reasons: First, since the dynamic range of the image is low, many spurious edges will be found that have the same strength as road edges. Second, the intensity of the road in reflectance images is very stable because it is insensitive to shadows and changes in illumination. This is to be compared with video images in which the appearance of the road region varies significantly, thus requiring the use of multiple classes of road and non-road regions [10]. Instead of extracting the road edges directly, a road region extractor identifies the pixels that are part of the road based on the road location and appearance predicted from a previous image.

The principle of the road region extractor is to keep the mean m_i and variance σ_i of the reflectance values inside the road region for each group i of four consecutive scanlines in the image. The statistics are computed on groups of scanlines instead of the entire image in order to account for the intensity attenuation at long range and for the presence of small markings on the road that would have an effect on a few scanlines only instead of propagating to the entire road region. The statistics are computed on the first image by selecting the road region interactively. The road region is extracted from the reflectance image by thresholding the pixels in swath i that are between $m_i+\sigma_i$ and $m_i-2*\sigma_i$, where (m_i, σ_i) are the values computed on the previous image. The resulting binary image is then processed to remove small isolated regions. The remaining large images are then searched for the most likely road region. The road must lie within the predicted bouds; have aproximately the correct width; and have approximately the right elongation. The most likely region is reported as the road location, and used to compute new m_i and σ_i values for each swath in order to process the next image. This algorithm is similar to [10,11], except that it uses only one road class, that is only one set of statistics, and that it computes and predicts the road appearance over small swaths instead of the entire image.

The final output of the road finder program is the direction of the center line of the road. Since the reflectance image is registered with a range image, all road region boundary points are represented by their full 3-D position. The road center line is computed by fitting two parallel lines to the left and right edges of the road polygon in the 3-D world coordinates. If the two lines are parametrized by a direction $\vec{v}$, which is the direction of the road common to both edges, and the signed distances of the lines to the origin d_l and d_r (Figure 4.1), the best fit is computed by minimizing:

$$\sum_{leftedge} (p_l.\vec{v} + d_l)^2 + \sum_{rightedge} (p_r.\vec{v} + d_r)^2 \tag{7.1}$$

The center line of the road is the middle line of $(\vec{v}, d_l)$ and $(\vec{v}, d_r)$, the width of the road is $w = |d_l - d_r|$. Figure 4.2 shows the result of the road following program on a typical reflectance image. The top part of Figure 4.2 shows the reflectance image, the bottom part shows the road edges and the center line of the road projected on the ground plane. The scale on the left side on the image is in meters; the road is found between five and fifteen meters in this case.

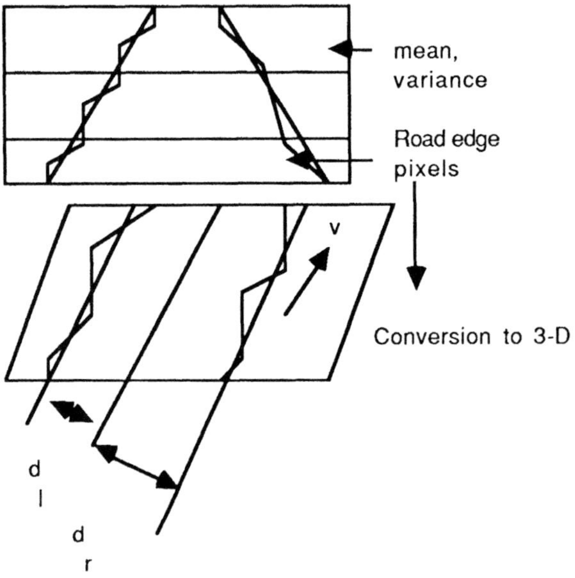

Figure 7.1: Road finding algorithm

In order to drive the vehicle, two points on the center line are sent to a local path planner. The path planner generates a sequence of circular arcs using a "pure pursuit" algorithm derived from [12]. The road following program successfully drove two vehicles, the CMU NAVLAB [7] and the Martin Marietta ALV, over several hundred meters at a speed of 40cm/s. In both cases, the road following is implemented on a Sun3 workstation. The average computation time is 3 seconds per reflectance image which allows for enough overlap between consecutive images.

7.3 Building maps from range and reflectance images

The map is built by matching features, such as road edges or trees, between consecutive images. The map is built incrementally as the vehicle follows the road autonomously using the system of Figure 4.3. This system using the road tracking algorithm to navigate, and a terrain analysis to generate a terrain description from each image. The details of the map building algorithms can be

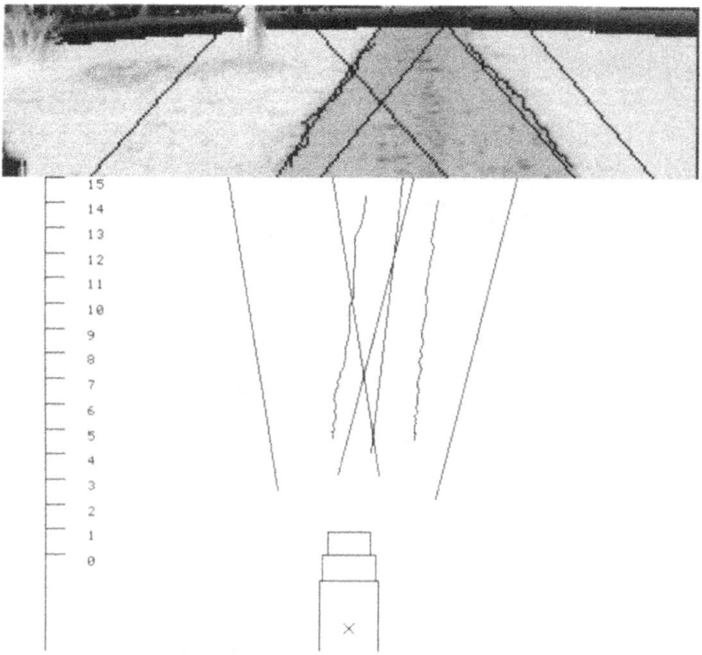

Figure 7.2: Road following on a reflectance image

found in Chapter 13. Even though a full terrain description can be stored in the map, we retain only the road edges and the detected objects for the purpose of navigation. Figure 4.4 shows a map built from twenty images over sixty meters. In this display, only the road edges, the center line of the road, and the discrete obstacles are shown.

7.4 Map-based road following

The last part of the system uses the map built from road following to traverse the same portion of the world. The map-based road following must proceed in three steps: computation of the starting position, path planning in the map, path execution. The first step is needed to avoid constraining the starting position and heading of the vehicle at the beginning of the traversal of the map to those used to initiate the map building stage. The position and heading of the vehicle with respect to the map are computed by matching the features, road edges and objects, observed in an image taken at the starting position with the features of the map that are predicted to be visible given a rough initial guess of starting position. The matching algorithm is basically the same as the one used for the map building except that in the current implementation, only road edges and discrete obstacles are used. For example, Figure 4.5 shows the initial guess of the starting position (marked by a cross) and the portion of the road and the obstacles that are used for the matching. The map features are predicted by

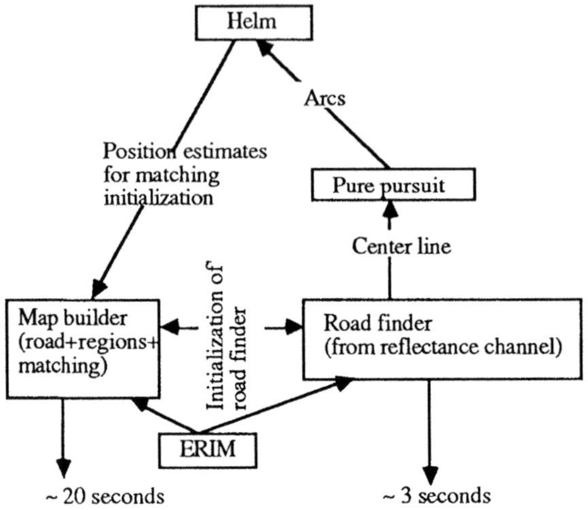

Figure 7.3: The map building/road following system

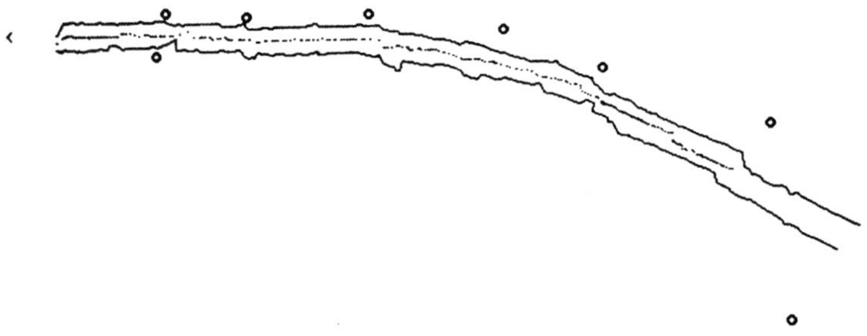

Figure 7.4: Complete map of a road scene

intersecting the sensor field of view with the map.

Figure 7.5: Estimation of the starting position and heading

Given the starting position, the second step is to compute a path that follows the road using the map. This step is the most straightforward in that any path planner that provides for smooth paths can be used. For example, Figure 4.6 shows a path composed of a sequence of circular arcs. The path is computed by dividing the center curve of the road into small segments over which the pure pursuit path planning algorithm of Section 4.2 is applied.

Once a path is computed, the vehicle is ready to follow the road based on the map. Ideally, the vehicle should be able to correctly execute the path without any perception at all. In practice, however, the vehicle will drift away from the ideal path due to wheel slippage, and the accumulation of small controller errors and numerical errors. Therefore, the position and heading of the vehicle with respect to the map must be recomputed periodically by comparing the features that are actually observed while executing the path and the features that are predicted from the map given the current estimate of the vehicle's position. The question now is how often should we make a position correction, that is take an image, extract road edges and objects, and match them with the map, in order to stay within reasonable bounds of the original path. This problem is the key to map-based navigation: If the corrections are performed too often we are back to the original road following approach and we lose the benefit of having a map. If, on the other hand, we do not perform enough corrections along the path, we may drift significantly far from the nominal path and eventually run off the road. Furthermore, the corrections should be meaningful in the sense that enough features should be present at the time of the correction to ensure that the newly computed position is indeed closer to the truth than the currently

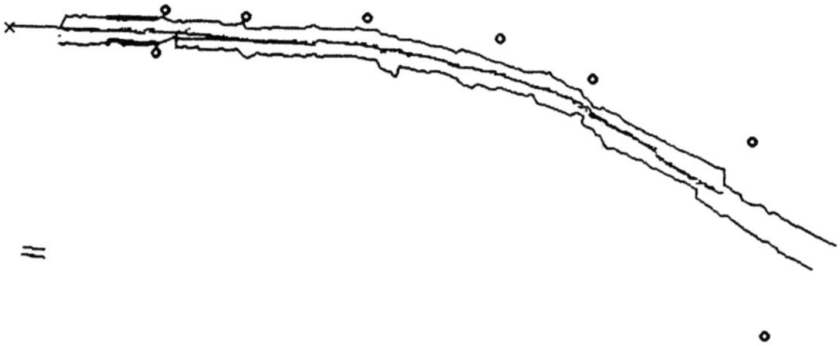

Figure 7.6: Path planned using a map

available estimate. Several strategies are possible to choose the locations at which corrections should be performed. An attractive strategy is to estimate the uncertainty on the position and heading as the vehicle moves, a new correction is requested whenever the uncertainty reaches a threshold that indicates that the vehicle is too far from its nominal path [9]. This approach guarantees that the distance between the vehicle's path and the nominal path always lies within preset bounds. It does not, however, guarantee that the images taken at the time at which a correction is needed contain enough features of interest. Another possible approach is to make a correction whenever the map predicts that features of interest may be observed from the current position. In our case, it is important to guarantee that the corrections are performed when objects are visible, since otherwise the correction would be computed on the basis of the road edges only and would therefore be ambiguous. A correction is therefore computed whenever at least one object is predicted to be visible from a position along the path. Matching the predicted objects and road edges from the map with the observed road and objects provides an unambiguous new estimate of the vehicle's position and heading. Figure 4.7 shows the locations at which new images are taken for computing the corrections along the path of Figure 4.6. The road edges and objects that are matched with the corresponding observed features are shown as bold segments of the road edges and dark circles respectively. The crosses along the path indicate the successive positions of the vehicle at regular intervals of one second (at a speed of 20 cm/s). The position is not displayed if an image is being processed, therefore the gaps in the stream of positions in this display illustrate the time spent in processing images while executing the initial path. (The gaps on this figure, from a simulated run, are somewhat larger than those during an actual run, due to different hardware configuration.)

Computing a correction gives an offset $\Delta = (\Delta x, \Delta y, \Delta\theta)$ between the nominal position and heading and the actual values at the time the image is taken. This offset must be used to correct the current course of the vehicle. This is achieved by shifting the path that has been executed while the image was being processed by Δ, by replanning from the current position as given by the shifted, and by replacing the pending set of motion commands by this new path. Figure 4.8 illustrates such a sequence of events: As the vehicle comes into view of the first objects, an image is taken and matched against the map, the new position is shown as a cross on the left of the initial path, a new path is planned that takes the vehicle back to its original course.

These results show that it is possible to use a map to efficiently guide the navigation of an autonomous vehicle. The main benefit is that considerably fewer images have to be processed while retraversing the map. For example, the map of Figure 4.7 requires seven images to be processed. Following the same road at the same speed without the support of a map would require at least 25 images for a displacement of two meters between consecutive images. The reason for the discrepancy is that even if the position of the road were computed perfectly from each individual image, the path planner would not have information far enough in front of the vehicle to plan a stable path that is guaranteed to remain

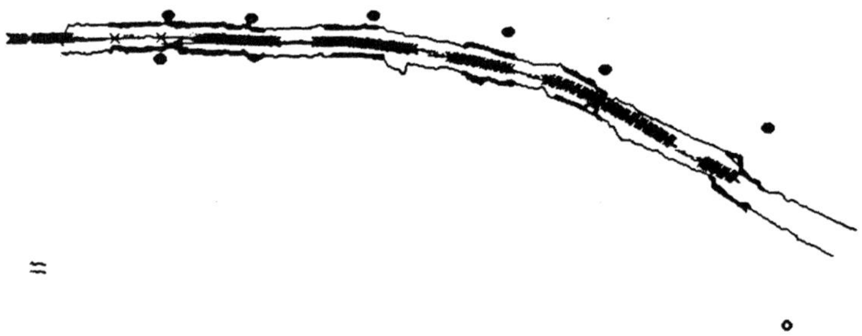

Figure 7.7: Locations at which images are taken along the path

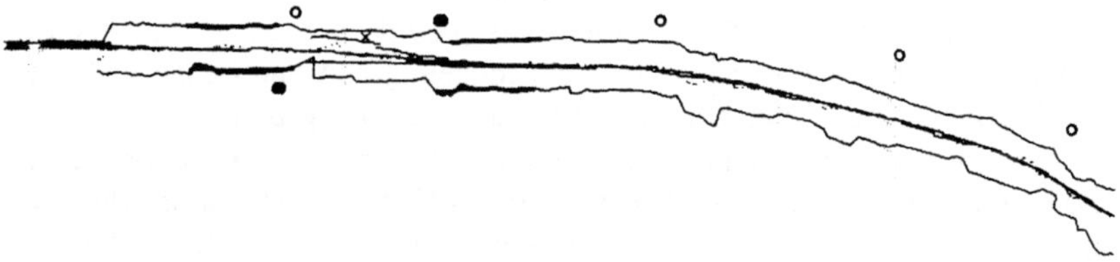

Figure 7.8: Corrected path

on the road. Although the same results could be obtained by using a map that is entered manually, it is important to note that the combination of map building from sensor data and map-based navigation results in a fully autonomous system that can learn its environment and use its new knowledge to navigate it.

7.5 Conclusion

The road following and map building system shows that road environments can be efficiently navigated and mapped using an active sensor such as a laser range finder. The map based navigation system shows that the information gathered during an initial traversal of the road can be used to improve the navigation over a portion of the road already explored. Specifically, using the map provides an initial path to follow, and a list of optimal locations at which visual data should be processed in order to correct the vehicle position that drifts over time. The combination of those three components provide a basis for autonomous navigation of roads including 3-D terrain modeling and knowledge gathering and utilization through map building and map based navigation.

We are currently extending the ideas used in those systems to the case of cross-country navigation and combined on road/off road navigation in which the map contains a representation of terrain regions in addition to the road model and the discrete obstacles. This type of information is currently extracted but it is not used for the map based navigation. The system presented here uses a simple path planner based on the pure pursuit control scheme. Our plan is to use the path planner described in [9] to take into account vehicle model and uncertainty, and to be able to apply this approach to cross-country navigation.

Acknowledgements

Mike Blackwell, James Frazier, and David Simon made the NAVLAB experiments possible. Chuck Thorpe provided the path planner used in this system. Keith Gremban ported and demonstrated the road following program on the Martin Marietta ALV.

7.6 References

[1] J. Beyer, C. Jacobus, and F. Pont. Autonomous Vehicle Guidance using Laser Range Imagery. In *SPIE Vol. 852, Mobile Robots II*, Cambridge, 1987.

[2] M. J. Daily, J. G. Harris, and K. Reiser. Detecting Obstacles in Range Imagery. In *Image Understanding Workshop*, Los Angeles, 1987.

[3] M.J. Daily, J.G. Harris, and K. Reiser. An Operational Perception System for Cross-Country Navigation. In *Proc. Image Understanding Workshop*, Cambridge, 1988.

[4] R. T. Dunlay and D. G. Morgenthaler. Obstacle Detection and Avoidance from Range Data. In *Proc. SPIE Mobile Robots Conference*, Cambridge, MA, 1986.

[5] K.D. Gremban, C.E. Thorpe, and T. Kanade. Geometric Camera Calibration Using Systems of Linear Equations. In *Proc. Image Understanding Workshop*, Cambridge, 1988.

[6] M. Hebert and T. Kanade. 3-D Vision for Outdoor Navigation by an Autonomous Vehicle. In *Proc. Image Understanding Workshop*, Cambridge, 1988.

[7] S. Shafer and W. Whittaker. *June 1987 Annual Report: Development of an Integrated Mobile Robot System at Carnegie Mellon*. Technical Report CMU-RI-TR-88-10, The Robotics Institute, Carnegie-Mellon University, 1988.

[8] U.K. Sharma and L.S. Davis. Road Following by an Autonomous Vehicle using Range Data. In *SPIE Vol. 727, Mobile Robots II*, Cambridge, 1986.

[9] T. Stentz. *The NAVLAB System for Mobile Robot Navigation*. PhD thesis, Carnegie-Mellon University, Fall 1988.

[10] C.E. Thorpe, M. Hebert, T. Kanade, and S.A. Shafer. Vision and Navigation for the Carnegie-Mellon Navlab. *PAMI*, 10(3), 1988.

[11] M.A. Turk, D.G. Morgenthaler, K.D. Gremban, and M. Marra. VITS- A Vision System for Autonomous Land Vehicle Navigation. *PAMI*, 10(3), may 1988.

[12] R. Wallace, K. Matsuzaki, Y. Goto, J. Crisman, J. Webb, and T. Kanade. Progress in robot road-following. In *IEEE International Conference on Robotics and Automation*, 1986.

[13] R. Watts, F. Pont, and D. Zuk. *Characterization of the ERIM/ALV Sensor - Range and Reflectance*. Technical Report , Environmental Research Institute of Michigan, Ann Arbor, MI, 1987.

[14] A.M. Waxman, J.J. LeMoigne, L.S. Davis, B. Srinivasan, T.R. Kushner, E. Liang, and T. Siddalingaiah. A Visual Navigation System for Autonomous Land Vehicles. *Journal of Robotics and Automation, Vol. 3*, 1987.

Chapter 8: 3-D Vision Techniques for Autonomous Vehicles

Martial Hebert, InSo Kweon, and Takeo Kanade

8.1 Introduction

There are many sensing strategies for perception for mobile robots, including single camera systems, sonars, passive stereo, and laser range finders. Here, we focus on perception algorithms for range sensors that provide 3-D data directly by active sensing. Using such sensors has the advantage of eliminating the calibration problems and computational costs inherent in passive techniques such as stereo. This Chapter describes some of the techniques we have developed in this area of research. The aim of our work is to produce models of the environment, which we call the *terrain*, for path planning and object recognition. [1]

We first describe the range sensor that we used in this work in Section 8.2. Even though we tested our algorithm on one specific range sensor, we believe that the sensor characteristics of Section 8.2 are fairly typical of a wide range of sensors [4]. The algorithms for building a terrain representation from a single sensor frame are discussed in Section 8.3 in which we introduce the concept of dividing the terrain representation algorithms into three levels depending on the sophistication of the path planner that would use the representation, and on the anticipated difficulty of the terrain. Since a mobile robot is by definition a dynamic system, it must process not one, but many observations along the course of its trajectory. The 3-D vision algorithms must therefore be able to reason about representations that are built from sensory data taken from different locations. We investigate this type of algorithm in Section 8.4 in which we propose algorithms for matching and merging multiple terrain representations. Finally, the 3-D vision algorithms that we propose are not meant to be used in isolation, but have to be eventually integrated in a system that includes other sensors. A typical example is the case of road following in which color cameras can track the road, while a range sensor can detect unexpected obstacles. Another example is a mission in which a scene must be interpreted in order to identify specific objects, in which case all the available sensors must contribute to the final scene analysis. We propose some algorithms for fusing 3-D representations with representations obtained from a color camera in Section 8.5. We also

[1] An earlier version of this paper appeared as part of CMU Robotics Institute report CMU-RI-TR-89-5. Reprinted by permission.

describe the application of this sensor fusion to a simple natural scene analysis program. Perception techniques for mobile robots have to be eventually validated by using real robots in real environments. We have implemented the 3-D vision techniques presented in this report on both the Terregator and the Navlab. The results presented in Sections 8.3.3 and 8.3.4 come from the 3-D vision module that we integrated in the Navlab system [40]. In addition, some of the techniques were used for the Ambler [2], a hexapod designed for planetary exploration. This vehicle is designed for navigation on very rugged terrain including high slopes, rocks, and wide gullies.

8.2 Active range and reflectance sensing

The basic principle of active sensing techniques is to observe the reflection of a reference signal (sonar, laser, radar, etc.) produced by an object in the environment in order to compute the distance between the sensor and that object. In addition to the distance, the sensor may report the intensity of the reflected signal which is related to physical surface properties of the object. In accordance with tradition, we will refer to this type of intensity data as "reflectance" data even though the quantity measured is not the actual reflectance coefficient of the surface.

Active sensors are attractive to mobile robot researchers for two main reasons: first, they provide range data without the computation overhead associated with conventional passive techniques such as stereo vision, which is important in time critical applications such as obstacle detection. Second, it is largely insensitive to outside illumination conditions, considerably simplifying the image analysis problem. This is especially important for images of outdoor scenes in which illumination cannot be controlled or predicted. For example, the active reflectance images of outside scenes do not contain any shadows from the sun. In addition, active range finding technology has developed to the extent that makes it realistic to consider it as part of practical mobile robot implementations in the short term [4].

The range sensor we used is a time-of-flight laser range finder developed by the Environmental Research Institute of Michigan (ERIM). The basic principle of the sensor is to measure the difference of phase between a laser beam and its reflection from the scene [44]. A two-mirror scanning system allows the beam to be directed anywhere within a 30° × 80° field of view. The data produced by the ERIM sensor is a 64 × 256 range image. Range is coded in eight bits from zero to 64 feet, which corresponds to a range resolution of three inches. All measurements are relative since the sensor measures differences of phase. That is, a range value is known *modulo* 64 feet. We have adjusted the sensor so that the range value 0 corresponds to the mirrors. In addition to range images, the sensor also produces active reflectance images of the same format (64 × 256 × 8 bits). The reflectance at each pixel encodes the energy of the reflected laser beam at each point. Figure 8.2 shows a pair of range and reflectance images of an outdoor scene. The next two Sections describe the range and reflectance data in more details.

8.2.1 From range pixels to points in space

The position of a point in a given coordinate system can be derived from the measured range and the direction of the beam at that point. We usually use the Cartesian coordinate system shown in Figure 8.1, in which case the coordinates of a point measured by the range sensor are given by the equations:

$$\begin{aligned} x &= D\sin\theta \\ y &= D\cos\phi\cos\theta \\ z &= D\sin\phi\cos\theta \end{aligned} \tag{8.1}$$

where ϕ and θ are the vertical and horizontal scanning angles of the beam direction. Note that the reference coordinate system is not the same as in [19] for consistency reasons. The two angles are derived from the row and column position in the range image, (r, c), by the equations:

$$\begin{aligned} \theta &= \theta_0 + c \times \Delta\theta \\ \phi &= \phi_0 + r \times \Delta\phi \end{aligned} \tag{8.2}$$

where θ_0 (respectively ϕ_0) is the starting horizontal (respectively vertical) scanning angle, and $\Delta\theta$ (respectively $\Delta\phi$) is the angular step between two consecutive columns (respectively rows). Figure 8.3 shows an overhead view of the scene of Figure 8.2, in which the coordinates of the points are computed using Eq. (8.3).

8.2.2 Reflectance images

A reflectance image from the ERIM sensor is an image of the energy reflected by a laser beam. Unlike conventional intensity images, this data provides us with information which is to a large extent independent of the environmental illumination. In particular, the reflectance images contain no shadows from outside illumination. The measured energy does depend, however, on the shape

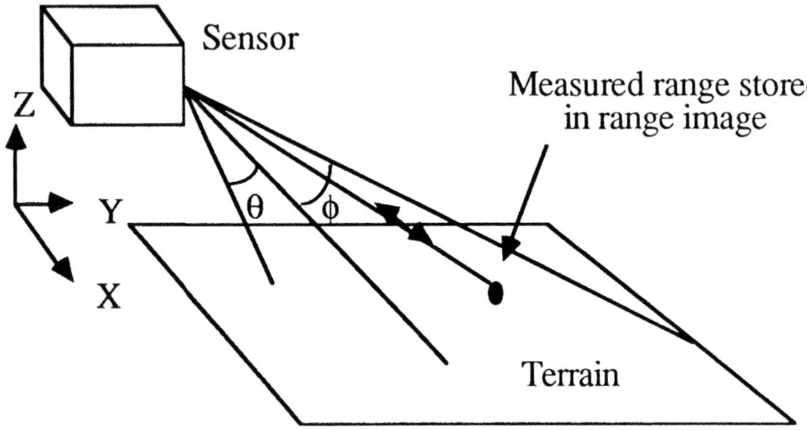

Figure 8.1: Geometry of the range sensor

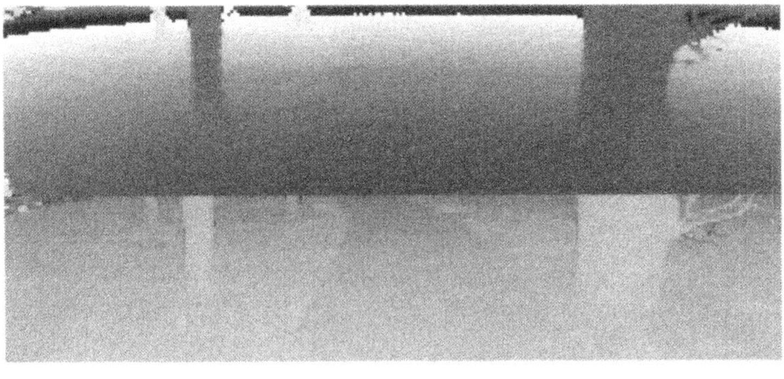

Figure 8.2: Range and reflectance images

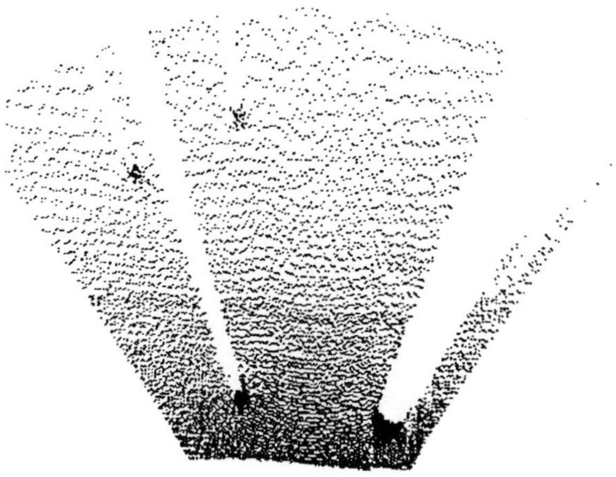

Figure 8.3: Overhead view

of the surface and its distance to the sensor. We correct the image so that the pixel values are functions only of the material reflectance. The measured energy, P_{return}, depends on the specific material reflectance, ρ, the range, D, and the angle of incidence, γ:

$$P_{return} = \frac{K\rho\cos\gamma}{D^2} \tag{8.3}$$

Due to the wide range of P_{return}, the value actually reported in the reflectance image is compressed by using a log transform. That is, the digitized value, P_{image} is of the form [42]:

$$P_{image} = A\log(\rho\cos\gamma) + B\log D \tag{8.4}$$

where A and B are constants that depend only on the characteristics of the laser, the circuitry used for the digitization, and the physical properties of the ambient atmosphere. Since A and B cannot be computed directly, we use a calibration procedure in which a homogeneous flat region is selected in a training image; we then use the pixels in this region to estimate A and B by least-squares fitting Eq. (8.4) to the actual reflectance/range data. Given A and B, we correct subsequent images by:

$$P_{new-image} = (P_{image} - B\log D)/A \tag{8.5}$$

The value $P_{new-image}$ depends only on the material reflectance and the angle of incidence. This is a sufficient approximation for our purposes since for smooth surfaces such as smooth terrain, the $cos\gamma$ factor does not vary widely. For efficiency purposes, the right-hand side of (8.5) is precomputed for all possible combinations (P_{image}, D) and stored in a lookup table. Figure 8.2 shows an example of an ERIM image, and Figure 8.4 shows the resulting corrected image.

Figure 8.4: Corrected reflectance image

8.2.3 Resolution and noise

As is the case with any sensor, the range sensor returns values that are measured with limited resolution, and which are corrupted by measurement noise. In the case of the ERIM sensor, the main source of noise is due to the fact that the laser beam is not a line in space but rather a cone whose opening is a 0.5° solid angle (the instantaneous field of view). The value returned at each pixel is actually the average of the range of values over a 2-D area, the *footprint*, which is the intersection of the cone with the target surface (Figure 8.5). Simple geometry shows that the area of the footprint is proportional to the square of the range at its center. The size of the footprint also depends on the angle θ between the surface normal and the beam as shown in Figure 8.5. The size of the footprint is roughly inversely proportional to $\cos\theta$ if we assume that the footprint is small enough and that θ is almost constant. Therefore, a first order approximation of the standard deviation of the range noise, σ is given by:

$$\sigma \propto \frac{D^2}{\cos\theta} \tag{8.6}$$

The proportionality factor in this equation depends on the characteristics of the laser transmitter, the outside illumination, and the reflectance ρ of the surface which is assumed constant across the footprint in this first order approximation. We validated the model of Equation 8.6 by estimating the RMS error of the range values on a sequence of images. Figure 8.6 shows the standard deviation with respect to the measured range. The Figure shows that σ roughly follows the D^2 behavior predicted by the first order model. The footprint affects all pixels in the image.

There are other effects that produce distortions only at specific locations in the image. The main effect is known as the "mixed point" problem and is illustrated in Figure 8.5 in which the laser footprint crosses the edge between two objects that are far from each other. In that case, the returned range value is some combination of the range of the two objects but does not have any physical meaning. This problem makes the accurate detection of occluding edges more difficult. Another effect is due to the reflectance properties of the observed surface; if the surface is highly specular then no laser reflection can be observed. In that case the ERIM sensor returns a value of 255. This effect is most noticeable on man-made objects that contain a lot of polished metallic surfaces. The noise characteristics of the ERIM sensor are fairly typical of the behavior of active range sensors [5].

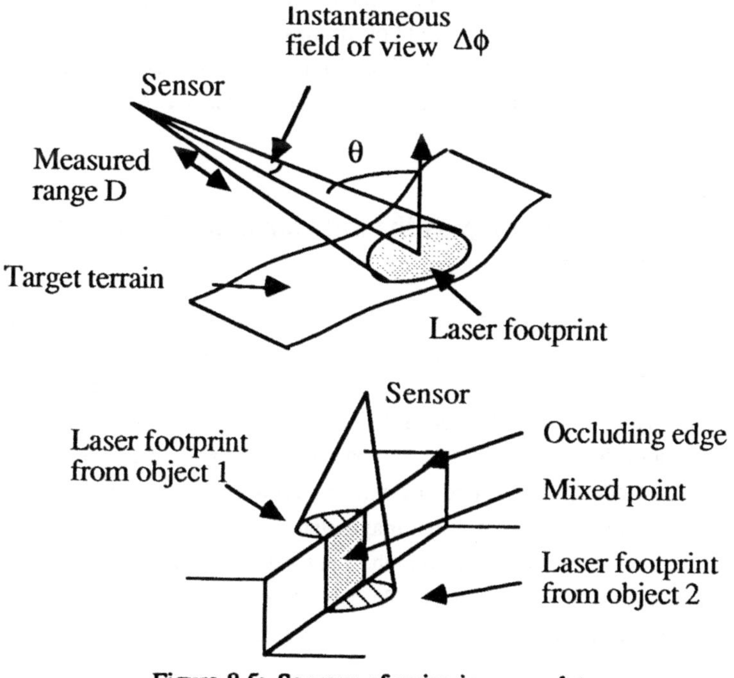

Figure 8.5: Sources of noise in range data

8.3 Terrain representations

The main task of 3-D vision in a mobile robot system is to provide sufficient information to the path planner so that the vehicle can be safely steered through its environment. In the case of outdoor navigation, the task is to convert a

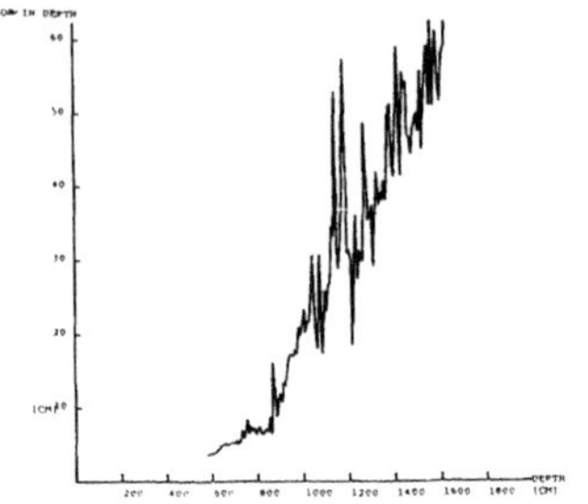

Figure 8.6: Noise in range data

range image into a representation of the terrain. We use the word "terrain" in a very loose sense in that we mean both the ground surface and the objects that may appear in natural environments (*e.g.* rocks or trees). In this Section we discuss the techniques that we have implemented for the Navlab and Ambler systems. We first introduce the concept of the elevation map as a basis for terrain representations and its relationship with different path planning techniques. The last four Sections spell out the technical details of the terrain representation algorithms.

8.3.1 The elevation map as the data structure for terrain representation

Even though the format of the range data is an image, this may not be the most suitable structuring of the data for extracting information. For example, a standard representation in 3-D vision for robot manipulators is to view a range image as a set of data points measured on a surface of the equation $z = f(x, y)$ where the $x-$ and $y-$axes are parallel to the axis of the image and z is the measured depth. This choice of axis is natural since the sensor is usually mounted overhead, and therefore the image plane is usually parallel to the plane of the scene. In our case, however, the "natural" reference plane is not the image plane but is the ground plane. In this context, "ground plane" refers to a plane that is horizontal with respect to the vehicle or to the gravity vector. The representation $z = f(x, y)$ is then the usual concept of an elevation map. To transform the data points into an elevation map is useful only if one has a way to access them. The most common approach is to discretize the (x, y) plane into a grid. Each grid cell (x_i, y_i) is the trace of a vertical column in space, its *field* (Figure 8.7). All the data that falls within a cell's field is stored in that cell. The description shown in Figure 8.7 does not necessarily reflect the actual implementation of an elevation map but is more of a framework in which we develop the terrain representation algorithms. As we shall see later, the actual implementation depends on the level of detail that needs to be included in the terrain description.

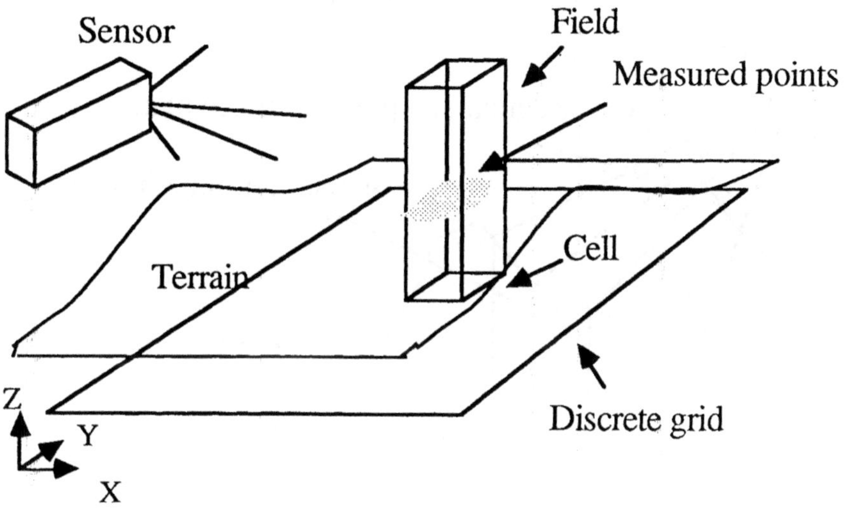

Figure 8.7: Structure of an elevation map

Although the elevation map is a natural concept for terrain representations, it exhibits a number of problems due to the conversion of a regularly sampled image to a different reference plane [24]. Although we propose solutions to these problems in Section 8.3.5, it is important to keep them in mind while we investigate other terrain representations. The first problem is the sampling problem illustrated in Figure 8.8. Since we perform some kind of image warping, the distribution of data points in the elevation map is not uniform, and as a result conventional image processing algorithms cannot be applied directly to the map. There are two ways to get around the sampling problem: We can either use a base structure that is not a regularly spaced grid, such as a Delaunay triangulation of the data points [32], or we can interpolate between data points to build a dense elevation map. The former solution is not very practical because of the complex algorithms required to access data points and their neighborhoods. We describe an implementation of the latter approach in Section 8.3.5. A second problem with elevation maps is the representation of the range shadows created by some objects (Figure 8.9). Since no information is available within the shadowed regions of the map, we must represent them separately so that no interpolation takes place across them and no "phantom" features are reported to the path planner. Finally, we have to convert the noise on the original measurements into a measure of uncertainty on the z value at each grid point (x, y).

This conversion is difficult due to the fact that the sensor's uncertainty is most naturally represented with respect to the direction of measurement (Figure 8.10) and therefore spreads across a whole region in the elevation map.

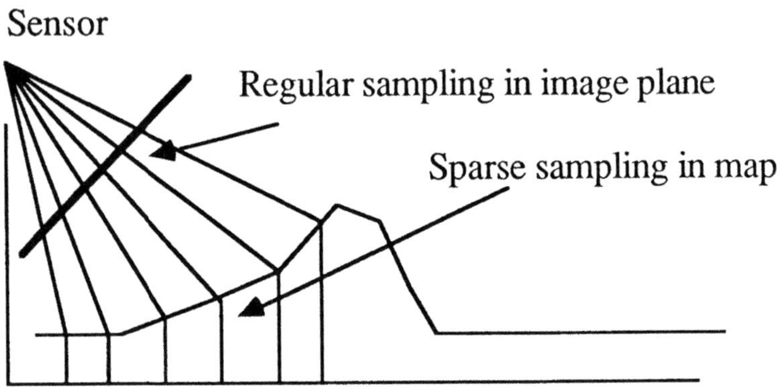

Figure 8.8: The sampling problem

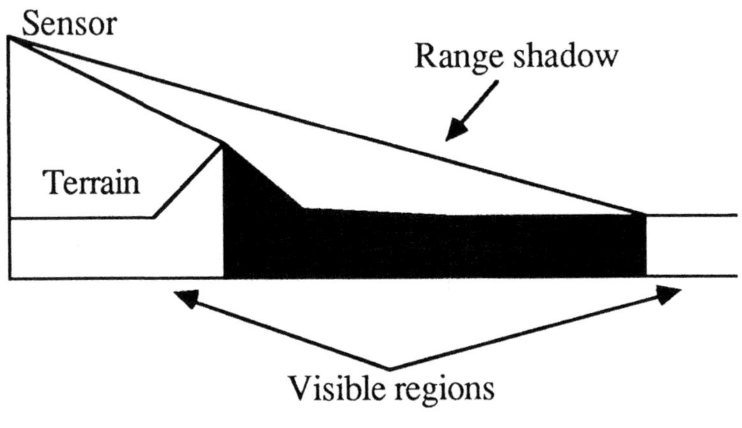

Figure 8.9: An example of a range shadow

8.3.2 Terrain representations and path planners

The choice of a terrain representation depends on the path planner used for actually driving the vehicle. For example, the family of planners derived from Lozano-Perez's approach [27] uses discrete obstacles represented by 2-D polygons. By contrast, planners that compare a vehicle model with the local terrain [9,36] use some intermediate representation of the raw elevation map. Furthermore, the choice of a terrain representation and a path planner in turn depend

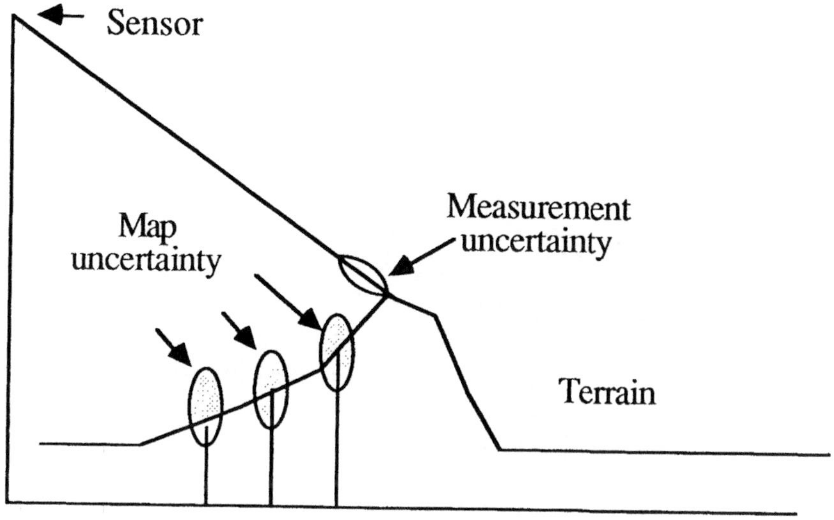

Figure 8.10: Representing uncertainty

on the environment in which the vehicle has to navigate. For example, representing only a small number of discrete upright objects may be appropriate if it is known in advance that the terrain is mostly flat, (*e.g.* a road) with a few obstacles (*e.g.* trees) while cross-country navigation requires a more detailed description of the elevation map. Generating the most detailed description and then extracting the relevant information is not an acceptable solution since it would significantly degrade the performance of the system in simple environments. Therefore, we need several levels of terrain representation corresponding to different resolutions at which the terrain is described (Figure 8.11). At the low resolution level we describe only discrete obstacles without explicitly describing the local shape of the terrain. At the medium level, we include a description of the terrain through surface patches that correspond to significant terrain features. At that level, the resolution is the resolution of the operator used to detect these features. Finally, the description with the highest resolution is a dense elevation map whose resolution is limited only by the sensor. In order to keep the computations involved under control, the resolution is typically related to the size of the vehicle's ground contacts. For example, the size of one foot is used to compute the terrain resolution in the case of a legged vehicle.

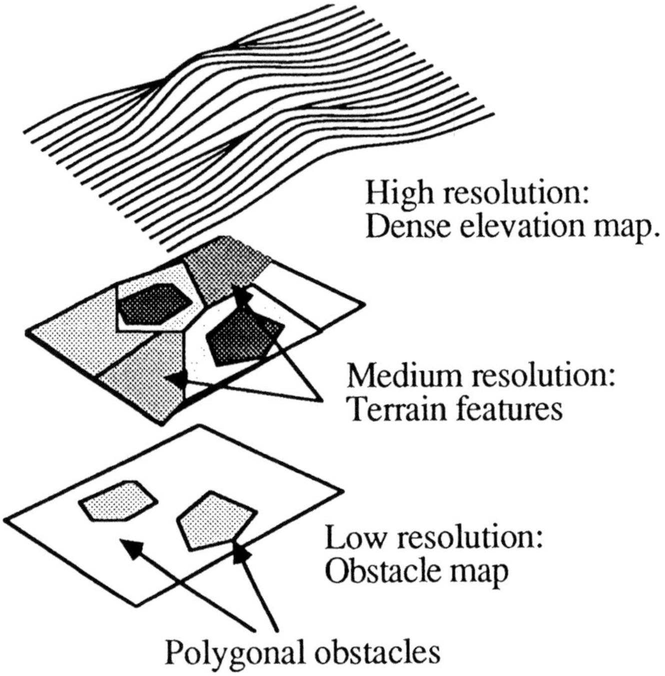

Figure 8.11: Levels of terrain representation

8.3.3 Low resolution: Obstacle map

The lowest resolution terrain representation is an obstacle map which contains a small number of obstacles represented by their trace on the ground plane. Several techniques have been proposed for obstacle detection. The Martin-Marietta ALV [10,11,41] detects obstacles by computing the difference between the observed range image and pre-computed images of ideal ground at several different slope angles. Points that are far from the ideal ground planes are grouped into regions that are reported as obstacles to a path planner. A very fast implementation of this technique is possible since it requires only image differences and region grouping. It makes, however, very strong assumptions on the shape of the terrain. It also takes into account only the positions of potential obstacles and not slopes, so it would be possible for adjacent points to be within the allowable elevation range, while forming a step that is too steep to climb.

Another approach proposed by Hughes AI group [8] is to detect the obstacles by thresholding the normalized range gradient, $\Delta D/D$, and by thresholding the radial slope, $D\Delta\phi/\Delta D$. The first test detects the discontinuities in range, while the second test detects the portion of the terrain with high slope. This approach has the advantage of taking a vehicle model into account when deciding whether a point is part of an obstacle.

In contrast to the Martin Marietta and Hughes groups, we used the terrain map paradigm to detect obstacles for the Navlab. Each cell of the terrain contains the set of data points that fall within its field (Figure 8.7). We can then estimate surface normal and curvatures at each elevation map cell by fitting a reference surface to the corresponding set of data points. Cells that have a high curvature or a surface normal far from the vehicle's idea of the vertical direction are reported as part of the projection of an obstacle. Obstacle cells are then grouped into regions corresponding to individual obstacles. The final product of the obstacle detection algorithm is a set of 2-D polygonal approximations of the boundaries of the detected obstacles that is sent to an A^*-type path planner (Figure 8.12). In addition, we can roughly classify the obstacles into holes or bumps according to the shape of the surfaces inside the polygons.

Figure 8.13 shows the result of applying the obstacle detection algorithm to a sequence of ERIM images. The Figure shows the original range images (top), the range pixels projected in the elevation map (left), and the resulting polygonal obstacle map (right). The large enclosing polygon in the obstacle map is the limit of the visible portion of the world. The obstacle detection algorithm does not make assumptions on the position of the ground plane in that it only assumes that the plane is roughly horizontal with respect to the vehicle.

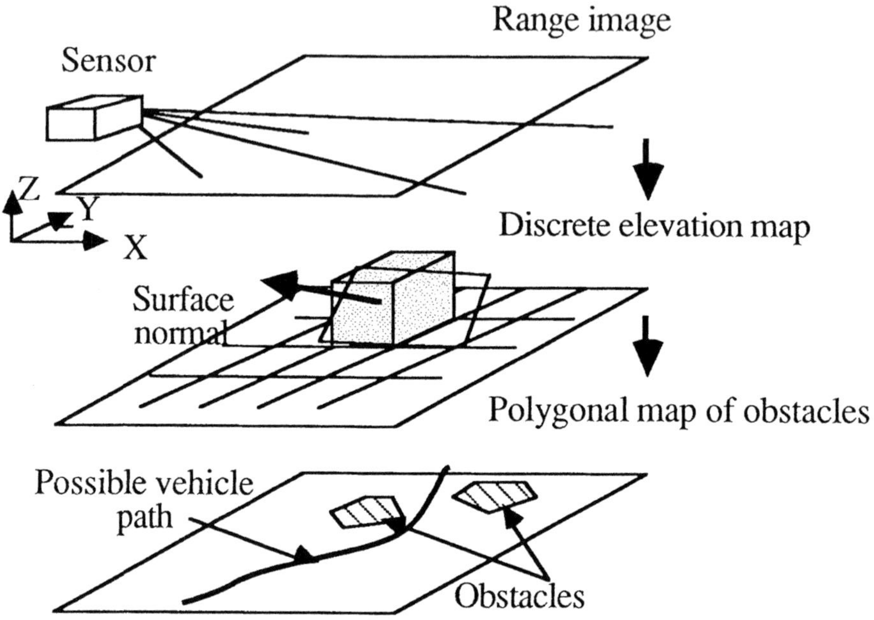

Figure 8.12: Building the obstacle map

Computing the slopes within each cell has a smoothing effect that may cause real obstacles to be undetected. Therefore, the resolution of the elevation map must be chosen so that each cell is significantly larger than the typical expected obstacles. In the case of Figure 8.13, the resolution is twenty centimeters. The size of the detectable obstacle also varies with the distance from the vehicle due to the sampling problem (Section 8.3.1).

One major drawback of our obstacle detection algorithm is that the computation of the slopes and curvatures at each cell of the elevation map is an expensive operation. Furthermore, since low-resolution obstacle maps are most useful for fast navigation through simple environments, it is important to have a fast implementation of the obstacle detection algorithm. A natural optimization is to parallelize the algorithm by dividing the elevation map into blocks that are processed simultaneously. We have implemented such a parallel version of the algorithm on a ten-processor Warp computer [43,20]. The parallel implementation reduced the cycle time to under two seconds, thus making it possible to use the obstacle detection algorithm for fast navigation of the Navlab. In that particular implementation, the vehicle was moving at a continuous speed of one meter per second, taking range images, detecting obstacles, and planning a path every four meters.

8.3.4 Medium resolution: Polygonal terrain map

Obstacle detection is sufficient for navigation in flat terrain with discrete obstacles, such as following a road bordered by trees. We need a more detailed description when the terrain is uneven as in the case of cross-country navigation. For that purpose, an elevation map could be used directly [9] by a path planner. This approach is costly because of the amount of data to be handled by the planner which does not need such a high resolution description to do the job in many cases (although we will investigate some applications in which a high resolution representation is required in Section 8.3.5). An alternative is to group smooth portions of the terrain into regions and edges that are the basic units manipulated by the planner. This set of features provides a compact representation of the terrain thus allowing for more efficient planning [36].

The features used are of two types: smooth regions, and sharp terrain discontinuities. The terrain discontinuities are either discontinuities of the elevation of the terrain, as in the case of a hole, or discontinuities of the surface normals, as in the case of the shoulder of a road [3]. We detect both types of discontinuities by using an edge detector over the elevation map and the surface normals map. The edges correspond to small regions on the terrain surface. Once we have detected the discontinuities, we segment the terrain into smooth regions. The segmentation uses a region growing algorithm that first identifies the smoothest locations in the terrain based on the behavior of the surface normals, and then grows regions around those locations. The result of the processing is a covering of the terrain by regions corresponding either to smooth portions or to edges.

The final representation depends on the planner that uses it. In our case, the terrain representation is embedded in the Navlab system using the path planner described in [36]. The basic geometric object used by the system is the three-dimensional polygon. We therefore approximate the boundary of each region by a polygon. The approximation is done in a way that ensures consistency between regions in that the polygonal boundaries of neighboring regions share common edges and vertices. This guarantees that no "gaps" exist in the resulting polygonal mesh. This is important from the point of view of the path planner since such gaps would be interpreted as unknown portions of the terrain. Each region is approximated by a planar surface that is used by the planner to determine the traversability of the regions. Since the regions are not planar in reality, the standard deviation of the parameters of the plane is associated with each region.

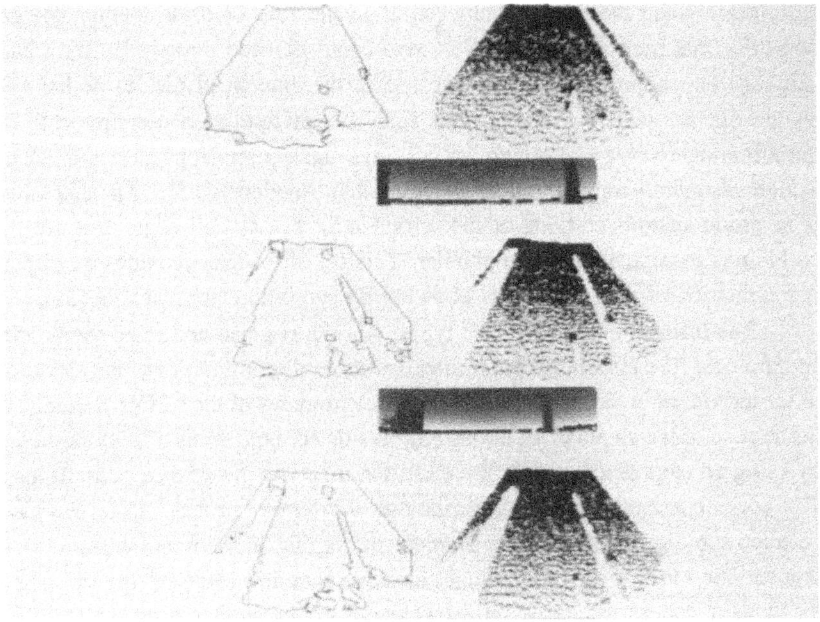

Figure 8.13: Obstacle detection on a sequence of images

Figure 8.14 shows the interpolated elevation map. Figure 8.15 shows the polygonal boundaries of the regions extracted from the image of Figure 8.14. In this implementation, the resolution of the elevation map is twenty centimeters. Since we need a dense map in order to extract edges, we interpolated linearly between the sparse points of the elevation map. This implementation of a medium resolution terrain representation is integrated in the Navlab system and will be part of the standard core system for our future mobile robot systems.

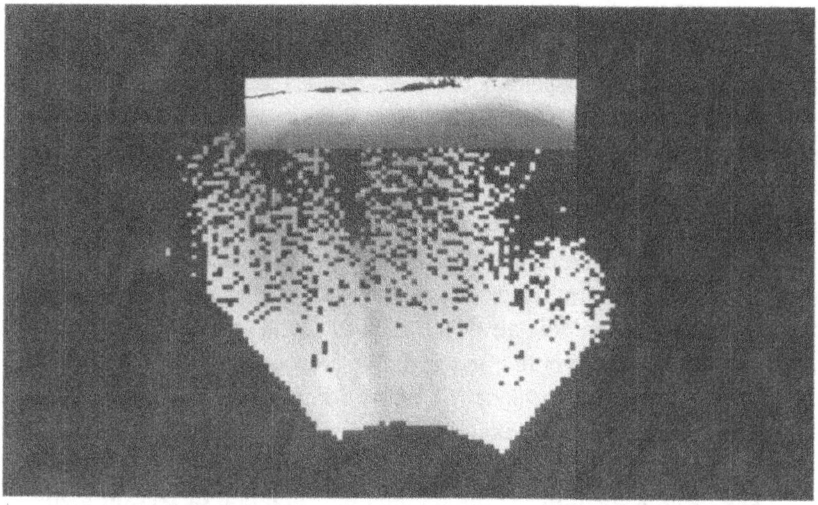

Figure 8.14: Range image and elevation map

8.3.5 High resolution: Elevation maps for rough terrain

The elevation map derived directly from the sensor is sparse and noisy, especially at greater distances from the sensor. Many applications, however, need a dense and accurate high resolution map. One way to derive such a map is to interpolate between the data points using some mathematical approximation of the surface between data points. The models that can be used include linear, quadratic, or bicubic surfaces [32]. Another approach is to fit a surface globally

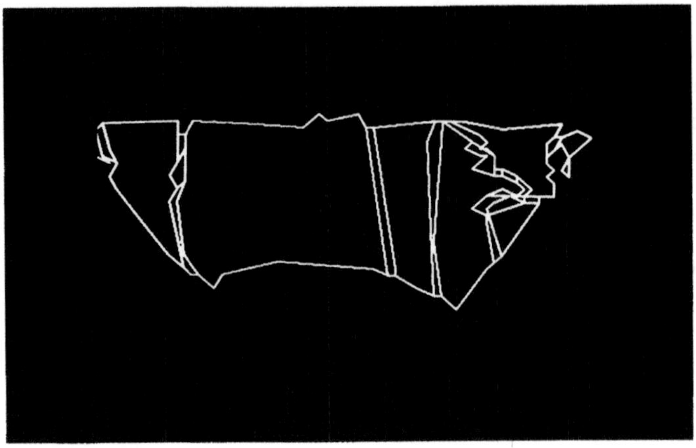

Figure 8.15: Polygonal boundaries of terrain regions

under some smoothness assumptions. This approach includes includes the family of regularization algorithms [6]. Two problems arise with both interpolation approaches: they make apriori assumptions on the local shape of the terrain which may not be valid (*e.g.* in the case of very rough terrain), and they do not take into account the image formation process since they are generic techniques independent of the origin of the data. In addition, the interpolation approaches depend heavily on the resolution and position of the reference grid. For example, they cannot compute an estimate of the elevation at an (x, y) position that is not a grid point without resampling the grid. We propose an alternative, the *locus* algorithm [24], that uses a model of the sensor and provides interpolation at arbitrary resolution without making any assumptions on the terrain shape other than the continuity of the surface.

The locus algorithm for the optimal interpolation of terrain maps.

The problem of finding the elevation z of a point (x, y) is equivalent to computing the intersection of the surface observed by the sensor and the vertical line passing through (x, y). The basic idea of the locus algorithm is to convert the latter formulation into a problem in image space (Figure 8.16). A vertical line is a curve in image space, the *locus*, whose equation as a function of ϕ is:

$$D = D_l(\phi) = \sqrt{\frac{y^2}{\cos^2\phi} + x^2} \tag{8.7}$$

$$\tag{8.8}$$

$$\theta = \theta_l(\phi) = \arctan\frac{x\cos\phi}{y}$$

where ϕ, θ, and D are defined as in Section 8.2. Equation (8.8) was derived by inverting Equation (8.2), and assuming x and y constant. Similarly, the range image can be viewed as a surface $D = I(\phi,\theta)$ in ϕ,θ, D space. The problem is then to find the intersection, if it exists, between a curve parametrized by ϕ and a discrete surface. Since the surface is known only from a sample of data, the intersection cannot be computed analytically. Instead, we have to search along the curve for the intersection point. The search proceeds in two stages: We first locate the two scanlines of the range image, ϕ_1 and ϕ_2, between which the intersection must be located, that is the two consecutive scanlines such that, $Diff(\phi_1) = D_l(\phi_1) - I(\phi_1, \hat{\theta}_l(\phi_1))$ and $Diff(\phi_2) = D_l(\phi_1) - I(\phi_2, \hat{\theta}_l(\phi_2))$ have opposite signs, where $\hat{\theta}_l(\phi)$ is the image column that is the closest to $\theta_l(\phi)$. We then apply a binary search between ϕ_1 and ϕ_2. The search stops when the difference between the two angles ϕ_n and ϕ_{n+1}, where $Diff(\phi_n)$ and $Diff(\phi_{n+1})$ have opposite signs, is lower than a threshold ϵ. Since there are no pixels between ϕ_1 and ϕ_2, we have to perform a local quadratic interpolation of the image in order to compute $\theta_l(\phi)$ and $D_l(\phi)$ for $\phi_1 < \phi < \phi_2$. The control points for the interpolation are the four pixels that surround the intersection point (Figure 8.17). The final result is a value ϕ that is converted to an elevation value by applying Equation (8.2) to $\phi, \theta_l(\phi), D_l(\phi)$. The resolution of the elevation is controlled by the choice of the parameter ϵ.

The locus algorithm enables us to evaluate the elevation at any point since we do not assume the existence of a grid. Figure 8.18 shows the result of applying the locus algorithm on range images of uneven terrain, in this case a construction site. The Figure shows the original range images and the map displayed as an isoplot surface. The centers of the grid cells are ten centimeters apart in the (x, y) plane.

Generalizing the locus algorithm.

We can generalize the locus algorithm from the case of a vertical line to the case of a general line in space. This generalization allows us to build maps using any reference plane instead of being restricted to the (x, y) plane. This is important when, for example, the sensor's (x, y) plane is not orthogonal to the gravity vector. A line in space is defined by a point $u = [u_x, u_y, u_z]^t$, and a unit vector $v = [v_x, v_y, v_z]^t$. Such a line is parametrized in λ by the relation $p = u + \lambda v$ if p is a point on the line. A general line is still a curve in image space that can be parametrized in ϕ if we assume that the line is not parallel to the (x, y) plane. The equation of the curve becomes:

$$D_l(\phi) = \sqrt{(v_x\lambda(\phi) + u_x)^2 + (v_y\lambda(\phi) + u_y)^2 + (v_z\lambda(\phi) + u_z)^2}$$

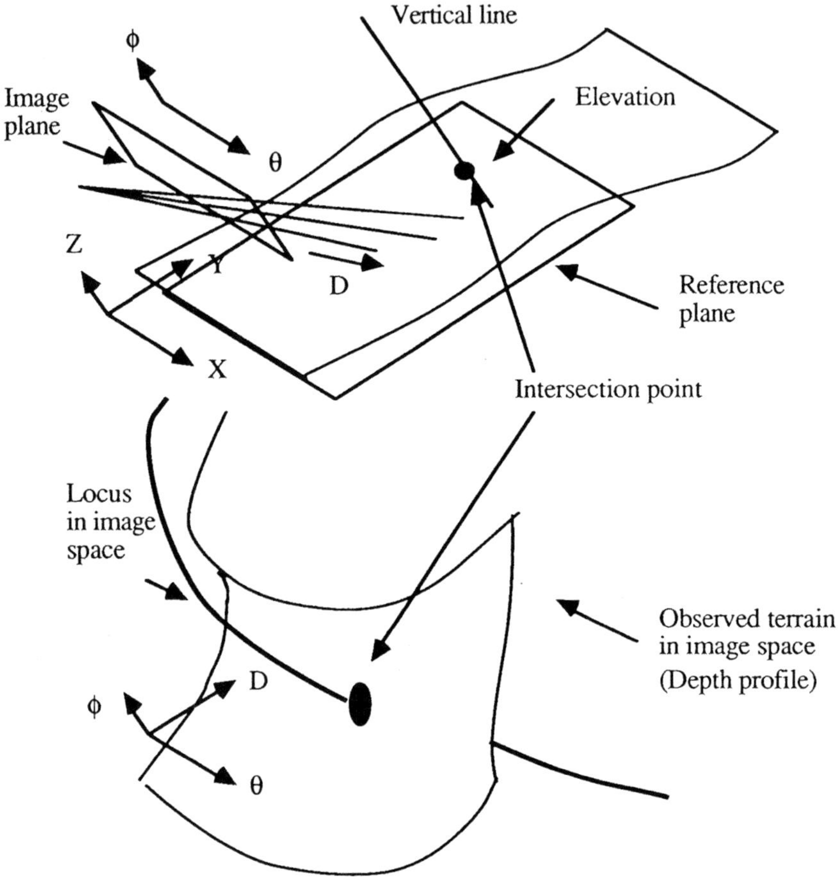

Figure 8.16: The locus algorithm for elevation maps

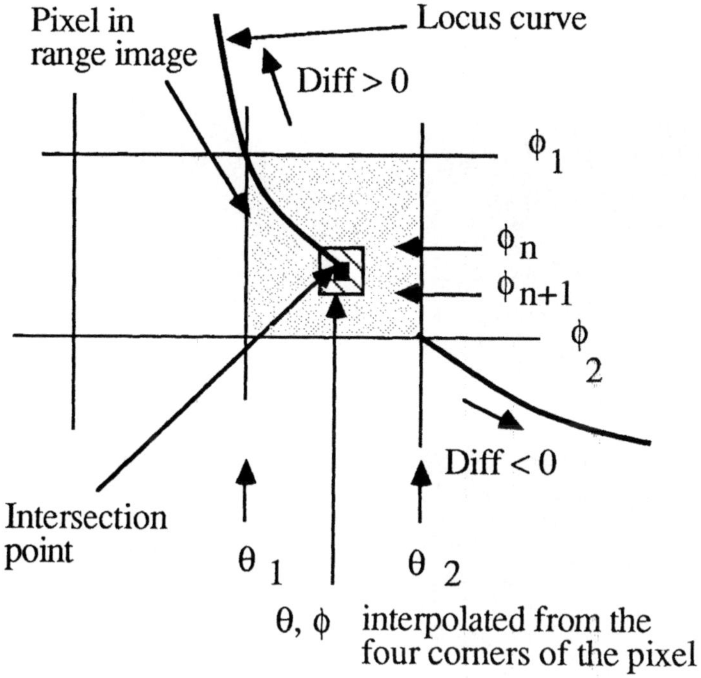

Figure 8.17: Image interpolation around the intersection point

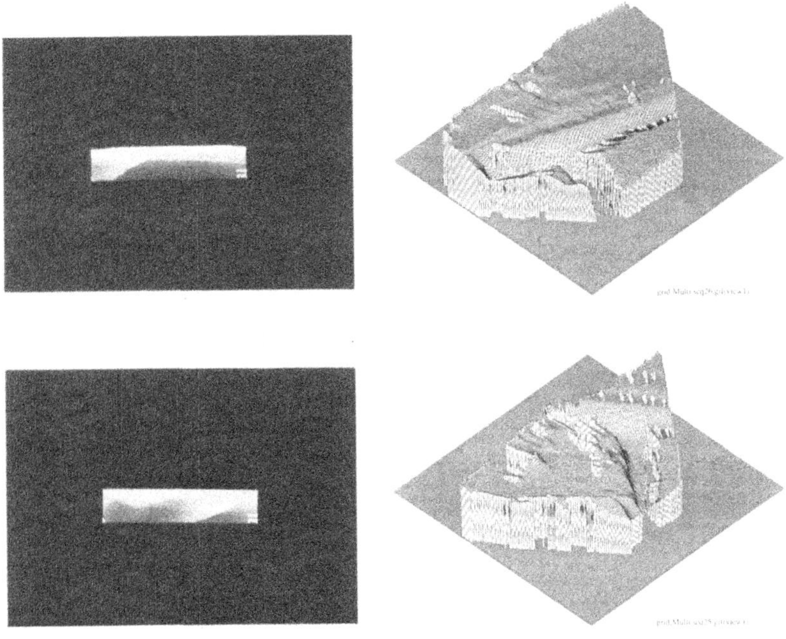

Figure 8.18: The locus algorithm on range images

$$\theta_l(\phi) = \arcsin \frac{v_x\lambda(\phi) + u_x}{D} \tag{8.9}$$

$$\lambda(\phi) = \frac{u_y \tan\phi - u_z}{v_z - v_y \tan\phi}$$

We can then compute the intersection between the curve and the image surface by using the same algorithm as before except that we have to use Equation (8.9) instead of Equation (8.8).

The representation of the line by the pair (u, v) is not optimal since it uses six parameters while only four parameters are needed to represent a line in space. For example, this can be troublesome if we want to compute the Jacobian of the intersection point with respect to the parameters of the line. A better alternative [21] is to represent the line by its slopes in x and y and by its intersection with the plane $z = 0$ (See [34] for a complete survey of 3-D line representations). The equation of the line then becomes:

$$x = az + p \tag{8.10}$$
$$y = bz + q$$

We can still use Equation (8.9) to compute the locus because we can switch between the (a, b, p, q) and (u, v) representations by using the Equations:

$$v = \begin{bmatrix} a \\ b \\ 1 \end{bmatrix}, u = \begin{bmatrix} p \\ q \\ 0 \end{bmatrix} \tag{8.11}$$
$$a = \frac{u_x}{u_z}, p = -\frac{v_x}{u_z}v_z$$
$$b = \frac{u_y}{u_z}, q = -\frac{v_y}{u_z}v_z$$

In the subsequent Sections, we will denote by $h(a, b, p, q)$ the function from R^4 to R^3 that maps a line in space to the intersection point with the range image.

Evaluating the locus algorithm.

We evaluate the locus algorithm by comparing its performance with the other "naive" interpolation algorithms on a set of synthesized range images of simple scenes. The simplest scenes are planes at various orientations. Furthermore, we add some range noise using the model of Section 8.2.3 in order to evaluate the robustness of the approach in the presence of noise. The performances of the algorithms are evaluated by using the mean square error:

$$E = \frac{\sum_{i=1}^{N}(h_i - \tilde{h}_i)^2}{N} \tag{8.12}$$

where h_i is the true elevation value and $\tilde{h}_i$ is the estimated elevation. Figure 8.19 plots E for the locus algorithm and the naive interpolation as a function of the slope of the observed plane and the noise level. This result shows that the locus algorithm is more stable with respect to surface orientation and noise level than the other algorithm. This is due to the fact that we perform the interpolation in image space instead of first converting the data points into the elevation map.

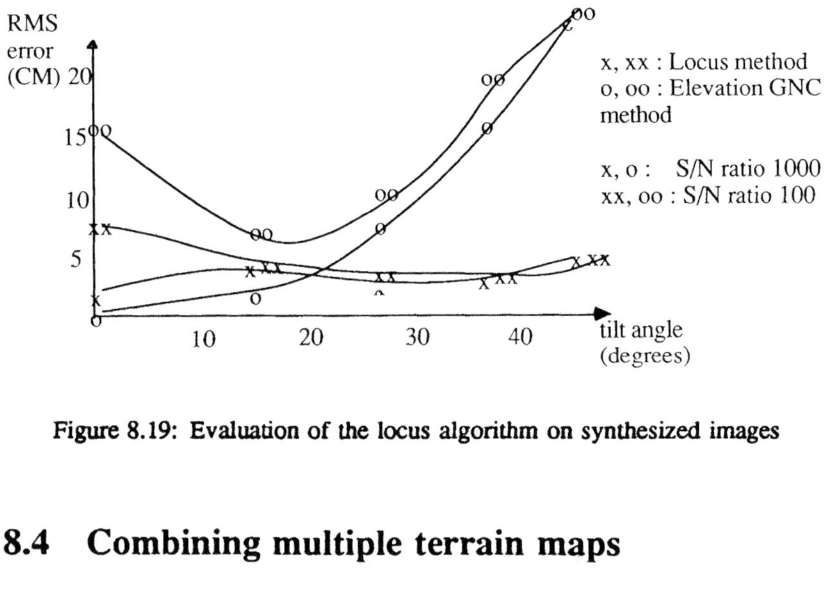

Figure 8.19: Evaluation of the locus algorithm on synthesized images

8.4 Combining multiple terrain maps

We have so far addressed the problem of building a representation of the environment from sensor data collected at one fixed location. For mobile robots, however, we have to deal with a stream of images taken along the vehicle's path. We could ignore this fact and process data from each viewpoint as if it were an entirely new view of the world, forgetting whatever information we may have extracted at past locations. This approach is not appropriate for mobile robot navigation; the representations computed from different vantage points must be combined into a coherent map. Although this has been observed first in the context of indoor mobile robots [13,15], the reasoning behind it holds true in our case. First of all, merging representations from successive viewpoints will produce a map with more information and better resolution than any of the individual maps. For example, a tall object observed by a range sensor creates an unknown area behind it, the range shadow, where no useful information can be extracted (Section 8.3.1). The shape and position of the range shadow change as we move to another location; merging images from several locations will therefore reduce the size of the shadow, thus providing a more complete description to the path planner (Figure 8.20). Another reason why merging maps increases the resolution of the resulting representation concerns the fact that the resolution of an elevation map is significantly better at close range. By merging maps, we can increase the resolution of the parts of the elevation map that were originally measured at a distance from the vehicle.

The second motivation for merging maps is that the position of the vehicle at any given time is uncertain. Even when using expensive positioning systems, we have to assume that the robot's idea of its position in the world will degrade in the course of a long mission. One way to solve this problem is to compute the position with respect to features observed in the world instead of a fixed coordinate system [35,29]. That requires the identification and fusion of common features between successive observations in order to estimate the displacement of the vehicle (Figure 8.21). Finally, combining maps is a mission requirement in the case of an exploration mission in which the robot is sent into an unknown territory to compile a map of the observed terrain.

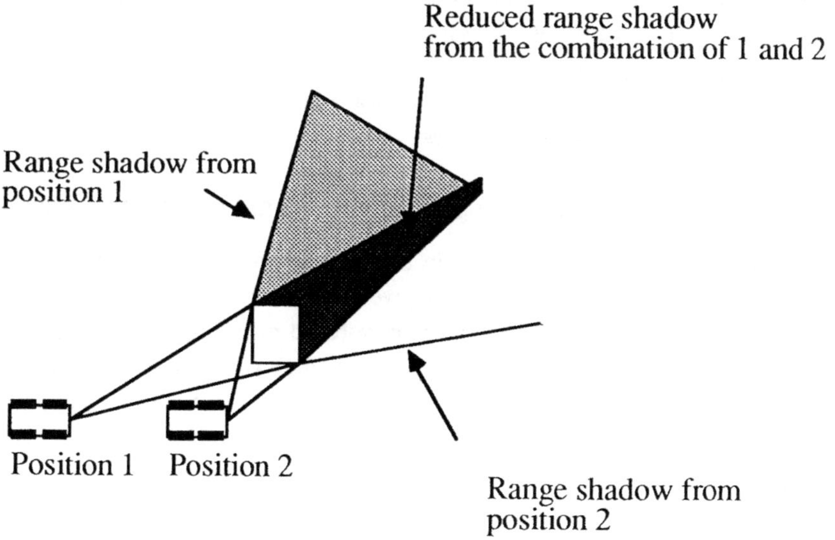

Figure 8.20: Reducing the range shadow

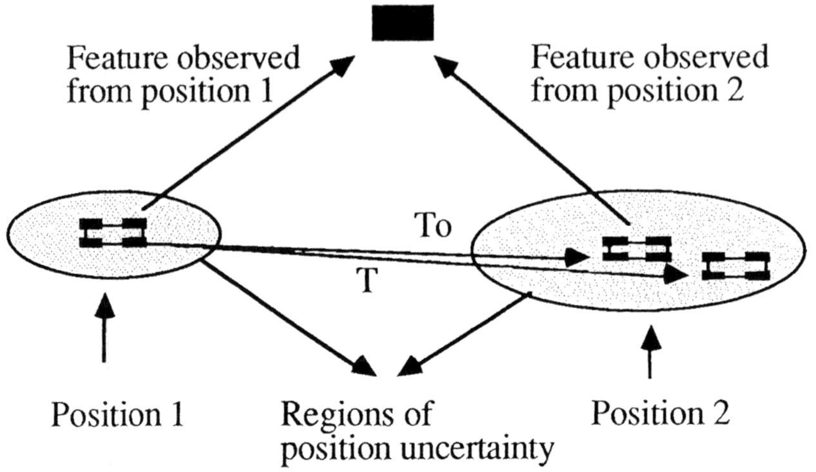

Figure 8.21: Matching maps for position estimation

Many new problems arise when combining maps: representation of uncertainty, data structures for combined maps, predictions from one observation to the next etc. We shall focus on the terrain matching problem, that is the problem of finding common features or common parts between terrain maps so that we can compute the displacement of the vehicle between the two corresponding locations and then merge the corresponding portions of the terrain maps. We always make the reasonable assumption that a rough estimate of the displacement is available since an estimate can always be computed either from dead reckoning or from past terrain matches.

8.4.1 The terrain matching problem: iconic *vs.* feature-based

In the terrain matching problem, as in any problem in which correspondences between two sets of data must be found, we can choose one of two approaches: feature-based or iconic matching. In feature-based matching, we first have to extract two sets of features (F_i^1) and (F_j^2) from the two views to be matched, and to find correspondences between features, $(F_{i_k}^1, F_{j_k}^2)$ that are globally consistent. We can then compute the displacement between the two views from the parameters of the features, and finally merge them into one common map. Although this is the standard approach to object recognition problems [5], it has also been widely used for map matching for mobile robots [13,22,29,7,1,39]. In contrast, iconic approaches work directly on the two sets of data points, P^1 and P^2 by minimizing a cost function of the form $F(T(P^2), P^1)$ where $T(P^2)$ is the set of points from view 2 transformed by a displacement T. The cost is designed so that its minimum corresponds to a "best" estimate of T in some sense. The minimization of F leads to an iterative gradient-like algorithm. Although less popular, iconic techniques have been successfully applied to incremental depth estimation [29,28] and map matching [38,12].

The proponents of each approach have valid arguments. The feature-based approach requires a search in the space of possible matches which may lead to a combinatorial explosion of the matching program. On the other hand, iconic approaches are entirely predictable in terms of computational requirements but are usually quite expensive since the size of the point sets P^i is typically on the order of several thousands. As for the accuracy of the resulting displacement T, the accuracy of iconic techniques can be better than the resolution of the sensors if we iterate the minimization of F long enough, while any feature extraction algorithm loses some of the original sensor accuracy. Furthermore, feature matching could in theory be used even if no a-priori knowledge of T, T_0, is available while iconic approaches require T_0 to be close to the actual displacement because of the iterative nature of the minimization of F.

Keeping these tenets in mind, we combine both approaches into one terrain matching algorithm. The basic idea is to use the feature matching to compute a first estimate $\hat{T}$ given a rough initial value T_0, and then to use an iconic technique to compute an accurate estimate $\hat{\hat{T}}$. This has the advantage of retaining the level of accuracy of iconic techniques while keeping the computation time of the iconic stage under control because the feature matching provides an estimate close enough to the true value. We describe in detail the feature-based and iconic stages in the next three sections.

8.4.2 Feature-based matching

Let F_i^1 and F_j^2 be two sets of features extracted from two images of an outdoor scene, I_1 and I_2. We want to find a transformation $\hat{T}$ and a set of pairs $C_k = (F_{i_k}^1, F_{j_k}^2)$ such that $F_{j_k}^2 \approx \hat{T}(F_{i_k}^1)$, where $T(F)$ denotes the transformation by T of a feature F. The features can be any of those discussed in the previous Sections: points or lines from the local feature extractor, obstacles represented by a ground polygon, or terrain patches represented by their surface equation and their polygonal boundaries. We first investigate the feature matching algorithm independently of any particular feature type so that we can then apply it to any level of terrain representation.

For each feature F_i^1, we can first compute the set of features F_{ij}^2 that could correspond to F_i^1 given an initial estimate T_0 of the displacement. The F_{ij}^2's should lie in a prediction region centered at $T_0(F_i^1)$. The size of the prediction region depends on the confidence we have in T_0 and in the feature extractors. For example, the centers of the polygonal obstacles of Section 8.3.4 are not known accurately, while high curvature points can be accurately located, for example. The confidence on the displacement T is represented by the maximum distance δ between a point in image I_1 and the transformation of its homologue in image I_2, $||Tp^2 - p^1||$, and by the maximum angle ϵ, between a vector in image I_2 and the transformation of its homologue in image I_1 by the rotation part of T. The prediction is then defined as the set of features that are at a Cartesian distance lower than δ, and at an angular distance lower than ϵ from $T_0(F_i^2)$. The parameters used to determine if a feature belongs to a prediction region depend on the type of that feature. For example, we use the direction of a line for the test on the angular distance, while the center of an obstacle is used for the test on the Cartesian distance. Some features may be tested only for orientation, such as lines, or only for position, such as point features. The features in each prediction region are sorted according to some feature distance $d(F_i^1, T_0(F_{ij}^2))$ that reflects how well the features are matched. The feature distance depends also on the type of the feature: for points we use the usual distance, for lines we use the angles between the directions, and for polygonal patches (obstacles or terrain patches) we use a linear combination of the distance between the centers, the difference between the areas, the angle between the surface orientations, and the number of neighboring patches. The features in image I_1 are also sorted according to an "importance" measure that reflects how important the features are for the matching. Such importance measures include the length of the lines, the strength of the point features (*i.e.* the curvature value) , and the size of the patches. The importance measure also includes the type of the features because some features such as obstacles are more reliably detected than others, such as point features.

Once we have built the prediction regions, we can search for matches between the two images. The search proceeds by matching the features F_i^1 to the features F_{ij}^2 that are in their prediction region starting at the most important feature. We have to control the search in order to avoid a combinatorial explosion by taking advantage of the fact that each time a new match is added both the displacement and the future matches are further constrained. The displacement is constrained by combining the current estimate T with the displacement computed from a new match (F_i^1, F_{ij}^2). Even though the displacement is described by six components, the number of components of the displacement that can be computed from one single match depends on the type of features involved: point matches provide only three components, line matches provide four components (two rotations and two translations), and region matches provide three components. We therefore combine the components of T with those components of the new match that can be computed. A given match prunes the search by constraining the future potential matches in two ways: if connectivity relations between features are available, as in the case of terrain patches, then a match (F_i^1, F_{ij}^2) constrains the possible matches for the neighbors of F_i^1) in that they have to be adjacent to F_{ij}^2. In the case of points or patches, an additional constraint is induced by the relative placement of the features in the scene: two matches, (F_i^1, F_{ij}^2) and $(F_{i'}^1, F_{i'j'}^2)$, are compatible only if the angle between the vectors $w^1 = \overline{F_{i'}^1 F_i^1}$ and $w^2 = \overline{F_{i'j'}^2 F_{ij}^2}$ is lower than π, provided the rotation part of T is no greater than π which is the case in realistic situations. This constraint means that the relative placement of the features remains the same from image to image which is similar to the classical ordering constraint used in stereo matching.

The result of the search is a set of possible matchings, each of which is a set of pairs $S = (F_{i_k}^1, F_{j_k}^2)_k$ between the two sets of features. Since we evaluated T simply by combining components in the course of the search, we have to evaluate T for each S in order to get an accurate estimate. T is estimated by minimizing an error function of the form:

$$E = \sum_k d(F_{i_k}^1 - T(F_{i_k}^2)) \tag{8.13}$$

The distance $d(.)$ used in Equation (8.13) depends on the type of the features involved: For point features, it is the usual distance between two points; for lines it is the weighted sum of the angle between the two lines and the distance between the distance vectors of the two lines; for regions it is the weighted sum of the distance between the unit direction vectors and the distance between the two direction vectors. All the components of T can be estimated in general by minimizing E. We have to carefully identify, however, the cases in which insufficient features are present in the scene to fully constrain the transformation. The matching S that realizes the minimum E is reported as the final match between the two maps while the corresponding displacement $\hat{T}$ is reported as the best estimate of the displacement between the two maps. The error $E(\hat{T})$ can then be used to represent the uncertainty in T.

This approach to feature based matching is quite general so that we can apply it to many different types of features, provided that we can define the distance $d(.)$ in Equation (8.13), the importance measure, and the feature measure. The approach is also fairly efficient as long as δ and ϵ do not become too large, in which case the search space becomes itself large. We describe two implementations of the feature matching algorithm in the next two Sections.

Example: Matching polygonal representations.

We have implemented the feature-based matching algorithm on the polygonal descriptions of Section 8.3.4 and 8.3.3. The features are in this case:

- The polygons describing the terrain parametrized by their areas, the equation of the underlying surface, and the center of the region
- The polygons describing the trace of the major obstacles detected (if any).
- The road edges found in the reflectance images if the road detection is reliable enough. The reliability is measured by how much a pair of road edges deviates from the pair found in the previous image.

The obstacle polygons have a higher weight in the search itself because their detection is more reliable than the terrain segmentation, while the terrain regions and the road edges contribute more to the final estimate of the displacement since their localization is better. Once a set of matches and a displacement T are computed, the obstacles and terrain patches that are common between the current map and a new image are combined into new polygons, the new features are added to the map while updating the connectivity between features.

This application of the feature matching has been integrated with the rest of the Navlab system. In the actual system, the estimates of the displacement T_0 are taken from the central database that keeps track of the vehicle's position. The size of prediction region is fixed with δ = one meter, and $\epsilon = 20°$. This implementation of the feature matching has performed successfully over the course of runs of several hundred meters. The final product of the matching is a map that combines all the observations made during the run, and a list of updated obstacle descriptions that are sent to a map module at regular intervals. Since errors in determining position tend to accumulate during such long runs, we always keep the map centered around the current vehicle position. As a result, the map representation is always accurate close to the current vehicle position. Figure 8.24 shows the result of the matching on five consecutive images separated by about one meter. The scene in this case is a road bordered by a few trees. Figure 8.22 shows the original sequence of raw range and reflectance images, Figure 8.23 shows perspective views of the corresponding individual maps, and Figure 8.24 is a rendition of the combined maps using the displacement and matches computed from the feature matching algorithm. This last display is a view of the map rotated by 45° about the x axis and shaded by the values from the reflectance image.

Example: Matching local features from high resolution maps.

Matching local features from high resolution maps provides the displacement estimate for the iconic matching of high resolution maps. The primitives

Figure 8.22: A sequence of range and reflectance images

used for the matching are high curvature points and lines. The initial matches are based on the similarity of the length of the lines and the similarity of the curvature strength of the points. The search among candidate matches proceeds as described in Section 8.4.2. Since we have dense elevation at our disposal in this case, we can evaluate a candidate displacement over the entire map by summing up the squared differences between points in one map and points in the transformed map. Figure 8.25 shows the result of the feature matching on a pair of maps. The top image shows the superposition of the contours and features of the two maps using the estimated displacement (about one meter translation and 4° rotation), while the bottom image shows the correspondences between the point and line features in the two maps. The lower map is transformed by T with respect to the lower right map. Figure 8.26 shows the result of the feature matching in a case in which the maps are separated by a very large displacement. The lower left display shows the area that is common between the two maps after the displacement. Even though the resulting displacement is not accurate enough to reliably merge the maps, it is close enough to the optimum to be used as the starting point of a minimization algorithm.

8.4.3 Iconic matching from elevation maps

The general idea of the iconic matching algorithm is to find the displacement T between two elevation maps from two different range images that minimizes an error function computed over the entire combined elevation map. The error

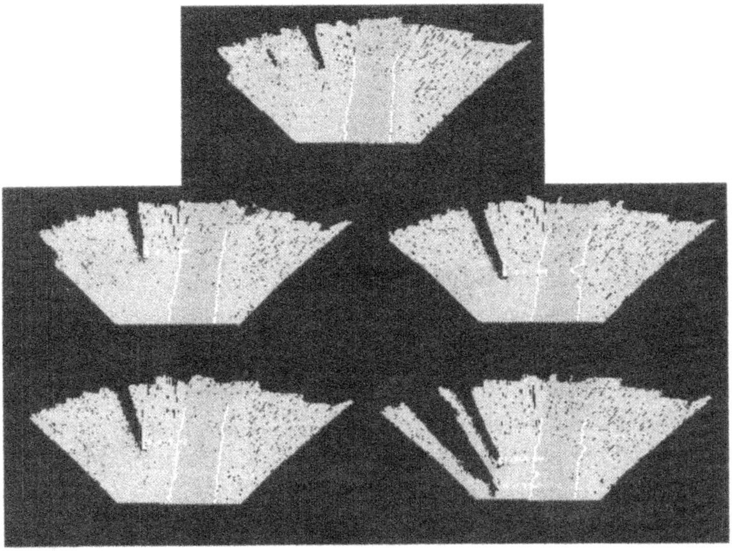

Figure 8.23: Individual maps

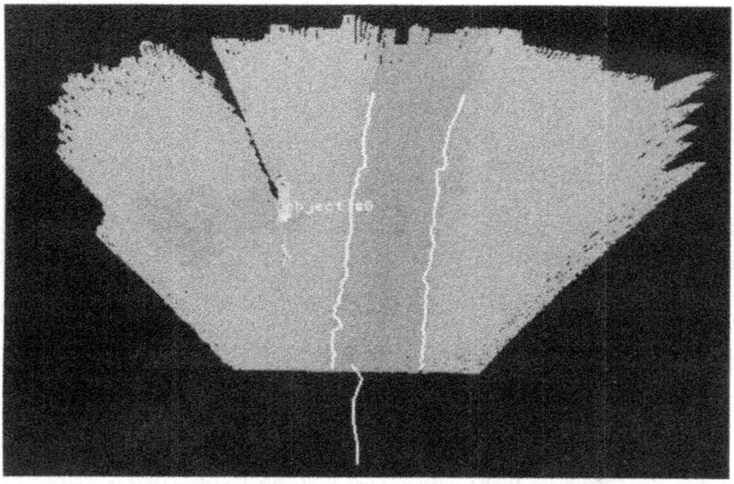

Figure 8.24: Perspective view of the combined map

Figure 8.25: Matching maps using local features

Figure 8.26: Matching maps using local features (large rotation component)

function E measures how well the first map and the transformation of the second map by T do agree. The easiest formulation for E is the sum of the squared differences between the elevation at a location in the first map and the elevation at the same location computed from the second map using T. To be consistent with the earlier formulation of the locus algorithm, the elevation at any point of the first map is actually the intersection of a line containing this point with the range image. We need some additional notations to formally define E: R and t denote the rotation and translation parts of T respectively, $f_i(u,v)$ is the function that maps a line in space, described by a point and a unit vector, to a point, by the generalized locus algorithm of Section 8.3.5 applied to image i. We have then:

$$E = \sum \|f_1(u,v) - g(u,v,T)\|^2 \tag{8.14}$$

where $g(u,v,T)$ is the intersection of the transformation of the line (u,v) by T with image I_2 expressed in the coordinate system of image I_1 (Figure 8.27). The summation in Equation (8.14) is taken over all the locations (u,v) in the first map where both $f_1(u,v)$ and $g(u,v,T)$ are defined. The lines (u,v) in the first map are parallel to the z-axis. In other words:

$$g(u,v,T) = T^{-1}(f_2(u',v')) = R'f_2(u',v') + t' \tag{8.15}$$

where $T^{-1} = (R',t') = (R^{-1}, -R^{-1}t)$ is the inverse transformation of T, and $(u',v') = (Ru + t, Rv)$ is the transformation of the line (u,v). This Equation demonstrates one of the reasons why the locus algorithm is powerful: in order to compute $f_2(Ru+t, Rv)$ we can directly apply the locus algorithm, whereas we would have to do some interpolation or resampling if we were using conventional grid-based techniques. We can also at this point fully justify the formulation of the generalized locus algorithm in Section 8.3.5: The transformed line (u', v') can be anywhere in space in the coordinate system of image I_2, even though the original line (u, v) is parallel to the z-axis, necessitating the generalized locus algorithm to compute $f_2(u', v')$.

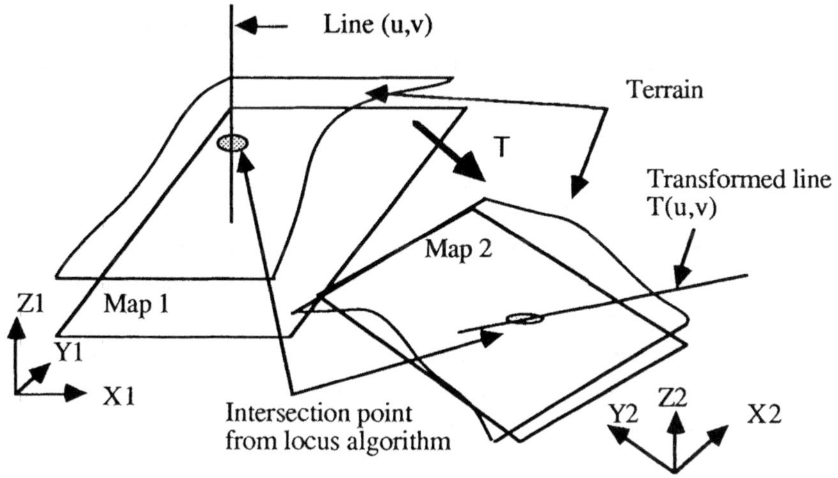

Figure 8.27: Principle of the iconic matching algorithm

We now have to find the displacement T for which E is minimum. If $\nu = [\alpha, \beta, \gamma, t_x, t_y, t_z]^t$ is the 6-vector of parameters of T, where the first three components are the rotation angles and the last three are the components of the translation vector, then E reaches a minimum when:

$$\frac{\partial E}{\partial \nu} = 0 \tag{8.16}$$

Assuming an initial estimate T_0, such a minimum can be found by an iterative gradient descent of the form:

$$\nu^{i+1} = \nu^i + k\frac{\partial E}{\partial \nu}(\nu^i) \tag{8.17}$$

where ν^i is the estimate of ν at iteration i. From Equation (8.14), the derivative of E can be computed by:

$$\frac{\partial E}{\partial \nu} = -2\sum (f_1(u,v) - g(u,v,T))\frac{\partial g}{\partial \nu}(u,v,T) \tag{8.18}$$

From Equation (8.15), we get the derivative of g:

$$\frac{\partial g}{\partial \nu}(u,v,T) = R'\frac{\partial f_2}{\partial \nu}(u',v') + \frac{\partial R'}{\partial \nu}f_2(u',v') + \frac{\partial t'}{\partial \nu} \tag{8.19}$$

The derivatives appearing in the last two components in Equation (8.19) are the derivatives of the transformation with respect to its parameters which can be computed analytically. The last step to compute the derivative of $g(u,v,T)$ is therefore to compute the derivative of $f_2(u',v')$ with respect to ν. We could write the derivative with respect to each component ν_i of ν by applying the chain rule directly:

$$\frac{\partial f_2}{\partial \nu_i}(u',v') = \frac{\partial f_2}{\partial u}\frac{\partial u'}{\partial \nu_i} + \frac{\partial f_2}{\partial v}\frac{\partial v'}{\partial \nu_i} \tag{8.20}$$

Equation (8.20) leads however to instabilities in the gradient algorithm because, as we pointed out in Section 8.3.5, the (u,v) representation is an ambiguous representation of lines in space. We need to use a non ambiguous representation in order to correctly compute the derivative. According to equation (8.12), we can use interchangeably the (u,v) representation and the unambiguous (a,b,p,q) representation. Therefore by considering f_2 as a function of the transform by T, $l' = (a',b',p',q')$, of a line $l = (a,b,p,q)$ in image 1, we can transform Equation (8.20) to:

$$\frac{\partial f_2}{\partial \nu_i}(l') = \frac{\partial f_2}{\partial l'}\frac{\partial l'}{\partial \nu_i} \tag{8.21}$$

Since the derivative $\partial f_2/\partial l'$ depends only on the data in image I_2, we cannot compute it analytically and have to estimate it from the image data. We approximate the derivatives of f_2 with respect to a, b, p, and q by differences of the type:

$$\frac{\partial f_2}{\partial a} = \frac{f(a+\Delta a,b,p,q) - f(a,b,p,q)}{\Delta a} \tag{8.22}$$

Approximations such as Equation (8.22) work well because the combination of the locus algorithm and the GNC image smoothing produces smooth variations of the intersection points.

The last derivatives that we have to compute to complete the evaluation of $\partial E/\partial \nu$ are the derivatives of l' with respect to each motion parameter ν_i. We start by observing that if $X = [x,y,z]^t$ is a point on the line of parameter l, and

$X' = [x', y', z']^t$ is the transformation of X by T that lies on a line of parameter l', then we have the following relations from Equation (8.12):

$$x = az + p, x' = a'z' + p' \quad (8.23)$$
$$y = bz + q, y' = b'z' + q'$$

By eliminating X and X' between Equation (8.23) and the relation $X' = RX + t$, we have the relation between l and l':

$$a' = \frac{R_x.V}{R_z.V}, \quad p' = R_x.U + t_x - a'(R_z.U + t_z) \quad (8.24)$$
$$b' = \frac{R_y.V}{R_z.V}, \quad q' = R_y.U + t_y - b'(R_z.U + t_z)$$

where R_x, R_y, R_z are the row vectors of the rotation matrix R, $A = [a, b, 1]^t$, $B = [p, q, 0]^t$. We now have l' as a function of l and T, making it easy to compute the derivatives with respect to ν_i from Equation (8.24).

In the actual implementation of the matching algorithm, the points at which the elevation is computed in the first map are distributed on a grid with 10 centimeter spacing. The lines (u, v) are vertical and pass through the centers of the grid cells. E is normalized by the number of points since the size of the overlap region between the two maps is not known in advance. We first compute the $f_1(u, v)$ for the entire grid for image 1, and then directly apply the gradient descent algorithm described above. The iterations stop either when the variation of error ΔE is small enough, or when E itself is small enough. Since the matching is computationally expensive, we compute E over an eight by eight meter window in the first image. The last test ensures that we do not keep iterating if the error is smaller than what can be reasonably achieved given the characteristics of the sensor. Figure 8.28 shows the result of combining three high resolution elevation maps. The displacements between maps are computed using the iconic matching algorithm. The maps are actually combined by replacing the elevation $f_1(u, v)$ by the combination:

$$\frac{\sigma_1 f_2 + \sigma_2 f_1}{\sigma_1 + \sigma_2} \quad (8.25)$$

where σ_1 and σ_2 are the uncertainty values computed. Equation (8.25) is derived by considering the two elevation values as Gaussian distributions. The resulting mean error in elevation is lower than ten centimeters. We computed the initial T_0 by using the local feature matching of Section 8.4.2. This estimate is sufficient to ensure the convergence to the true value. This is important because the gradient descent algorithm converges towards a local minimum, and it is therefore important to show that T_0 is close to the minimum. Figure 8.29 plots the value of the ν_i's with respect to the number of iterations. These curves show that E converges in a smooth fashion. The coefficient k that controls the rate of convergence is very conservative in this case in order to avoid oscillations about the minimum.

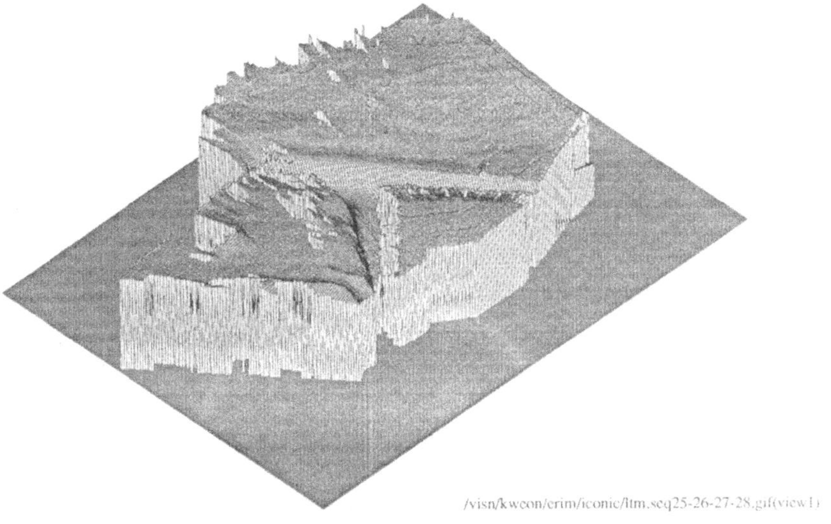

Figure 8.28: Combining four maps by the iconic matching algorithm

8.5 Combining range and intensity data

In the previous Section we have concentrated on the use of 3-D vision as it relates solely to the navigation capabilities of mobile robots. Geometric accuracy was

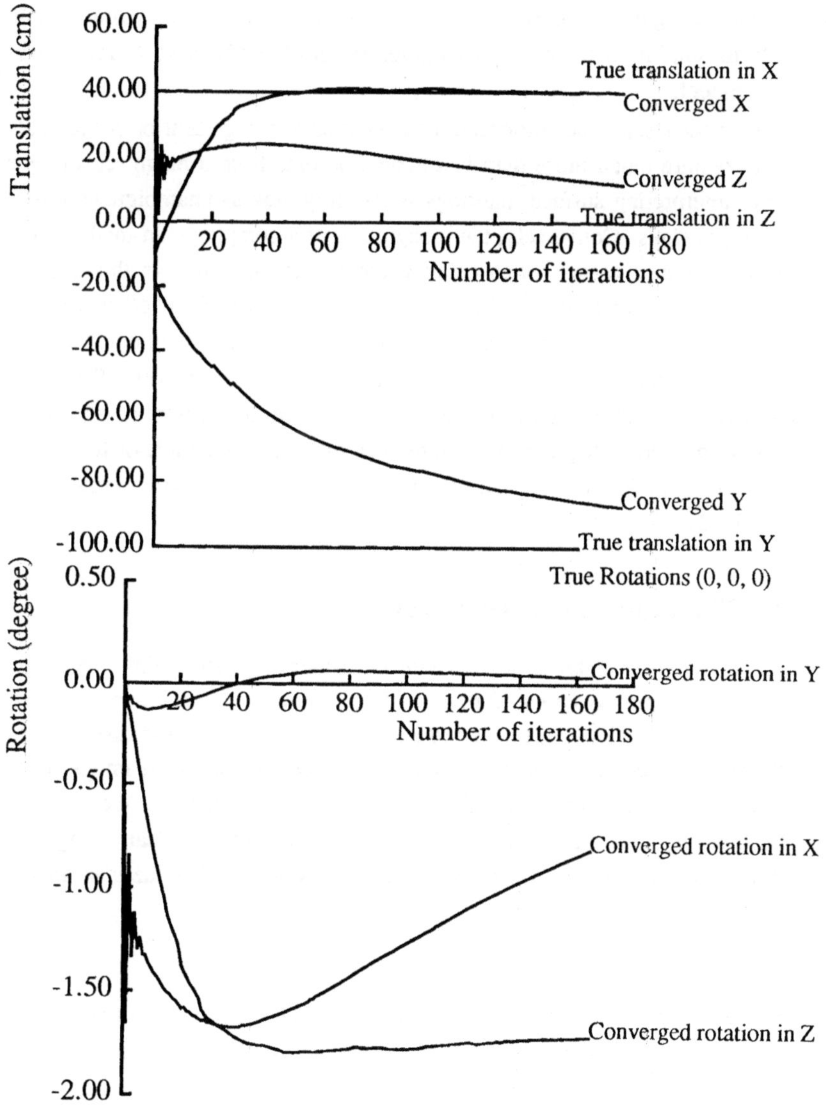

Figure 8.29: Convergence rate of the matching algorithm

the deciding factor in the choice of representations and algorithms while we gave very little attention to the extraction of semantic information. A mobile robot needs more than just navigation capabilities, however, since it also must be able to extract semantic descriptions from its sensors. For example, we will describe a landmark recognition algorithm in Section 5. In that case, the system is able not only to build a geometric representation of an object but also to relate it to a stored model.

Extracting semantic information for landmark recognition or scene analysis may require much more than just geometric data from a range sensor. For example, interpreting surface markings is the only way to unambiguously recognize traffic signs. Conversely, the recognition of a complex man-made object of uniform color is easiest when using geometric information. In this Section we address the problem of combining 3-D data with data from other sensors. The most interesting problem is the combination of 3-D data with color images since these are the two most common sensors for outdoor robots. Since the sensors have different fields of view and positions, we first present an algorithm for transforming the images into a common frame. As an example of the use of combined range/color images, we describe a simple scene analysis program in Section 8.5.3.

8.5.1 The geometry of video cameras

The video camera is a standard color vidicon camera equipped with wide-angle lenses. The color images are 480 rows by 512 columns, and each band is coded on eight bits. The wide-angle lens induces a significant geometric distortion in that the relation between a point in space and its projection on the image plane does not obey the laws of the standard perspective transformation. We alleviate this problem by first transforming the actual image into an "ideal" image: if (R, C) is the position in the real image, then the position (r, c) in the ideal image is given by:

$$r = f_r(R, C), c = f_c(R, C) \quad (8.26)$$

where f_r and f_c are third order polynomials. This correction is cheap since the right-hand side of (8.26) can be put in lookup tables. The actual computation of the polynomial is described in [30] The geometry of the ideal image obeys the laws of the perspective projection in that if $P = [x, y, z]^t$ is a point in space, and (r, c) is its projection in the ideal image plane, then:

$$r = fx/z, c = fy/z \quad (8.27)$$

where f is the focal length. In the rest of the paper, row and column positions will always refer to the positions in the ideal image, so that perspective geometry is always assumed.

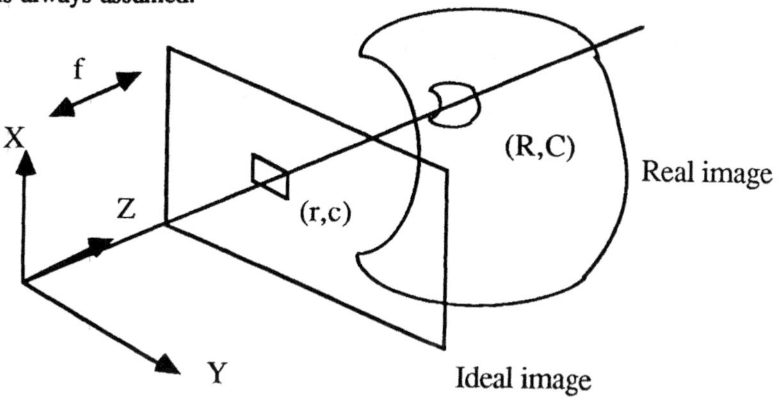

Figure 8.30: Geometry of the video camera

8.5.2 The registration problem

Range sensors and video cameras have different fields of view, orientations, and positions. In order to be able to merge data from both sensors, we first have to estimate their relative positions, known as the calibration or registration problem (Figure 8.31). We approach the problem as a minimization problem in which pairs of pixels are selected in the range and video images. The pairs are selected so that each pair is the image of a single point in space as viewed from the two sensors. The problem is then to find the best calibration parameters given these pairs of points and is further divided into two steps: we first use a simple linear least-squares approach to find a rough initial estimate of the parameters, and then apply a non-linear minimization algorithm to compute an optimal estimate of the parameters.

The calibration problem as a minimization problem.

Let P_i be a point in space, with coordinates P_i^e with respect to the range sensor, and coordinates P_i^c with respect to the video camera. The relationship between the two coordinates is:

$$P_i^c = RP_i^e - T \qquad (8.28)$$

where R is a rotation matrix, and T is a translation vector. R is a non-linear function of the orientation angles of the camera: pan (α), tilt (β), and rotation (γ). P_i^e can be computed from a pixel location in the range image. P_i^c is not

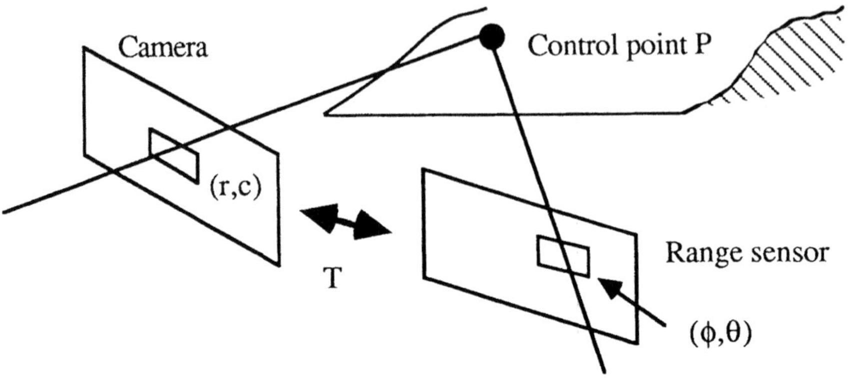

Figure 8.31: Geometry of the calibration problem

completely known, but is related to the pixel position in the video image by the perspective transformation:

$$z_i^c r_i = f x_i^c \tag{8.29}$$

$$z_i^c c_i = f y_i^c \tag{8.30}$$

where f is the focal length. Substituting (8.28) into (8.29) and (8.30) we get:

$$R_z P_i^e r_i - T_z r_i - f R_x P_i^e + T_x' = 0 \tag{8.31}$$

$$R_z P_i^e c_i - T_z c_i - f R_y P_i^e + T_y' = 0 \tag{8.32}$$

where R_x, R_y, and R_z are the row vectors of the rotation matrix R, and $T_y' = fT_y$, $T_x' = fT_x$.

We are now ready to reduce the calibration problem to a least-squares minimization problem. Given n points P_i, we want to find the transformation (R,T) that minimizes the left-hand sides of equations (8.31) and (8.32). We first estimate T by a linear least-squares algorithm, and then compute the optimal estimate of all the parameters.

Initial estimation of camera position.

Assuming that we have an estimate of the orientation R, we want to estimate the corresponding T. The initial value of R can be obtained by physical measurements using inclinometers. Under these conditions, the criterion to be minimized is:

$$C = \sum_{i=1}^{n} [(A_i - T_z B_i - fC_i + T_x')^2 + (D_i - T_z E_i - fF_i + T_y')^2] \tag{8.33}$$

where $A_i = R_z P_i^e r_i$, $B_i = r_i$, $C_i = R_x P_i^e$, $D_i = R_z P_i^e c_i$, $E_i = c_i$, and $F_i = R_y P_i^e$ are known and T_z, T'_x, T'_y, f are the unknowns.

Equation (8.33) can be put in matrix form:

$$C = ||U - AV||^2 - ||W - BV||^2 \tag{8.34}$$

where $V = [T'_x, T'_y, T_z, f]^t$, $U = [A_1, .., A_n]^t$, $W = [D_1, .., D_n]^t$, $A = \begin{bmatrix} B_i & 0 & -1 & C_i \\ . & . & . & \\ B_n & 0 & -1 & C_n \end{bmatrix}$, and $B = \begin{bmatrix} E_i & -1 & 0 & F_i \\ . & . & . & \\ E_n & -1 & 0 & F_n \end{bmatrix}$. The minimum for the criterion of Equation (8.34) is attained at the parameter vector:

$$V = (A^t A + B^t B)^{-1}(A^t U + B^t W) \tag{8.35}$$

Optimal estimation of the calibration parameters.

Once we have computed the initial estimate of V, we have to compute a more accurate estimate of (R, T). Since R is a function of (α, β, γ), we can transform the criterion from equation (8.33) into the form:

$$C = \sum_{i=1}^{n} ||I_i - H_i(S)||^2 \tag{8.36}$$

where I_i is the 2-vector representing the pixel position in the video image, $I_i = [r_i, c_i]^t$, and S is the full vector of parameters, $S = [T'_x, T'_y, T_z, f, \alpha, \beta, \gamma]^t$. We cannot directly compute C_{min} since the functions H_i are non-linear; instead we linearize C by using the first order approximation of H_i [26]:

$$C \approx \sum_{i=1}^{n} ||I_i - H_i(S_0) - J_i \Delta S||^2 \tag{8.37}$$

where J_i is the Jacobian of H_i with respect to S, S_0 is the current estimate of the parameter vector, and $\Delta S = S - S_0$. The right-hand side of (8.37) is minimized when its derivative with respect to ΔS vanishes, that is:

$$\sum_{i=1}^{n} J_i^t J_i \Delta S + J_i^t \Delta C_i = 0 \tag{8.38}$$

where $\Delta C_i = I_i - H_i(S_0)$. Therefore, the best parameter vector for the linearized criterion is:

$$\Delta S = -\sum_{i=1}^{n} (J_i^t J_i)^{-1} J_i^t \Delta C_i \tag{8.39}$$

Equation (8.39) is iterated until there is no change in S. At each iteration, the estimate S_0 is updated by: $S_0 \leftarrow S_0 + \Delta S$.

Implementation and performance.

The implementation of the calibration procedure follows the steps described above. Pairs of corresponding points are selected in a sequence of video and range images. We typically use twenty pairs of points carefully selected at interesting locations in the image (*e.g.* corners). An initial estimate of the camera orientation is $(0, \beta, 0)$, where β is physically measured using an inclinometer. The final estimate of S is usually obtained after less than ten iterations. This calibration procedure has to be applied only once, as long as the sensors are not displaced.

Once we have computed the calibration parameters, we can merge range and video images into a colored-range image. Instead of having one single fusion program, we implemented this as a library of fusion functions that can be divided in two categories:

1. Range $\rightarrow$ video: This set of functions takes a pixel or a set of pixels (r^a, c^a) in the range image and computes the location (r^c, c^c) in the video image. This is implemented by directly applying Equations (8.31) and (8.32).
2. Video $\rightarrow$ range: This set of functions takes a pixel or a set of pixels (r^c, c^c) in the video image and computes the location (r^a, c^a) in the range image. The computed location can be used in turn to compute the location of a intensity pixel in 3-D space by directly applying Equation (8.3). The algorithm for this second set of functions is more involved because a pixel in the video image corresponds to a line in space (Figure 8.30) so that Equations (8.31) and (8.32) cannot be applied directly. More precisely, a pixel (r^c, c^c) corresponds, after transformation by (R, T), to a curve C in the range image. C intersects the image at locations (r^a, c^a), where the algorithm reports the location (r^a, c^a) that is the minimum among all the range image pixels that lie on C of the distance between (r^c, c^c) and the projection of (r^a, c^a) in the video image (using the first set of functions). The algorithm is summarized in Figure 8.32.

Figure 8.33 shows the colored-range image of a scene of stairs and sidewalks. The image is obtained by mapping the intensity values from the color image onto the range image. Figure 8.34 shows a perspective view of the colored-range image. In this example [16], we first compute the location of each range pixel (r^a, c^a) in the video image, and then assign the color value to the 64×256 colored-range image. The final display is obtained by rotating the range pixels, the coordinates of which are computed using Equation (8.3).

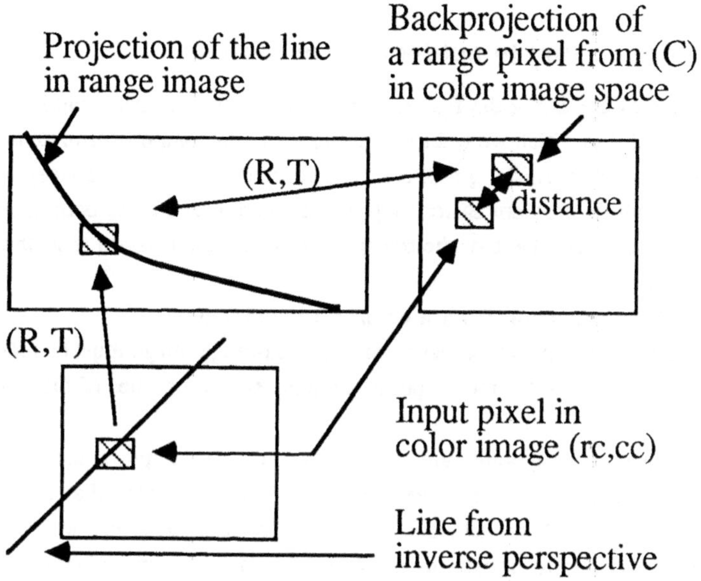

Figure 8.32: Geometry of the "video → range" transformation

Figure 8.33: Colored-range image of stairs

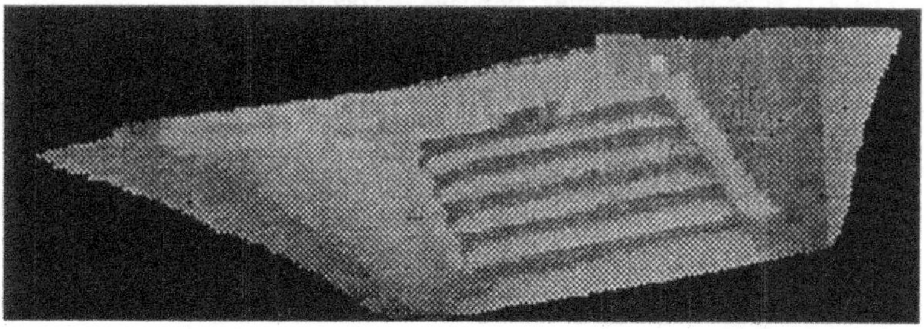

Figure 8.34: Perspective view of registered range and color images

8.5.3 Application to outdoor scene analysis

An example of the use of the fusion of range and video images is outdoor scene analysis [19,25] in which we want to identify the main components of an outdoor scene, such as trees, roads, grass, etc. The colored-range image concept makes the scene analysis problem easier by providing data pertinent to both geometric information (*e.g.* the shape of the trees) and physical information (*e.g.* the color of the road).

Feature extraction from a colored-range image.

The features that we extract from a colored-range image must be related to two types of information: the shapes and the physical properties of the observed surfaces.

The geometric features are used to describe the shape of the objects in the scene. We propose to use two types of features: regions that correspond to smooth patches of surface, and edges that correspond either to transitions between regions, or to transitions between objects (occluding edges). Furthermore, we must be able to describe the features in a compact way. One common approach is to describe the regions as quadric patches, and the edges as sets of tri-dimensional line segments. More sophisticated descriptions are possible [5], such as bicubic patches or curvature descriptors. We use simpler descriptors since the range data is relatively low resolution, and we do not have the type of accurate geometric model that is suited for using higher order geometric descriptors. The descriptors attached to each geometric feature are:

- The parameters describing the shape of the surface patches. That is the parameters of the quadric surface that approximate each surface patch.
- The shape parameters of the surface patches such as center, area, and elongations.
- The 3-D polygonal description of the edges.
- The 3-D edge types: convex, concave, or occluding.

The surface patches are extracted by fitting a quadric of equation $X^tAX + B^tX + C = 0$ to the observed surfaces, where X is the Cartesian coordinate vector computed from a pixel in the range image. The fitting error,

$$E(A,B,C) = \sum_{X_i \in patch} [X_i^t A X_i + B^t X_i + C]^2 \tag{8.40}$$

is used to control the growing of regions over the observed surfaces. The parameters A, B, C are computed by minimizing $E(A,B,C)$ as in [14].

The features related to physical properties are regions of homogeneous color in the video image, that is regions within which the color values vary smoothly. The choice of these features is motivated by the fact that an homogeneous region is presumably part of a single scene component, although the converse is not true as in the case of the shadows cast by an object on an homogeneous patch on the ground. The color homogeneity criterion we use is the distance $(X-m)^t\Sigma^{-1}(X-m)$ where m is the average mean value on the region, Σ is the covariance matrix of the color distribution over the region, and X is the color value of the current pixel in $(red, green, blue)$ space. This is a standard approach to color image segmentation and pattern recognition. The descriptive parameters that are retained for each region are:

- The color statistics (m, Σ).
- The polygonal representation of the region border.
- Shape parameters such as center or moments.

The range and color features may overlap or disagree. For example, the shadow cast by an object on a flat patch of ground would divide one surface patch into two color regions. It is therefore necessary to have a cross-referencing mechanism between the two groups of features. This mechanism provides a two-way direct access to the geometric features that intersect color features. Extracting the relations between geometric and physical features is straightforward since all the features are registered in the colored-range image.

An additional piece of knowledge that is important for scene interpretation is the spatial relationships between features. For example, the fact that a vertical object is connected to a large flat plane through a concave edge may add evidence to the hypothesis that this object is a tree. As in this example, we use three types of relational data:

- The list of features connected to each geometric or color feature.
- The type of connection between two features (convex/concave/occluding) extracted from the range data.
- The length and strength of the connection. This last item is added to avoid situations in which two very close regions become accidentally connected along a small edge.

Scene interpretation from the colored-range image.

Interpreting a scene requires the recognition of the main components of the scene such as trees or roads. Since we are dealing with natural scenes, we cannot use the type of geometric matching that is used in the context of industrial parts recognition [5]. For example, we cannot assume that a given object has specific quadric parameters. Instead, we have to rely on "fuzzier" evidence such as the verticality of some objects or the flatness of others. We therefore implemented the object models as sets of properties that translate into constraints on the surfaces, edges, and regions found in the image. For example, the description encodes four such properties:

- $P1$: The color of the trunk lies within a specific range $\Longrightarrow$ constraint on the statistics (m, Σ) of a color region.
- $P2$: The shape of the trunk is roughly cylindrical $\Longrightarrow$ constraint on the distribution of the principal values of the matrix A of the quadric approximation.
- $P3$: The trunk is connected to a flat region by a concave edge $\Longrightarrow$ constraint on the neighbors of the surface, and the type of the connecting edge.
- $P4$: The tree has two parallel vertical occluding edges $\Longrightarrow$ constraint on the 3-D edges description.

Other objects such as roads or grass areas have similar descriptions. The properties P_{ij} of the known object models M_j are evaluated on all the features F_k extracted from the colored-range image. The result of the evaluation is a score S_{ijk} for each pair (P_{ij}, F_k). We cannot rely on individual scores since some may not be satisfied because of other objects, or because of segmentation problems. In the tree trunk example, one of the lateral occluding edges may itself be occluded by some other object, in which case the score for $P4$ would be low while the score for the other properties would still be high. In order to circumvent this problem, we first sort the possible interpretations M_j for a given feature F_k according to all the scores $(S_{ij})_i$. In doing this, we ensure that all the properties contribute to the final interpretation and that no interpretations are discarded at this stage while identifying the most plausible interpretations.

We have so far extracted plausible interpretations only for individual scene features F_k. The final stage in the scene interpretation is to find the interpretations (M_{j_k}, F_k) that are globally consistent. For example, property $P3$ for the tree implies a constraint on a neighboring region, namely that this has to be a flat ground region. Formally, a set of consistency constraints C_{mn} is associated with each pair of objects (M_m, M_n). The C_{mn} constraints are propagated through the individual interpretations (M_{j_k}, F_k) by using the connectivity information stored in the colored-range feature description. The propagation is simple considering the small number of features remaining at this stage.

The final result is a consistent set of interpretations of the scene features, and a grouping of the features into sets that correspond to the same object. The last result is a by-product of the consistency check and the use of connectivity data. Figure 8.35 shows the color and range images of a scene which contains a road, a couple of trees, and a garbage can. Figure 8.36 shows a display of the corresponding colored-range image in which the white pixels are the points in the range image that have been mapped into the video image. This set of points is actually sparse because of the difference in resolutions between the two sensors, and some interpolation was performed to produce the dense regions of Figure 8.36.

Only a portion of the image is registered due to the difference in field of view between the two sensors (60° for the camera versus 30° in the vertical direction for the range sensor). Figure 8.37 shows a portion of the image in which the edge points from the range image are projected on the color image. The edges are interpreted as the side edges of the tree and the connection between the ground and the tree. Figure 8.38 shows the final scene interpretation. The white dots are the main edges found in the range image. The power of the colored-range image approach is demonstrated by the way the road is extracted. The road in this image is separated into many pieces by strong shadows. Even though the shadows do not satisfy the color constraint on road region, they do perform well on the shape criterion (flatness), and on the consistency criteria (both with the other road regions, and with the trees). The shadows are therefore interpreted as road regions and merge with the other regions into one road region. This type of reasoning is in general difficult to apply when only video data is used unless one uses stronger models of the objects such as an explicit model of a shadowed road region. Using the colored-range image also makes the consistency propagation a much easier task than in purely color-based scene interpretation programs [31].

8.6 Conclusion

We have described techniques for building and manipulating 3-D terrain representations from range images. We have demonstrated these techniques on real images of outdoor scenes. Some of them (Sections 8.3.3, 8.3.4, and 8.4.2) were integrated in a large mobile robot system that was successfully tested in the field. We expect that the module that manipulates and creates these terrain representations will become part of the standard core system of our outdoor mobile robots, just as a local path planner or a low-level vehicle controller are standard modules of a mobile robot system independent of its application. This work will begin by combining the polygonal terrain representation of Section 8.3.4 with the path planner of [36] in order to generate the basic capabilities for an off-road vehicle.

Many issues still remain to be investigated. First of all, we must define a uniform way of representing and combining the uncertainties in the terrain maps.

Figure 8.35: Color and range images of an outdoor scene

Figure 8.36: A view of the corresponding colored-range image

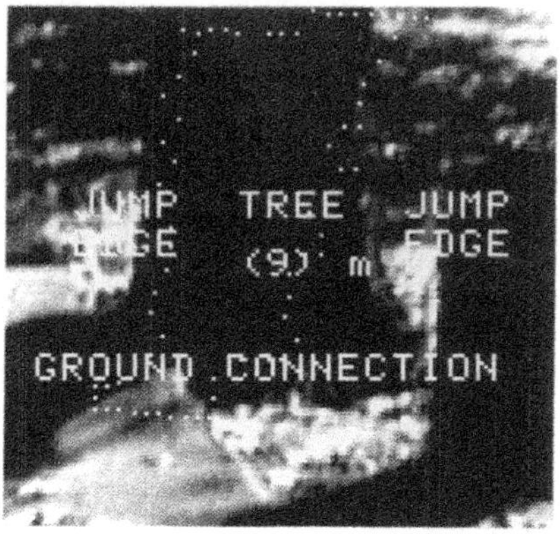

Figure 8.37: Edge features from the colored-range image

Figure 8.38: Final scene interpretation

Currently, the uncertainty models depend heavily on the type of sensor used and on the level at which the terrain is represented. Furthermore, the displacements between terrain maps are known only up to a certain level of uncertainty. This level of uncertainty must be evaluated and updated through the matching of maps, whether iconic or feature-based. Regarding the combination of the 3-D representations with representations from other sensors, we need to define an algorithm for sensor registration that is general enough for application to a variety of situations. The algorithms presented in Section 8.5 are still very dependent on the sensors that we used, and on the intended application. Registration schemes such as [17] would enable us to have a more uniform approach to the problem. An added effect of using such a registration algorithm is that we could explicitly represent errors caused by the combination of the sensors, which we did not do in Section 8.5. Another issue concerns our presentation of the three levels of terrain representation, the matching algorithms, and the sensor combination algorithms as separate problems. We should define a common perceptual architecture to integrate these algorithms in a common representation that can be part of the core system of a mobile robot. Finally, we have tackled the terrain representation problems mainly from a geometrical point of view. Except in Section 8.5, we did not attempt to extract semantic interpretations from the representations. A natural extension of this work is to use the 3-D terrain representations to identify known objects in the scene. Another application along these lines is to use the terrain maps to identify objects of interest, such as terrain regions for sampling tasks for a planetary explorer [23]. Although we have performed some preliminary experiments in that respect [18,2], extracting semantic information from terrain representations remains a major research area for outdoor mobile robots.

8.7 References

[1] M. Asada. Building a 3-D World Model for a Mobile Robot from Sensory Data. In *Proc. IEEE Robotics and Automation*, Philadelphia, 1988.

[2] J. Bares and W. Whittaker. Configuration of an Autonomous Robot for Mars Exploration. In *Proc. World Conference on Robotics*, 1988.

[3] A. Bergman and C. K. Cowan. Noise-Tolerant Range Analysis for Autonomous Navigation. In *IEEE Conf. on Robotics and Automation*, San Francisco, 1986.

[4] P. Besl. *Range Imaging Sensors*. Technical Report GMR-6090, General Motors Research Lab, Warren, MI, March 1988.

[5] P. J. Besl and R. C. Jain. Three-dimensional Object Recognition. *ACM Comp. Surveys*, 17(1), march 1985.

[6] A. Blake and A. Zisserman. *Visual Reconstruction*. MIT Press, Cambridge, MA, 1987.

[7] R. Brooks. Aspects of Mobile Robot Visual Map Making. In *Second International Robotics Research Symposium*, MIT press, 1985.

[8] M. J. Daily, J. G. Harris, and K. Reiser. Detecting Obstacles in Range Imagery. In *Image Understanding Workshop*, Los Angeles, 1987.

[9] M.J. Daily, J.G. Harris, and K. Reiser. An Operational Perception System for Cross-Country Navigation. In *Proc. Image Understanding Workshop*, Cambridge, 1988.

[10] R. T. Dunlay and D. G. Morgenthaler. Obstacle Detection and Avoidance from Range Data. In *Proc. SPIE Mobile Robots Conference*, Cambridge, MA, 1986.

[11] T. Dunlay. Obstacle Avoidance Perception Processing for the Autonomous Land Vehicle. In *Proc. IEEE Robotics and Automation*, Philadelphia, 1988.

[12] A. Elfes. Sonar-Based Real-World Mapping and Navigation. *Journal of Robotics and Automation, Vol. 3*, 1987.

[13] O.D. Faugeras, N. Ayache, and B. Faverjon. Building Visual Maps by Combining Noisy Stereo Measurements. In *Proc. IEEE Conf. on Robotics and Automation*, 1986.

[14] O.D. Faugeras and M. Hebert. The Representation, Recognition, and Locating of 3-D Objects. *International Journal of Robotics Research*, 5(3), 1986.

[15] G. Giralt, R. Chatila, and M. Vaisset. An Integrated Navigation and Motion Control System for Autonomous Multisensory Mobile Robots. In *Proc. 1st International Symposium Robotics Research*, Cambridge, 1984.

[16] Y. Goto, K. Matsuzaki, I. Kweon, and T. Obatake. CMU Sidewalk Navigation System: a Blackboard-Based Outdoor Navigation System Using Sensor Fusion with Colored-Range Images. In *Proc. First Joint Computer Conference*, Dallas, 1986.

[17] K. Gremban, C.E. Thorpe, and T. Kanade. Geometric Calibration Using Systems of Linear Equations. In *Proc. IEEE Robotics and Automation Conf.*, Philadelphia, 1988.

[18] M. Hebert and T. Kanade. 3-D Vision for Outdoor Navigation by an Autonomous Vehicle. In *Proc. Image Understanding Workshop*, Cambridge, 1988.

[19] M. Hebert and T. Kanade. First Results on Outdoor Scene Analysis. In *Proc. IEEE Robotics and Automation*, San Francisco, 1985.

[20] T. Kanade and J.A. Webb. *End of Year Report for Parallel Vision Algorithm Design and Implementation.* Technical Report CMU-RI-TR-87-15, The Robotics Institute, Carnegie-Mellon University, 1987.

[21] M.G. Kendall and P.A.P. Moran. *Geometrical Probabilities.* Hafner Publishers, New York, 1963.

[22] D. Kriegman, E. Triendl, and T.O. Binford. A Mobile Robot: Sensing, Planning and Locomotion. In *Proc. IEEE Conf. on Robotics and Automation*, 1987.

[23] I. Kweon. Modeling Rugged 3-D Terrain from Multiple Range Images for Use by Mobile Robots. 1988. PhD thesis proposal.

[24] I. Kweon, M. Hebert, and T. Kanade. Perception for Rough Terrain Navigation. In *Proc. SPIE Mobile Robots*, Cambridge, MA, 1988.

[25] I. Kweon, M. Hebert, and T. Kanade. Sensor Fusion of Range and Reflectance Data for Outdoor Scene Analysis. In *Proc. Space Operations Automation and Robotics*, Cleveland, 1988.

[26] D.G. Lowe. Solving for the Parameters of Object Models from Image Descriptions. In *ARPA Image Understanding Workshop*, 1980.

[27] T. Lozano-Perez. An Algorithm for Planning Collision Free Paths among Polyhedral Obstacles. *Communications of the ACM*, October 1979.

[28] B.D. Lucas. *Generalized Image Matching by the Method of Differences.* Technical Report CMU-CS-85-160, Carnegie-Mellon University, 1985.

[29] L. Matthies and S.A. Shafer. Error Modeling in Stereo Navigation. *Journal of Robotics and Automation, Vol. 3*, 1987.

[30] H.P. Moravec. *Obstacle Avoidance and Navigation in the Real World by a Seeing Robot Rover.* Technical Report CMU-RI-TR-3, Carnegie-Mellon University, 1980.

[31] Y. Ohta. *Knowledge-based Interpretation of Outdoor Natural Color Scenes.* Pittman Publishing, Inc., 1984.

[32] D.J. Orser and M. Roche. The Extraction of Topographic Features in Support of Autonomous Underwater Vehicle Navigation. In *Proc. Fifth International Symposium on Unmanned Untethered Submersible*, University of New Hampshire, 1987.

[33] J. Ponce and M. Brady. Toward a Surface Primal Sketch. In *IEEE International Conference on Robotics and Automation*, St Louis, 1985.

[34] K.S. Roberts. A New Representation for a Line. In *Proc. Computer Vision and Patter Recognition*, Ann Arbor, MI, 1988.

[35] R.C. Smith and P. Cheeseman. On the Representation and Estimation of Spatial Uncertainty. *International Journal of Robotics Research*, 1986.

[36] T. Stentz. *The NAVLAB System for Mobile Robot Navigation*. PhD thesis, Carnegie-mellon University, Fall 1988.

[37] R. Szeliski. *Bayesian Modeling of Uncertainty in Low-Level Vision*. PhD thesis, Carnegie-mellon University, Computer Science Departement, July 1988.

[38] R. Szeliski. Estimating Motion from Sparse Range Data without Correspondance. In *International Conf. on Computer Vision*, Tarpon Springs, Florida, December 1988.

[39] C.E. Thorpe. *The CMU Rover and the FIDO Vision and Navigation System*. PhD thesis, Carnegie-mellon University, 1984.

[40] C.E. Thorpe, M. Hebert, T. Kanade, and S.A. Shafer. Vision and Navigation for the Carnegie-Mellon Navlab. *PAMI*, 10(3), 1988.

[41] M.A. Turk, D.G. Morgenthaler, K.D. Gremban, and M. Marra. VITS- A Vision System for Autonomous Land Vehicle Navigation. *PAMI*, 10(3), may 1988.

[42] R. Watts, F. Pont, and D. Zuk. *Characterization of the ERIM/ALV Sensor - Range and Reflectance*. Technical Report ???, Environmental Research Institute of Michigan, Ann Arbor, MI, 1987.

[43] J.A. Webb and T. Kanade. Vision on a Systolic Array Machine. In L. Uhr, editor, *Evaluation of multicomputers for image processing*, Academic Press, 1986.

[44] D. Zuk, F. Pont, R. Franklin, and V. Larrowe. *A System for Autonomous Land Navigation*. Technical Report IR-85-540, Environmental Research Institute of Michigan, Ann Arbor MI, 1985.

Chapter 9: The CODGER System for Mobile Robot Navigation

Anthony Stentz

9.1 Introduction

Most sophisticated mobile robot systems are large and complex [7, 6]. For this reason, such systems are usually developed by a team of researchers, rather than an individual. Furthermore, in order for these systems to execute in a reasonable amount of time, the computations must be parallelized to some degree. These two characteristics pose two problems:

- What software engineering tools are needed for developing a large mobile robot system?
- What software support is needed for successfully distributing the computation across a number of processors?

In the case of the Navlab and its development laboratory, the software system consists of five to ten large sensing, planning, control, and graphics modules running on a mixture of Suns and Vaxes interconnected via an EtherNet. Both Lisp and C are available for programming. From a software engineering standpoint, we would like to support module development in multiple languages running on multiple machine architectures, all tied together in a single system. Furthermore, we would like one module programmer to be able to modify or extend one part of the system without requiring redesign or recompilation of the rest of the system. Concerning multiprocessing support, we need a common data format to all modules, regardless of programming language or machine type. We need a mechanism for moving data between modules, and more importantly, synchronizing this exchange. For geometric data, which is used by nearly every module in the system, we must guarantee that it is consistent, so that all modules agree as to the location of the robot and all objects in its environment. A number of systems exist for integrating various modules into a single system and coordinating their activities. Many of these systems are grouped together under the name of *blackboards* [4, 8, 9]. While these systems do permit many modules (or *knowledge sources*) to be integrated into a single system, traditionally multiprocessing is not supported as modules are invoked one at a time. Furthermore, modules are scheduled for execution according to a *static* priority scheme. This mechanism is too restrictive for the Navlab system; a *dynamic* scheme is needed that changes from one context to another. Other systems [1, 2] do provide adequate multiprocessing support, but fall short in supporting geometric data, which is central to robotic systems. We have developed a system named CODGER (for COmmunications Database with GEometric Reasoning) that provides ample support for module development and multiprocessing support. In this paper we describe CODGER and illustrate how it addresses these issues.

9.2 Overview of the CODGER System

The computing resources for the Navlab consist of three to five Sun microcomputers interconnected with an EtherNet. The CODGER system was designed to facilitate communication between system modules for a static mapping of modules to processors in this arrangement. The basic structure of CODGER is illustrated in Figure 9.1. CODGER consists of a central database, called the *local map*, a program to manage the database, called the *local map builder* (LMB), and a library of subroutines for communicating with the LMB, called the *LMB Interface*. The other boxes in the figure are user modules. These modules run as separate programs in the system. Each module is linked with an LMB Interface, through which the module can store and retrieve information from the database. The LMB Interface handles all communication and synchronization with the LMB over the network.

Note that the system configuration is "star-shaped", that is, all data passed between modules must pass through the central database. There are several advantages to this arrangement. First, during development of the system, the communication paths between modules are frequently changed as the system grows and ideas are tested. A star-shaped configuration makes data re-routing easy without requiring new communication channels. Second, data created in one module is often used by many other modules. Figure 9.2 illustrates the "interconnectivity" of a typical CODGER application. In this example, the system contains five large modules that exchange data of 28 different types (tokens). A box is darkened in this grid if the module corresponding to the row reads and/or writes the data type corresponding to the column. Figure 9.3 illustrates the relative activity of each data type (in total transactions), and thus the activity of each communication channel. The CODGER configuration precludes the need for a full clique of communication channels. Third, the CODGER communications scheme is very extensible in that it supports anonymity between modules. We can add a new module that taps into a communication line via the database (for example, a graphics module that displays data in the system), without making changes to the source or destination modules in the line.

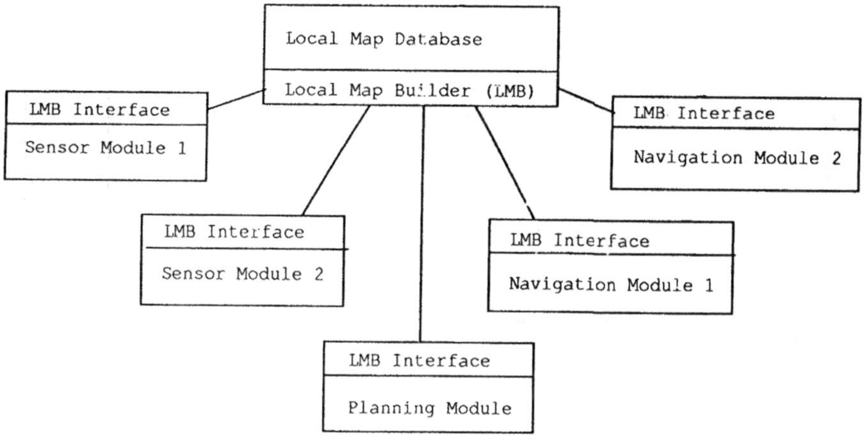

Figure 9.1: CODGER System Structure

CODGER is similar to a traditional blackboard system, in that it has a central database, a database manager, and a pool of modules that can read and write the database. Such systems are *heterarchical* because the "knowledge" in the system is distributed among a number of modules (often called *knowledge sources*), each of which can read or write any piece of the data in the database. Each knowledge source (KS) has a set of preconditions, or predicates defined over the database, which must be satisfied in order for the KS to execute. In a traditional blackboard, the database manager determines which preconditions are satisfied, selects one KS from the eligible set using a fixed priority scheme, invokes the KS as a subroutine, and repeats.

A traditional blackboard functions as a problem-solving framework. At any point in time, the database represents the current state of the solution. The knowledge sources use techniques such as *forward-* or *backward-chaining* to transform the database into a solution state. The database manager relies on a priority scheme to focus the activity of the KS's on those search paths likely to lead to a solution. This problem-solving framework is not needed for our navigation tasks, as the tasks are more algorithmic in nature. Instead, a system is needed that can effectively distribute the computation across multiple processors to maximize parallelism. The serialized control and execution

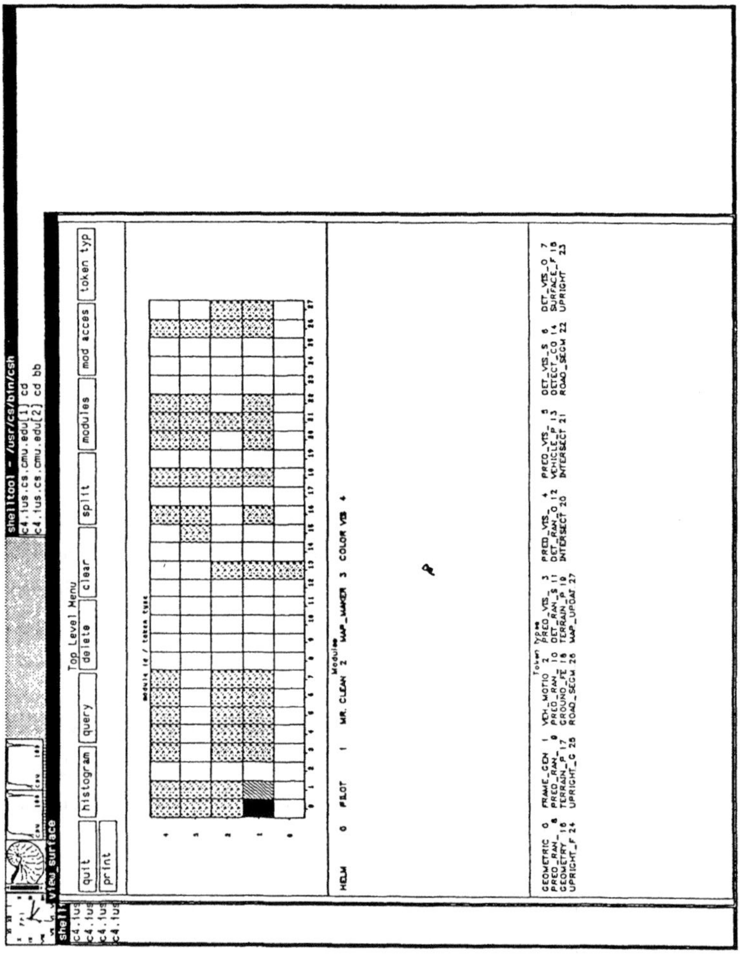

Figure 9.2: Interconnectivity of Typical CODGER Application

scheme of a traditional blackboard is unsuited for a distributed computing environment. CODGER differs from these blackboards in a number of ways. The modules in the CODGER system run continuously and in parallel. Modules communicate by storing and retrieving data from the database. A module is able to run as soon as its precondition is satisfied, regardless of the state of execution of the other modules. Synchronization is achieved by suspending execution within a module until the requested data appears in the database. CODGER even offers a mechanism whereby a module can perform one computation and be interrupted when a precondition is met to perform another. Thus, parallel processors can be effectively utilized. Preconditions in CODGER are not compiled prior to execution but can be changed by the modules dynamically during execution. This feature frees CODGER from a fixed control scheme, allowing it to change dynamically in response to different computing requirements. Furthermore, CODGER has no fixed priority scheme for activating modules. The LMB matches preconditions to the database in using a FIFO scheme. Therefore, priority scheduling is left to the modules, to be encoded in the data itself. Overall, the system can be viewed as data-driven, where the data flow pattern can change during execution.

The ability of the system to change its control scheme and data flow dynamically is of special importance to the Navlab system. Certain modules in the system (particularly perception and planning modules) require different amounts of time depending on the Navlab's environment (e.g., road following, intersection navigation, off-road driving). By observing the behavior of these modules during execution, the Pilot acts as a "meta-module" and adjusts parameters in the data that flows between the other modules to maximize parallelism (see [5]). This control scheme cannot be compiled into the system a priori.

9.3 Data Storage and Transfer

Since the Navlab development environment is diverse, it is important to have a data format that can be understood and manipulated by all modules, regardless of the machine architecture or programming language. The fundamental unit of data in CODGER, the *token*, exhibits this property. As the system evolves, so does the data format used by the modules. CODGER isolates the impact of this restructuring, requiring only those modules directly affected by the change to be modified. In order to effectively utilize a multiprocessing environment, primitives must exist for routing the data and synchronizing the transfer between modules. CODGER employs pattern-matching for routing and a

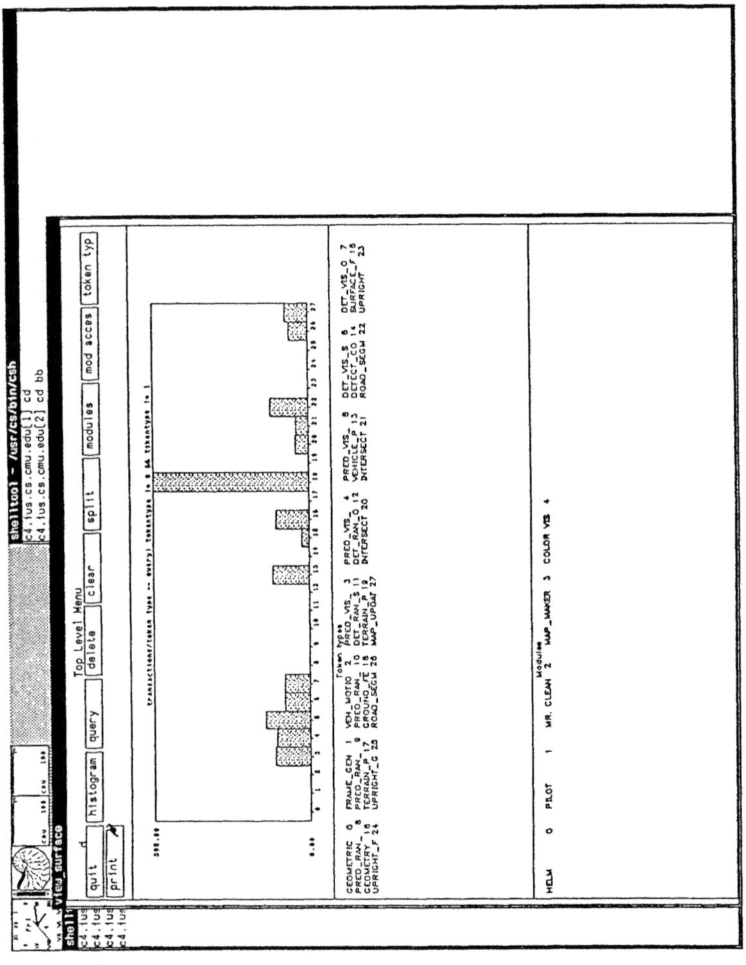

Figure 9.3: Relative Activities of each Data Type

blocking/interrupting scheme for synchronizing transfer. Both the data format and transfer mechanisms are described in greater detail in this section.

9.3.1 Database Tokens

CODGER tokens consist of classical *attribute-value pairs*. Attribute names are strings and their values can be scalars (integer, float, boolean, enumerated types, and strings), arrays (including arrays of arrays), pointers (to other tokens), or geometric location (described in the next section). Tokens are the basic unit of data in CODGER and can be used to represent, physical objects, predictions, commands, status information, etc. Modules create, write, read, and delete tokens in the global database by calling functions in the LMB Interface. The LMB Interface takes care of transforming the data format between different hardware architectures and programming languages.

Each token *type* (set of attribute names and data types) is defined in a *template* file. At startup time, the LMB parses the template file to determine the format of all tokens in the system. The binding of attribute names (strings) to fields in the token occurs in each module when it connects to the LMB. Functions in the LMB Interface perform type checking during execution to ensure that data types are used correctly. Since operations are dynamic, the work needed to modify the system is minimized. Attributes can be added, deleted, or changed in a token type without requiring re-compilation of any modules except those that use the attribute. This feature is of particular importance since typical CODGER applications have consisted of over 100,000 lines of code.

The attributes themselves fall into three classes: internal, local, and global. Internal attributes are automatically included in every token type and specify information such as the token's type, unique ID number, creation time, most recent modification time, version number, and creator. These attributes are managed by CODGER but can be examined by the user modules. Local attributes are defined by the user modules in the template file. The scope of these attributes is restricted to the token type in which they are declared. Global attributes are also defined by the user but can be used in a number of token types. They are intended to have the same semantics (universal) regardless of the token type in which they are used. The real power of global attributes is that they permit modules which understand a given global attribute to manipulate token types that it has not been pre-programmed to handle. The module can accomplish this task without needing to "understand" the other local attributes of the token.

9.3.2 Synchronization Primitives

Modules can retrieve tokens from the central database by direct address (request by ID number) or through the use of a pattern called a *specification*. Specifications are boolean expressions defined across token attribute values. If a token in the database satisfies a specification, the token is sent to the querying module. For example:

if surface **equals** navigable **and** boundary **equals** perceivable

In this example, "surface" and "boundary" are global attribute names defined in the template file. The strings "navigable" and "perceivable" are scalar constants from an enumerated type, and "**equals**" and "**and**" are function names embedded in the specification language. This specification matches any tokens (possibly representing roads, intersections, or off-road patches) over which the robot can drive and the boundary of which it can perceive. One possible use of such a specification is for path planning. Note that above specification matches tokens of *any* type, provided they have the above global attributes with the appropriate values. In this way new token types can be added to the system without needing to modify the path planner. Routing via pattern matching is more powerful than a simple routing table because it channels data based on semantic interpretation, rather than data format type.

In addition to routing, the transfer of data between modules must be synchronized. Whenever a module sends a specification to the LMB for matching, it also selects one of the following synchronization modes:

- *Immediate Request:* As soon as the LMB receives such a request, it matches the specification against the database. The calling module blocks execution during this time. If one or more tokens match, the tokens are sent to the calling module and it resumes execution. If there is no match, one of the following actions happens at the module's discretion:
    - *Non-Blocking:* The LMB responds that there are no matching tokens, and the module resumes execution.
    - *Blocking:* The module remains blocked until a token is deposited that matches the specification, at which point the

token is sent to the module and it resumes execution.

- *Standing Request:* The calling module passes the name of a function (interrupt handler) along with the specification to the LMB Interface and resumes execution. Whenever an incoming token matches the specification, the LMB sents the token to the calling module. The LMB Interface in the module generates an interrupt and calls the interrupt handler passing the token as an argument.

An immediate request facilitates synchronous communication between modules. Such requests are typically used for modules expecting a certain type of data at a certain point in time or sequence. The non-blocking option allows the module to take action if "expected" data is not resident in the database (e.g., to prevent deadlocking). A standing request facilitates asynchronous communication. Such requests are typically used for modules which function as servers but cannot afford to block, such as a vehicle-control module which must monitor signals from the vehicle while fielding steering requests from other modules.

Because all of the modules have access to all of the data, CODGER employs a locking scheme to ensure that the database is consistent. Modules that wish to change tokens in the database must lock them upon retrieval. This mechanism ensures that two modules do not modify a token in parallel. If a module sends a specification to the LMB with a locking request, the matching tokens are not returned until they are unlocked. Thus, an immediate, blocking specification will block until a matching token is unlocked. Our approach is that the database should be consistent so that it appears to be "local memory" to the user module. Note that our locking scheme does not prevent deadlock; it is the responsibility of the modules to avoid or detect such a condition.

9.4 Geometric Representation and Reasoning

The ability to represent, manipulate, and index geometric data is important to any navigation system. Perception modules must be able to represent the shape and location of detected objects. Planning modules must be able to search a map to find the best path for the robot, and control modules must be able to field requests to sense data at particular location and to steer the robot along a trajectory. Since nearly every module makes use of geometry, CODGER support must be distributed. However, in order to guarantee that the geometry used in each module is globally consistent, the support must include a centralized component. In this section, we describe CODGER's facility for representing and indexing geometric data and for ensuring global consistency.

9.4.1 Geometric Data and Indexing

Geometry in CODGER is a separate data *type*, called a *location*. A location consists of a basic shape and a coordinate frame in which the shape is expressed. The primitive shapes that CODGER supports are: points, edges, arcs, ribbons, and polygons. Although these shapes are of no higher dimension than two, by using the appropriate coordinate frames, a three-dimensional object (such an a polyhedron) can be constructed from a set of locations.

Because geometry is a data type in CODGER, it can be embedded in specifications for geometric indexing. The specification language supports a number of indexing functions, including Euclidian distance, polygonal intersection, inclusion test, and minimum bounding rectangle (MBR). Geometric indexing is used by planning and perception modules for operations like searching a map database to find the best sequence of road segments to a goal or for determining which map objects should be visible to the robot. A simple example of the latter operation is shown below:

if objecttype **equals** mapobject **and** (tlocation
poly-intersect viewframe) **equals** true

In this example, "objecttype" is an attribute which is set to "mapobject" if the object is part of the map database. The string "tlocation" is an internal attribute of type "location" which holds the shape and location of the map object. The string "viewframe" is a constant of type "location" which defines (using a polygon) the robot's field of view projected onto the ground. The function "poly-intersect" returns "true" if its two arguments intersect. Thus, the above specification returns all map object tokens that appear in "viewframe".

It should be noted that the geometric indexing functions supported in CODGER are two-dimensional. For typical Navlab scenarios, the ground around the robot can be assumed to be approximately planer. Thus, for local planning and perception operations, two-dimensional indexing is sufficient.

9.4.2 Frames and Frame Generators

As described in the previous section, geometric data consists of two parts: a shape and a coordinate *frame*. In CODGER, a frame consists of a base frame and a homogeneous transformation from the base frame to the frame itself.

There are two system-defined coordinate frames: the WORLD frame affixed to the ground, and the VEHICLE frame affixed to the robot. User modules are able to define their own coordinate frames and to use them for expressing geometric data. The advantage of this feature is that a module can define coordinate frames that are convenient for expressing its data. For example, a perception module can define a coordinate frame affixed to the camera (relative to the VEHICLE) for representing detected objects. This is certainly more convenient than expressing detected objects in a world-based coordinate frame, especially since the robot is moving. CODGER provides a facility to allow the user to express any piece of data in any coordinate frame known to the system. This feature is useful for a planning module that needs to express perception data from multiple sources or times into a single coordinate frame for trajectory planning. When performing geometric indexing, CODGER automatically transforms all geometric data accessed into the same coordinate frame.

Because the robot moves in its environment, the transformation between some frames varies over time. CODGER provides *frame generators* for representing time-varying transformations. A frame generator is a function that takes a time parameter as input and returns a homogeneous transformation as output, to be interpreted as the transform between the frame generator's base and object frames. CODGER supports one system-defined frame generator, namely the one between the WORLD and VEHICLE coordinate frames. In order to generate transformations between the robot and its environment, CODGER must be supplied with the sequence of arcs and vehicle speeds (called the *history list*) specifying the robot's trajectory relative to the world.

Frame generators can be included in location data structures in place of a coordinate frame, provided a time parameter is provided somewhere in the chain of frames. For example, assume that an object is sighted by a perception module at time t. The module creates a location data object representing the object based on the camera frame at time t. The camera frame has been previously defined as fixed relative to the VEHICLE. When CODGER is called upon to determine the distance between the object and a landmark relative to the WORLD (possibly to determine if they are the same object), it uses the time parameter t to "select" a transformation between the VEHICLE and WORLD. Thus, CODGER is able to transform the object from the camera frame into the WORLD frame for the distance calculation.

9.4.3 Geometric Consistency and Affixment Groups

Provided that each frame in the system has only one base frame, the set of frames forms a tree, as illustrated in Figure 9.4. In this figure, the nodes are coordinate frames and the arcs are transformations between the frames. In this example, the central "spine" of arcs is the robot's history list. Arcs emanating from this spine are object sightings from the robot. The remaining arcs are landmarks in the map. Because of the tree structure, there exists a single and unique path between any two coordinate frames in the system. In some cases, it is desirable to have a coordinate frame with more than one base frame and transformation. Such a case arises when an object is sighted multiple times or when a sighted object is identified as a landmark in the map. In such cases, the tree becomes a graph, and more than one path exists between pairs of frames. Unless all measurements are perfect, the paths will yield different transformations, and the graph is globally inconsistent. If we have a model for the discrepancy, the transformations can be adjusted to make the graph consistent [10, 3]. CODGER retains the tree structure (and thus global consistency) by discarding all but the most recent base frame for each coordinate frame. While this strategy is more flexible, it is less powerful if an error model is known. In general this strategy works well because the most recent measurements reflect the best estimate of a frame's transformation (e.g., the robot is closer to the object). See Stentz [11] for a further discussion of this subject. When an object's position is recorded relative to some frame at time t, a mechanism is needed for specifying how the transformation should be handled at times other than t. For example, assume an object is sighted from the robot at time t with transformation T. If we can determine that the object is not moving relative to the WORLD, at time t' the transformation will be T'. However, if the object is not moving relative to the VEHICLE, the transformation at time t' will still be T. In order to disambiguate these cases, in addition to a time value the user can specify an *affixment object* when defining a coordinate frame. An affixment object is a coordinate frame such that the transformation to this frame from the new frame is constant over *all* time. Coordinate frames which are "tied" to the same affixment object are called an *affixment group*. If no affixment object is specified in a frame definition, then the transformation is assumed to valid only at the given time.

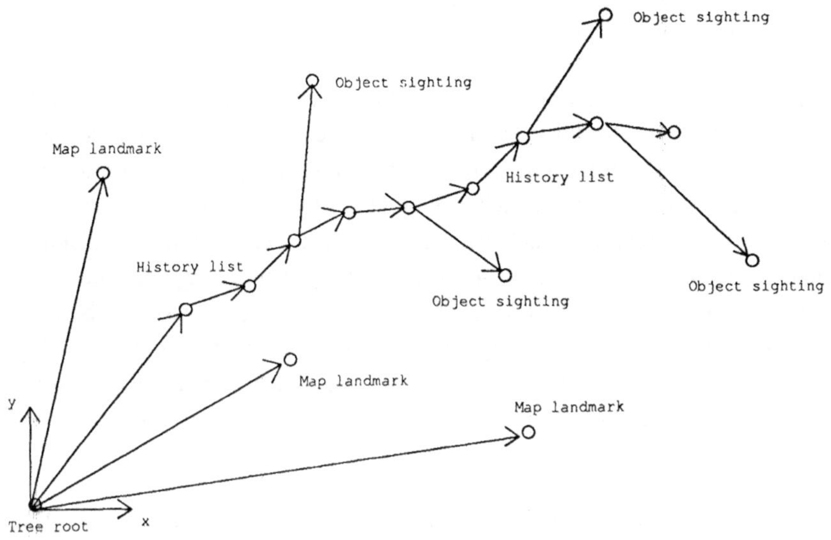

Figure 9.4: Tree of Coordinate Frames in CODGER

9.5 Conclusions

The CODGER system has been shown to be a good tool for developing a large robot navigation system. The "star-shaped" architecture allows any module to communicate with any other module. Although in the final system configuration the communication channels are typically less than clique-like, CODGER was designed to be flexible by allowing the data-routing to change as the system evolves. The data format is flexible enough to represent a broad range of objects needed in a navigation system, including map information, physical objects, status reports, hypotheses, and commands. The system was designed so that changes in the data format affect only those modules which depend on the changes, thus avoiding recompilation of the entire system. This feature is especially important since large navigation systems typically include over one hundred thousand lines of code. Of special importance to navigation systems is geometric data. Sensing, planning, and control modules all need to use this type of data. CODGER provides mechanisms for representing and querying such data. It is able to automatically handle these operations, while accounting for the fact that such data is typically acquired at different times while the robot is moving.

CODGER is able to decrease the overall run-time for a navigation system by providing a means for distributing the computation across multiprocessors. Modules in the system are able to run in parallel, invoking a set of primitives in CODGER to synchronize the exchange of data between them. This mechanism is especially important for balancing the load between sensing and planning modules at the local navigation level, as used by the Driving Pipeline ([5]). CODGER provides a locking mechanism to ensure data consistency between modules. A tree-based hierarchy of frames and frame generators is used to ensure consistency for geometric data. These mechanisms are needed to ensure that concurrently executing modules in the system operate with an updated and consistent model of the world at all times.

CODGER has a number of limitations. First, it is not a real-time system. Due to the relative long latencies in TCP/IP message passing and the varying execution patterns of processes in a UNIX time-sharing environment, data transfer cannot be guaranteed within given time bounds. As discussed in the next chapter, this limitation is acceptable given the types of navigation scenarios we have addressed. Second, the low data transfer bandwidth of CODGER precludes the transfer of iconic or image-level data through the database. Again, the systems we have developed decompose the problem in such a way that high bandwidth is not required. Finally, the mapping of KS's or modules to processors in CODGER is static. CODGER is unable to redistribute modules on the processors to balance the load; this configuration must be set by the user a priori. We are able to compensate for this deficiency by adjusting the execution times of the various modules in the system to balance the load.

9.6 References

[1] Alleva, F., Bisiani, R., Forin, A., Lerner, R.
AGORA Distributed, Interactive Speech Kernel.
1985.

[2] Bjornson, R., Carriero, N., Gelernter, D., Leichter, J.
Linda, the Portable Parallel.
Technical Report, Yale University, February, 1987.
YALEU/DCS/TR-615.

[3] Durrant-Whyte, H.
Integration, Coordination and Control of Mult-Sensor Robot Systems.
PhD thesis, University of Pennsylvania, 1986.

[4] Erman, L. D., Hayes-Roth, F., Lesser, V. R., and Reddy, D. R.
The Hearsay-II Speech-Understanding System: Integrating Knowledge to Resolve Uncertainty.
ACM Computing Surveys 12(2), June, 1980.

[5] Goto, Y., Shafer, S., Stentz, A.
The Driving Pipeline: A Driving Control Scheme for Mobile Robots.
International Journal of Robotics and Automation 4(1), 1989.

[6] Harmon, S. Y.
Implementation of Complex Robot Subsystems on Distributed Computing Resources.
Technical Report, Naval Ocean Systems Center, 1986.

[7] McTamaney, L.
Real-Time Intelligent Control.
IEEE Expert 2(4), Winter, 1987.

[8] Nii, H. Penny.
Blackboard Systems: The Blackboard Model of Problem Solving and the Evolution of Blackboard Architectures, Part One.
AI Magazine 7(2):38-53, Summer, 1986.

[9] Nii, H. Penny.
Blackboard Systems: The Blackboard Model of Problem Solving and the Evolution of Blackboard Architectures, Part Two.
AI Magazine 7(3), Fall, 1986.
Forthcoming.

[10] Smith, R., Self, M., Cheeseman, P.
Estimating Uncertain Spatial Relationships in Robotics.
In *AAAI Workshop on Uncertainty.* 1986.

[11] Stentz, A.
The NAVLAB System for Mobile Robot Navigation.
PhD thesis, CMU School of Computer Science, November, 1989.

Chapter 10: The Driving Pipeline: A Driving Control Scheme for Mobile Robots

Yoshimasa Goto, Steven A. Shafer, Anthony Stentz

Abstract

Mobile robot vehicles must control the execution of numerous perception and planning processes to navigate successfully in complex environments. In the past, most mobile robot systems have utilized "stop-and-go" control schemes that avoid addressing the driving control problem, or have used fixed control schemes that do not allow for the changing environment and field of view of the vehicle. This paper presents a new architecture for mobile robot control called the "Driving Pipeline", that integrates multiple perception and planning processes and provides continuous motion with adaptive control. The Driving Pipeline has been implemented and tested on numerous versions of two vehicles: the Terregator and the Navlab. It has proven to be a flexible and powerful mechanism for building integrated software for mobile robot perception and planning.

10.1 Introduction

This paper describes a driving control scheme for a mobile robot that drives the robot vehicle outdoors, avoiding obstacles, and keeping the vehicle within a navigable area. As illustrated by Figure 10.1, the driving control scheme takes a high-level navigation plan from planning modules and sensor data from sensors, and generates vehicle motion commands, performing the necessary computations including perception, environment modeling, path planning, and vehicle control. We have developed a scheme for the coordination of these tasks, which we call the *Driving Pipeline*. This paper describes the Driving Pipeline, the various processes that it coordinates, and the experiments in which the Driving Pipeline has been successfully used for building mobile robot systems.

Our objective is to build an autonomous mobile robot working in the real world in real-time, so we adopted the following design goals:

- **Flexibility:** Other systems have been developed that perform a single navigation task well; however, these systems are not easily extended to handle a broad range of tasks.
- **Continuous Vehicle Motion:** Continuous motion is more desirable

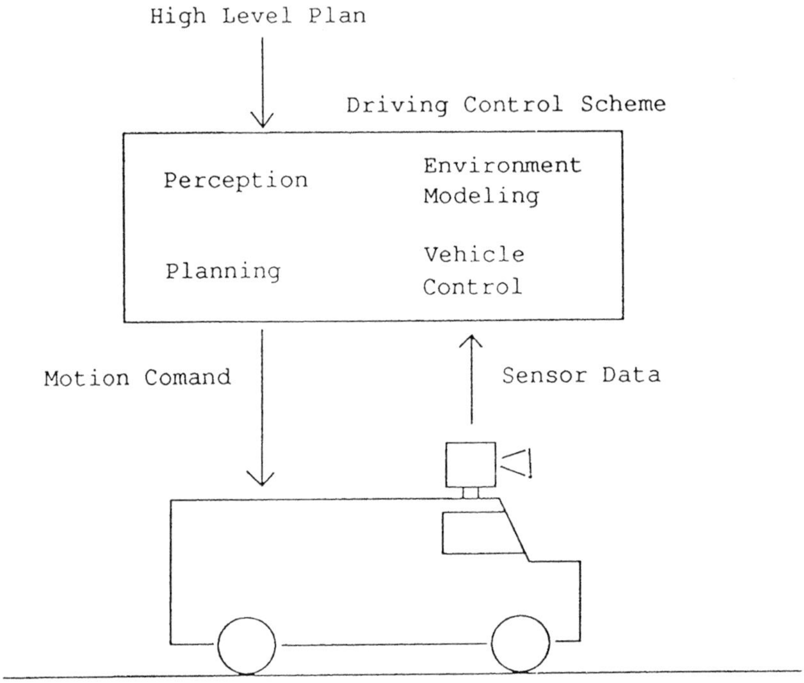

Figure 10.1: Driving Control Scheme

than stop-and-go motion, because it produces higher vehicle speeds and smoother control.

- **Adaptive Control:** Driving control must be adaptive to the environment and to the internal condition of the robot vehicle. For example, the vehicle should be able to drive faster using less sensor data on a flat broad ground than on a winding narrow road. The driving control scheme must adjust its computation and maintain effective coordination among numerous perception and planning processes.

- **Parallel Execution:** For real-time motion, driving control requires a large amount of computation in a variety of different procedures. For this end, parallel computing is the most practical solution. In addition to small-grain parallelism such as parallel machines for signal data processing, large-grain parallelism can be used to coordinate the various tasks involved in driving. Parallel computing can take advantage of two kinds of parallelism: parallelism in processing steps and parallelism in data to be processed.

In order to achieve these goals, we developed the **Driving Pipeline**. A *pipeline* is a form of parallelism in which the computation is decomposed into a sequence of processing steps, called *stages*, to be executed in a fixed order. Typically, each stage is a separate processor receiving input data from the previous stage and providing output data to the next stage. A stage commences execution whenever data arrives from the input. A pipeline is used for performing the same computation over a number of different data sets. Since the pipeline can begin processing a second data set before the first has finished, the stages run in parallel. The pipeline processes data sets at the rate of one per *cycle time*. The cycle time is the longest stage time. The total time required to process a given data set (called the *job time*) is the sum of individual stage times. The construction of our pipeline is based on two key ideas:

- **The Driving Unit:** We divide the area in which the vehicle navigates (road, hillside, etc.) into a sequence of small areas called *driving units* so that it can process each driving unit separately. Each processing module for perception and planning will operate successively on each driving unit in turn.
- **Execution Pipeline:** The Driving Pipeline allocates the primitive processing steps along a pipeline so each one can work independently, receiving input data from the previous processing step and passing data to the following processing step.

These two key ideas enable the pipelined execution of the primitive processing steps on the sequence of driving units, which provides enough throughput to allow continuous vehicle motion. As the vehicle encounters changes in the road configuration, it can place driving units with different sizes and intervals by adjusting the sensor view frames, execution intervals, and vehicle speed.

Although several mobile robot systems have been built in the past, they did not address driving control scheme very deeply. Stop-and-go motion, although it does incorporate all of the primitive processing steps, deliberately avoids the problem of continuous motion control [2, 4, 7, 10]. Waxman et al. mentioned the necessity for vehicle speed adjustment using knowledge, but didn't show any method for doing so [11]. Brooks developed a layered control structure that drives a vehicle continuously [1]. However, it does not have the ability to adapt the control to meet the changing needs of perception. Dickmanns and Zapp developed a system for high-speed navigation on the German Autobahn [3]. This system tracks simple visual features (e.g., white lines bordering the road) and cannot be easily extended to handle more difficult perceptual scenarios.

To solve these problems, we have developed the concept of the Driving Pipeline and verified it in two experimental mobile robot systems: the Terregator and the Navlab. This paper describes the Driving Pipeline, including the component concepts of the Driving Unit and the Execution Pipeline, and describes our experiments with these vehicles.

10.2 Processing Steps and Driving Unit

We divide the computation necessary for driving control into the following primitive processing steps:

- The **Prediction** step plans the area that the vehicle will move into next.
- The **Perception** step detects navigable area boundaries and obstacles using sensor data.
- The **Environment Modeling** step makes a description of the vehicle environment and updates the estimate of the vehicle position.
- The **Local Path Planning** step plans the vehicle trajectory.
- The **Vehicle Control** step drives the vehicle mechanism.

These steps must each execute in turn to process each area of terrain that the vehicle will traverse.

We developed the concept of the *driving unit* to indicate the area that each primitive step will process once in each execution cycle. The vehicle's entire route is divided into driving units which are passed, one at a time, to each of the primitive processing steps. In this way, planning and perception are synchronized to provide driving control.

10.2.1 Prediction and the Driving Unit

The Prediction step works as the manager of the Driving Pipeline. It receives the high-level plan from the map navigation level of the system, predicts the next chunk of area into which the robot vehicle should move, and indicates it by defining a new driving unit. Because the driving units are placed in the order that the vehicle travels, the sequence of driving units forms the vehicle passage, which outlines the planned path of the vehicle (Figure 10.2).

The parameters for placing the driving units are:

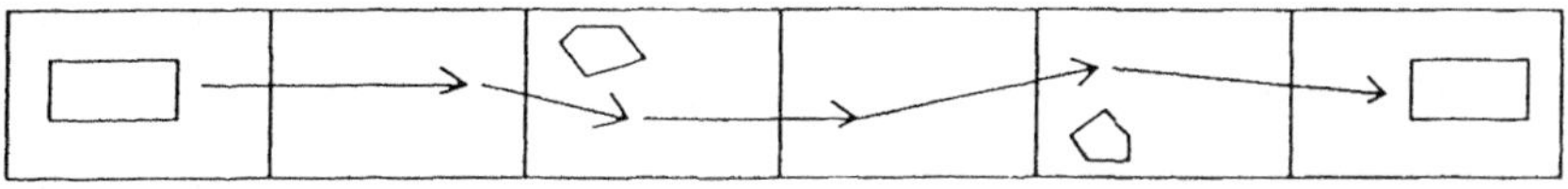

Figure 10.2: Sequence of Driving Units

- location of the driving unit
- type of the driving unit : such as *on-road, open-terrain.*
- size of the driving unit : the width and length of the driving unit
- interval of driving units : the distance between the centers of consecutive driving units along the vehicle trajectory.

The driving unit location is determined based on the high-level plan derived from the navigational map, combined with the vehicle's current position estimate. The type of driving unit can be road or intersection, depending also on the map and the vehicle position. The factors that determine the size and the interval area are discussed in the following sections.

10.2.2 Perception and Driving Unit

The Perception step scans a driving unit with sensors to determine the key objects within it. Perception results will be used by the Environment Modeling step both for determining navigable areas and for updating the vehicle position estimate.

Two parameters, the driving unit and a *scanning position*, direct the Perception step. The driving unit, which is given by the Prediction step, indicates the area that the Perception should see. Because sensor data must cover the driving unit, the sizes of sensor view frames give the upper limit of the driving unit sizes.

The scanning position is the position at which the Perception step should scan the driving unit. Two factors determine the scanning position: the required accuracy of the visual measurement, and the need for specific vehicle position information. The required accuracy of the visual measurement is important because of the reduced accuracy as distance increases. Thus, the vehicle should be close enough to the driving unit to satisfy the accuracy needs of the Environmental Modeling step. The need for specific vehicle position information also constrains the scanning position. The vehicle position estimation is updated with both the perceptual results and dead reckoning from the control system. In general, the perception result gives a more accurate vehicle position estimate. The vehicle position estimated with the perception result will, of course, be a scanning position. Therefore, when the mobile robot system needs an accurate vehicle position estimation at a specific position, this position should be the scanning position.

Once the driving unit and the scanning position are determined, the Perception step can calculate the sensor view frame relative to the vehicle and aim the sensors. This enables Perception to aim the sensors adaptively.

10.2.3 Environment Modeling and the Driving Unit

By analyzing the perception results, the Environment Modeling step produces an environment description that indicates a navigable area from the current vehicle position toward the end of the last scanned driving unit.

The Environment Modeling step also updates the vehicle position estimation. Because the vehicle is traveling continuously and the the scanning positions are discrete, the Modeling step merges the perception result and the dead reckoning updates to estimate the vehicle positions between the scanning positions and beyond the last scanning position.

10.2.4 Local Path Planning and the Driving Unit

The Local Path Planning step determines the physical vehicle trajectory within the navigable area determined by the Modeling step, from the current vehicle position to the end of the last scanned driving unit.

As shown in Figure 10.3, the local path plan restricts the minimum size of a driving unit, because the driving unit must be large enough to allow the vehicle to maneuver and avoid obstacles.

The Driving Pipeline includes two levels of path planning: the driving passage from the Prediction step and the trajectory from the Local Path Planning step. If the map database is complete, the driving passage can be planned before navigation by consulting the map data. If not, it is determined gradually based on perception results from the previous driving units. This is the reason why we

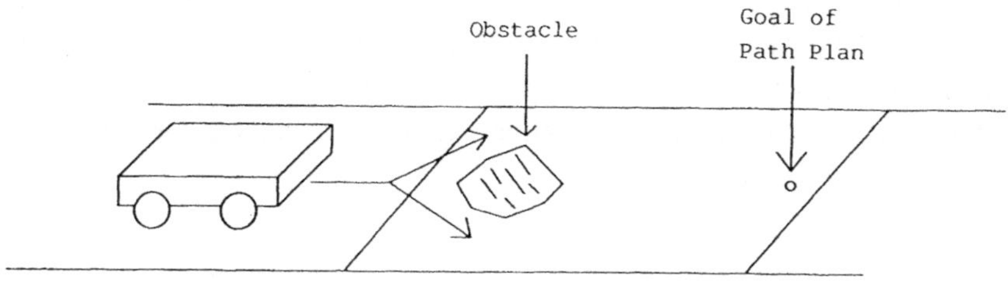

Figure 10.3: Driving Unit Size for Vehicle Maneuvering

include planning the vehicle passage in the the Driving Pipeline level of the system rather than in a higher level.

10.2.5 Vehicle Control and the Driving Unit

The Vehicle Control step drives the physical vehicle. It generates a set of motion commands for the vehicle mechanism from the trajectory plan given by the Local Path Planning step. Because the trajectory plan ends at the far edge of the last scanned driving unit, the vehicle never moves into an unscanned area. Also, this step adjusts the vehicle speed to be optimal unless the Local Path Planning step gives commands on speeds (such as stopping at a specific place). The details will be described in Section 10.3.4.

10.3 Continuous Motion, Adaptive Control, and the Driving Pipeline

The simplest control structure for implementing the Driving Unit concept would be for the vehicle to stop at the end of each driving unit, process the next one through each of the primitive steps, then drive across the next driving unit and stop, repeating this cycle over and over. This paradigm is known as the "stop-and-go" model of vehicle control, and it produces very jerky motion as well as being far below the optimum vehicle speed. To remedy these problems, we apply the concept of pipelined execution of the primitive steps to form the Driving Pipeline.

10.3.1 Pipelined Execution for Continuous Motion

In order to drive the robot vehicle continuously, the Vehicle Control step should work on one driving unit after another without stopping the vehicle. To accomplish this, the Prediction step, the Perception step, the Modeling step, and the Local Path Planning step must have finished processing the next driving unit before the Vehicle Control step finishes the current driving unit. This is the reason that continuous vehicle motion needs a Driving Pipeline to process multiple driving units in parallel.

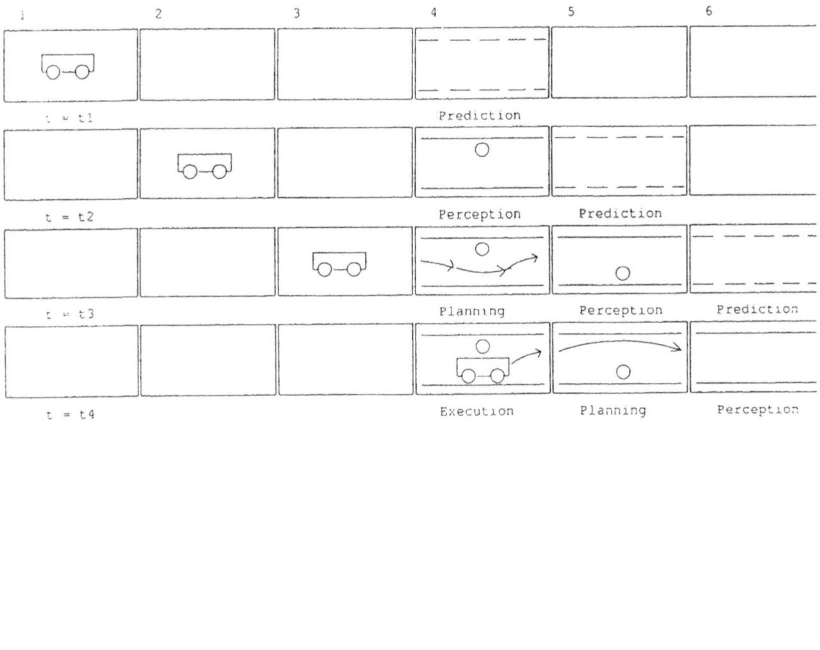

Figure 10.4: Pipelined Execution of the Driving Pipeline

The Driving Pipeline supports continuous vehicle motion by using *pipelined execution*. As described in Section 10.2, the processing steps are allocated along the pipeline, and the Driving Pipeline executes the processing steps in parallel by passing a sequence of the driving units through this pipeline. Figure 10.4 illustrates the pipeline execution of the Driving Pipeline as follows:

1. When the vehicle is on Driving Unit 1, the Prediction step places a new prediction for Driving Unit 4.
2. When the vehicle is on Driving Unit 2, the Perception step works on Driving Unit 4. At the same time, the Prediction step places the next driving unit, Driving Unit 5.
3. When the vehicle is on Driving Unit 3, the Modeling step determines the vehicle passage and the Local Path Planning step plans the path to the end of Driving Unit 4. In parallel, the Prediction step defines Driving Unit 6 and the Perception step works on Driving Unit 5.
4. When the the Vehicle control step drives the vehicle on Driving Unit 4, the Prediction step is defining Driving Unit 7, Perception is working on Driving Unit 6, and the Modeling and the Local Path Planning step are working on Driving Unit 5.

Several key features of the Driving Pipeline make the pipelined execution possible. First is the concept of the driving unit, which is critical because it allows the route ahead of the vehicle to be partitioned into individual units for processing by the successive steps. Because each driving unit specifies an area on which one processing step works, the Driving Pipeline may assign the different processing steps to different areas along the vehicle passage.

The second is the constant flow of the driving units through the processing steps in a prearranged sequence. Each driving unit is created at the Prediction step and is passed through the following steps from one step to the next step ending with the Vehicle Control Step, thus forming the data flow through the processing steps. This flow is always one way and in the same direction; no driving unit skips any processing step or goes back to the previous steps. Therefore, the order of execution of the primitive processing steps can be "hard-wired" into the system without the need for symbolic reasoning to decide what to do next.

The third necessary feature is the independent computation of the processing steps. The computation for driving control is divided into processing steps in such a way that each processing step performs a different function. Each step requires as input only the outputs of the previous steps. Therefore, each step can only work on a driving unit after the previous steps have completed their processing on that driving unit.

The fourth feature is the order of the driving units themselves. Since the driving units are created as the vehicle travels and are placed along the vehicle passage, the order of their generation is always the same as the order in which they are processed by the processing steps. Therefore, the Driving Pipeline can feed the driving units to the processing steps continuously.

Finally, the ability of the sensors to look ahead of the vehicle more than one driving unit's distance is necessary. This permits Perception to be working at a distance beyond the next driving unit. This ultimately limits the distance over which pipelining can be effective.

The existence of all of these features allows pipelined execution in both of the necessary aspects, the processing and the data. The name "Driving Pipeline" comes from the pipeline of processing steps, the sequence of driving units, and the pipelined execution. The following sections provide a more detailed examination of the pipelined execution.

10.3.2 Execution Intervals of the Driving Pipeline

The "execution interval" of the driving control system refers to how often the mobile robot system executes the cycle of the primitive processing steps. Adjusting the execution interval to be optimal is essential for an autonomous mobile robot system, because the necessary execution intervals depend on driving conditions such as the width, flatness, and curvature of the road. Execution intervals that are too long may cause unstable vehicle motion, because the vehicle position and the path plan are updated only once in each interval. On the other hand, execution intervals that are too short consume unnecessary computation and slow down the vehicle speed because the amount of computation in each interval is roughly constant.

To provide the optimal vehicle speed control, the driving control scheme needs a way to compute and change the execution intervals. In the Driving Pipeline the sizes of the consecutive driving units determine the execution intervals, because each execution cycle works on one driving unit and the number of driving units per unit trajectory length is equal of the number of the execution cycles. Therefore, the Driving Pipeline is able to adjust the execution intervals by changing the driving unit intervals.

If the vehicle could be controlled to exactly follow the planned path, the driving units could be made as long as the range of the effective field of view of the sensors. Unfortunately, the actual vehicle trajectory may differ from the local path plan because of many reasons, particularly the error in the control mechanism and the inaccuracy of dead reckoning. The cumulative error in the control of vehicle motion and the allowed error tolerance in the vehicle position are the factors used to determine the driving unit intervals.

The error in the vehicle position and direction, which grows as the vehicle travels, must be canceled by the execution of the driving pipeline before it surpasses an error tolerance. Therefore, if the accumulated error increases very rapidly, the intervals of the driving pipeline must be shorter. If the accumulated error increases slowly, they can be longer. For example, because errors in the vehicle direction can produce a larger accumulated error in the vehicle position than errors in the vehicle displacement, the interval must be shorter in turning than in moving straight.

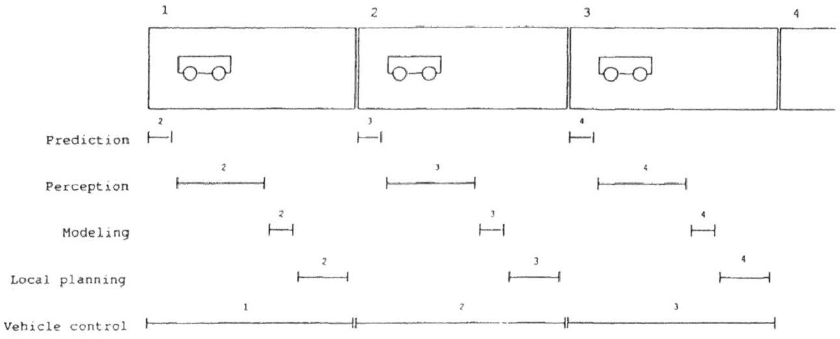

Figure 10.5: Badly-Balanced Execution of the Driving Pipeline

As mentioned in Section 10.2.4, vehicle maneuverability restricts the minimum size of a driving unit. If a driving unit interval is shorter than a driving unit length, adjacent driving units overlap.

10.3.3 Parallelism in the Driving Pipeline

Although the pipelined execution allows the processing steps to work in parallel, it does not ensure a high degree of parallelism. Figure 10.5 illustrates an extreme example in which parallel execution is not well maintained. In this figure, the vehicle speed is too high. This brings the vehicle to the end of the local path plan before the next plan is produced by the Local Path Planning step. The vehicle then has to stop at the end of the current driving unit to wait for the new path plan to be completed. In this example, the Prediction step, the Perception step, the Environment Modeling step, and the Local Path Plan step must work serially without any parallelism. In this section and the next we discuss the parallelism in the Driving Pipeline and a mechanism for keeping it high. This section discusses parallel execution among the Prediction, Perception, Environment Modeling, and Local Path Planning steps. The next section discusses parallelism between these steps and the Vehicle Control step.

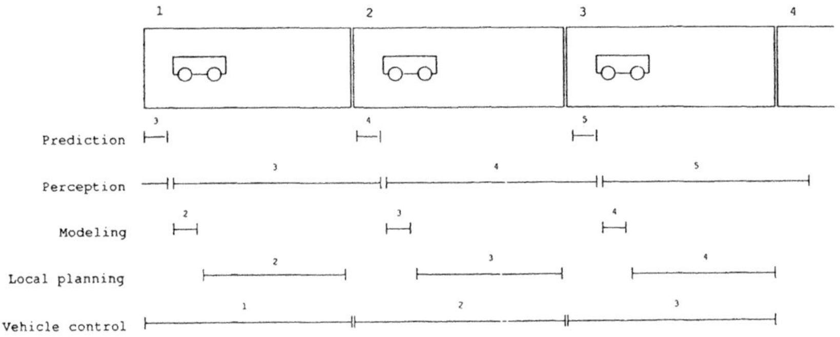

Figure 10.6: Parallel Execution Pattern in the Map Navigation Mode

The Prediction, Perception, Environment Modeling, and Local Path Planning steps generally work on each driving unit sequentially, with their execution times overlapping each other on consecutive driving units due to the execution pipeline. However, the parallelism among these steps depends on whether or not there exists a sufficiently rich map database. When such a map exists, we call this the *map navigation mode*; if not, the vehicle drives in the *map building mode*. The timing of the start of pipelined execution varies in these these two modes. In the map navigation mode, the map database can offer enough information so that the Prediction step is able to place a new driving unit without using the perception results from the preceding driving unit, relying instead on the map database and the perception results from earlier driving units. Therefore, the Prediction step can work on the next driving unit before the Perception and the Environment Modeling steps finish the current driving unit. This produces the execution pattern illustrated in Figure 10.6. In this case, since all processing steps are ready to work on the next driving unit just after finishing the current one, complete pipelined execution is achieved.

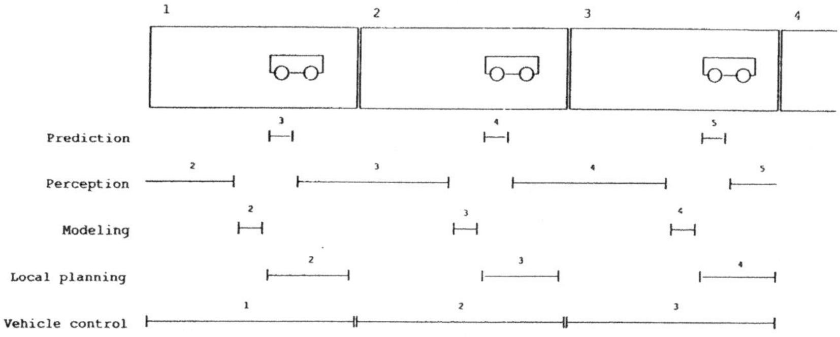

Figure 10.7: Parallel Execution Pattern in the Map Building Mode

In the map building mode, the map database does not have enough information about the unscanned areas, so the Prediction step needs the perception result on the current driving unit in order to place the next driving unit. In this case, the Prediction step has to wait until the Perception step and the Environment Modeling step finish the current driving unit. The resulting execution pattern is illustrated in Figure 10.7. Consecutive execution cycles overlap less in the map building mode than the map navigation mode.

The difference between the map navigation and map building modes explains one reason that a rich map database is able to produce the higher vehicle speed than the poor map database. In addition, a rich map database allows perception to potentially be faster and more accurate, thus reducing the processing time and/or allowing larger driving units.

In both execution modes, the scanning position is a key factor in maintaining these parallel execution patterns because it regulates the execution patterns. The Environment Modeling step, the Local Path Plan step, and the Vehicle Control step start just after the previous step finishes. The Prediction step starts just after the Perception step finishes in the map building mode, and may start any time in the map navigation mode. So, all of these steps can start at a time independent of the actual vehicle progress. On the other hand, the Perception step can start working only when the vehicle reaches the desired scanning position. The scanning positions that produce the highest parallelism, illustrated in Figures 10.6 and 10.7, are given by the following equation:

$$\textit{scanning distance} = \frac{T_p}{T_c} \cdot L_i \tag{10.1}$$

where

L_i = *driving unit interval*

T_p = *total job time of Perception, Modeling and Planning*

T_c = *cycle time of Driving Pipeline*

In this equation, the "scanning distance" is the distance from the scanning position to the driving unit to be scanned. The "cycle time" is the time between consecutive execution cycles, which is the time taken for the vehicle to travel one driving unit. In the map navigation mode, the cycle time is determined as:

$$T_c = T_m \tag{10.2}$$

whereas in the map building mode, the cycle time is:

$$T_c = \mathit{Max}\,(\,T_m\,,\,T_t\,) \tag{10.3}$$

where

T_m = *job time of the most time consuming step*

T_t = *total job time of Prediction, Perception and Modeling*

In the map navigation mode, if the most time consuming processing step works in the whole cycle time, the execution pattern will be the most condensed and will exhibit the highest degree of parallelism. In this execution pattern, the Perception, Environment Modeling, and Local Path Planning steps must work after the vehicle passes the scanning position. That is the derivation of the above equation for the map navigation mode. In the map building mode, the processing for the sequence of the Prediction, Perception, and Modeling steps can not overlap with the processing of this sequence for consecutive driving units. Therefore, this execution sequence behaves like one individual processing step. That is the reason for the above equation for the map building mode.

10.3.4 Vehicle Speed and Driving Pipeline

The Vehicle Control step must take into account the execution time of all the processing steps in order to achieve the optimum vehicle speed. Too high a vehicle speed requires the vehicle to stop at the end of each driving unit, as described in the previous section. In this section, we discuss the highest possible vehicle speed and the method to achieve it.

Because the distance that the vehicle moves in one cycle time is equals to the interval of the driving unit, the highest vehicle speed is described by the following equation:

$$\textit{vehicle speed} < \frac{L_i}{T_c} \tag{10.4}$$

The maximum vehicle speed is less than the driving unit interval divided by the cycle time because distance must be allocated for decelerating the vehicle in the event that some stage of the pipeline requires more time than expected.

If the scanning position is adjusted as described in Section 10.3.3, the cycle time is given by Equations (10.2) and (10.3). Then the above equation can be rewritten as follows:

in the map navigation mode,

$$\textit{vehicle speed} = \frac{L_i}{T_m} \tag{10.5}$$

and in the map building mode,

$$\textit{vehicle speed} = \frac{L_i}{Max(T_m, T_t)} \tag{10.6}$$

These equations are based on the highest degree of parallelism among the processing steps and therefore give the highest achievable vehicle speed.

The vehicle speeds given by these equations are possible only when the scanning position is optimally adjusted. The scanning position, however, may be determined by other factors as described in Section 10.2.2. For example, the scanning distance may be shorter than the distance given by Equation (10.1) because the Perception step requires a closer distance for more accurate measurement. If the scanning distance is shorter than the distance given by Equation (10.1), the speed of the Driving Pipeline is given by the following equation:

$$\textit{vehicle speed} = \frac{D_s}{T_p} \tag{10.7}$$

where

D_s = *scanning distance*

These equations (Equation (10.4) - (10.7)) describing the vehicle speeds explain the following vehicle behavior patterns, which demonstrate the adaptive control capabilities of the Driving Pipeline:

- The most time consuming processing step limits the highest vehicle speed. The Driving Pipeline is capable of adjusting the vehicle speed to be as high as the processing times will allow.
- Longer driving unit intervals produce a higher vehicle speed. If the robot vehicle drives in easy driving conditions such as a broad, flat, straight road, then the Prediction step may define driving units with large intervals. The vehicle speed will then be adjusted to be higher.
- Likewise, shorter scanning distances produces a slower vehicle speed. If the Perception step has to look at objects from a closer distance, the vehicle slows down. This behavior is similar to a human driver looking around carefully.

These behaviors need not be explicitly programmed into the system. They arise naturally as a result of the operation of the Driving Pipeline and the calculation of each driving unit interval based on the geometry of the road, the vehicle, and the sensor field of view.

Although Equations (10.4)~ (10.7) assume that each processing step always requires a constant execution time, the actual requirements may vary from time to time and place to place. In such a case, the Driving Pipeline calculates the vehicle speed with the following equation, which is a modified version of Equation (10.7):

$$vehicle\ speed = \frac{D_r}{T_r} \qquad (10.8)$$

D_r = *remaining distance of local path plan*

T_r = *remaining job time*

In this equation, D_r is the distance from the current vehicle position to the end of the path plan in the *current* driving unit, and T_r is an estimate of the total remaining execution time for the Prediction, Perception, Modeling, and Local Path Planning steps working on the *next* driving unit. The initial value of T_r is a predicted execution time for these processing steps. Whenever these processing steps finish processing a driving unit, T_r and D_r are recalculated and the vehicle speed is updated. This allows the vehicle speed to adaptively respond to the changing requirements for its own computation time.

10.4 The Driving Pipeline in Action: Experimental Results

10.4.1 Implementing the Driving Pipeline

We have developed and tested the Driving Pipeline through building several experimental mobile robot systems, called *Sidewalk System 2*, *Sidewalk System 3*, and the *Park System*, [5] [6] [9]. The Sidewalk 2 and the Sidewalk 3 systems drive an experimental vehicle called the Terregator on the network of sidewalks on the campus of Carnegie Mellon University. The Park system drives the Navlab, a computer-controlled van, on a road in Schenley Park adjacent to Carnegie Mellon. Figure 10.8 shows these vehicles, which are both equipped with color TV cameras and a laser range scanner made by ERIM. While the Terregator is linked to several SUN-3 workstations in the laboratory with radio communication and cables, the Navlab carries four SUN-3s on board. In the remainder of this chapter, we will describe primarily Sidewalk System 3 because

Figure 10.8: Terregator and Navlab

it demonstrates the Driving Pipeline most clearly.

Figure 10.9 shows the module structure of Sidewalk System 3. The processing steps are implemented as individual programs and are linked through the CODGER distributed database, a system-building tool written at Carnegie Mellon to support large-grain parallelism for mobile robot navigation [8]. CODGER makes it relatively easy to build the Driving Pipeline because of its capability to support parallel processing among multiple computers. All of the systems mentioned above use CODGER in this way.

10.4.2 Processing Steps and Driving Units

Figure 10.10 shows a diagram of the primitive processing steps working on one driving unit in approaching an intersection. Figure 10.10(a) shows the driving unit placed by the Prediction step. In Figure 10.10(b), the trapezoid is the sensor view frame aimed by the Perception step to cover the driving unit. Figure 10.10(c) shows the vehicle position estimated by the Modeling step. The

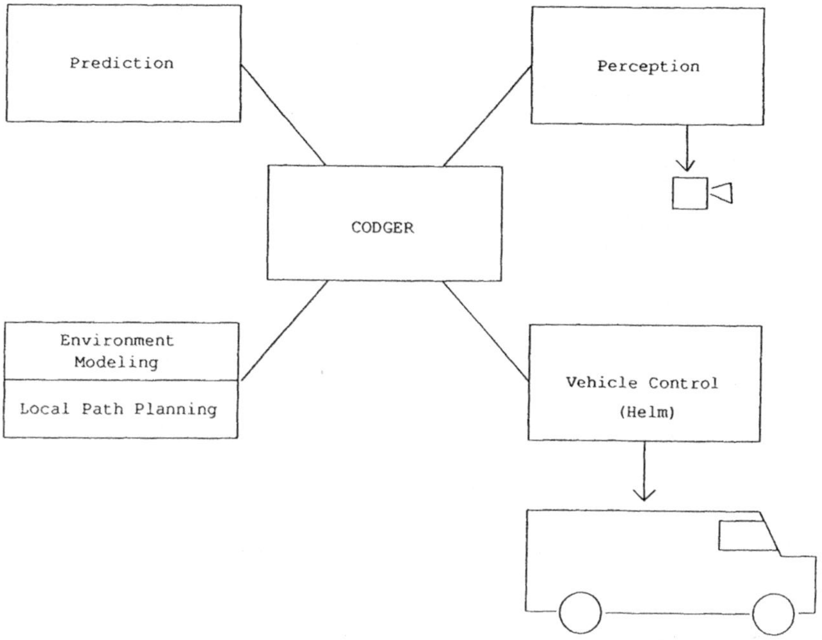

Figure 10.9: Module Structure

Vehicle Control step drove the vehicle as illustrated in Figure 10.10(d).

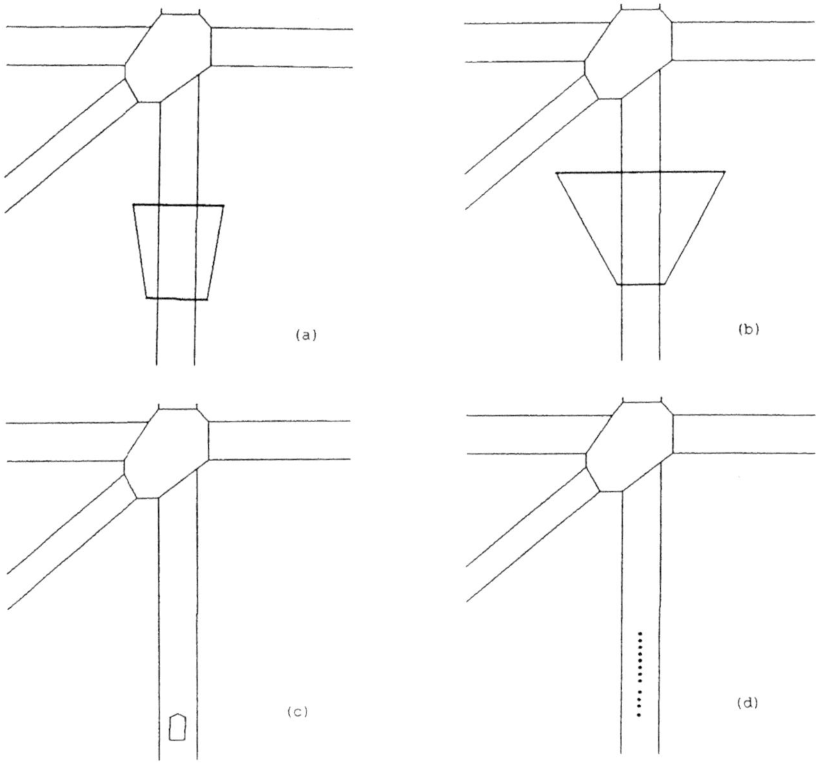

Figure 10.10: Processing Steps

10.4.3 Pipeline Execution and Parallelism

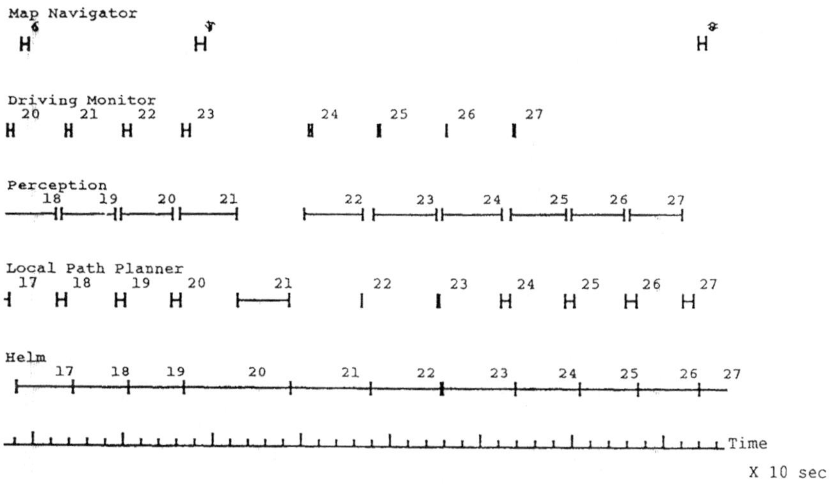

Figure 10.11: Timing Diagram of the Processing Steps

Figure 10.11 is a recorded timing diagram of the processing steps. The bars in the figure indicate the time during which each step is processing a driving unit. The driving unit number appears next to the bar. Because Sidewalk System 3 has a complete pre-stored map database, the Prediction step does not need to wait for the Perception step for placing a new driving unit and the consecutive pipeline executions overlap completely. This is the "map navigation" mode described above. Because the scanning position and the vehicle speed were adjusted as described in Section 10.3.3, the most time consuming step (Perception) was the limiting factor in the cycle time of the system.

10.4.4 Execution Intervals

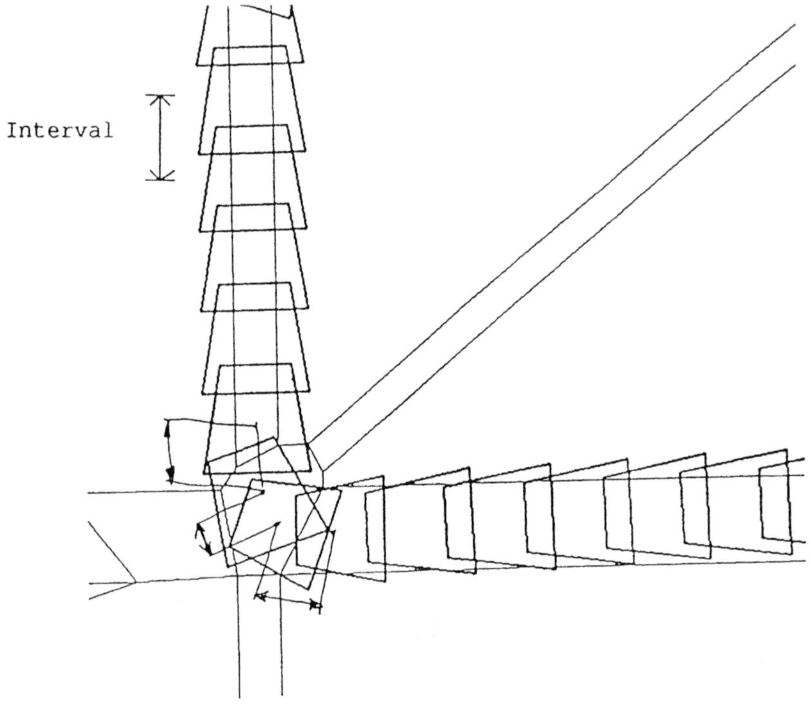

Figure 10.12: Driving Unit Intervals

Because turning at intersections requires more accurate vehicle position estimation than following sidewalks, and because the Terregator vehicle makes larger dead reckoning errors in turning than in straight motion, the Prediction step uses a shorter driving unit interval while the vehicle is turning. Figure 10.12 shows the driving unit intervals around the intersection and the straight sidewalks. On the other hand, Sidewalk System 2 used constant driving unit intervals and had unstable turning because of the large dead reckoning error. Sidewalk System 3, however, did not have such unstable motion thanks to the adjustment of the driving unit intervals.

10.4.5 Vehicle Speed

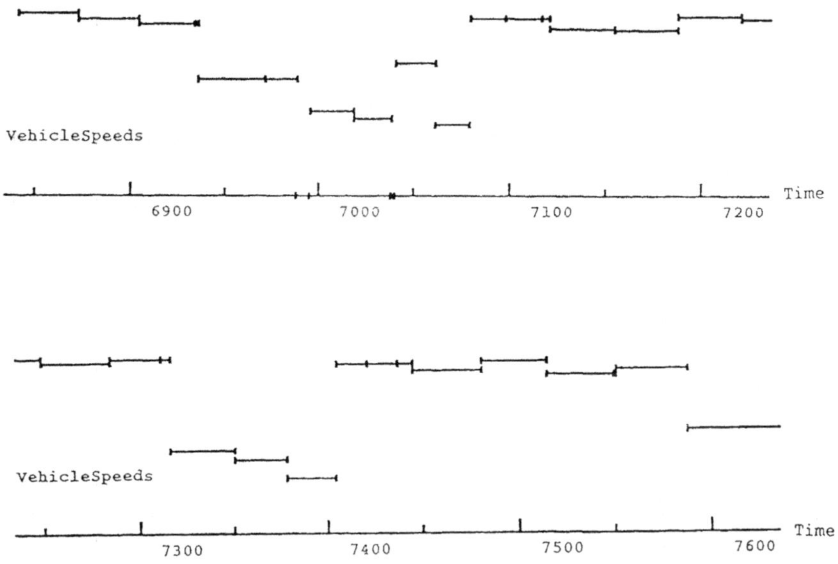

Figure 10.13: Control of the Vehicle Speed

Figure 10.13 shows a recorded vehicle speed that was adjusted according to Equation (10.8). The vehicle speed was recalculated whenever the processing steps were done. The vehicle slowed down around the intersection where the driving unit intervals were shorter and went back to a high speed on the straight road where the driving unit intervals were longer. Because of the hardware limitations of the Terregator vehicle, the vehicle speed could not be changed frequently; this is the reason that the recorded vehicle speed is not smooth.

10.4.6 Sensor Aiming

Our experiments on the Carnegie Mellon campus test site showed the necessity for adaptive sensor aiming. The fixed sensor view frame created a problem in turning at the intersections, because the vehicle had to turn through a large angle and the fixed sensor view frame could not cover the destination sidewalk while the vehicle was turning. To remedy this problem, the sensor view frame has to be aimed so that it covers the vehicle's destination. In addition, the scanning distance must be different in following straight sidewalks and in turning through intersections. In turning through an intersection, the vehicle position estimation must be accurate in both the vehicle's heading direction and the direction perpendicular to the vehicle's heading. Therefore, the scanning distance must be short. During straight travel, however, the vehicle position estimation along the vehicle's heading direction does not need to be so accurate and the scanning distance may be longer.

Figure 10.14 shows the sensor view frames and the scanning positions. The scanning positions were calculated using Equation (10.1) and the local path plan that was produced in the previous execution cycle. The scanning distance varied at the intersection and on the sidewalks.

To aim the TV camera into the predicted driving units, *pan* and *tilt* mechanisms are needed. This can present a very challenging timing problem if mechanical pan and tilt mechanisms are used. To avoid this, the Terregator vehicle was equipped with two cameras and switched between them instead of using a mechanical pan. The TV cameras had wide angle lenses and covered broad areas. The Perception step processed the desired rows of the image in place of a mechanical tilt. This "software pan/tilt" is very fast and simple to program, as opposed to a mechanical pan/tilt which is relatively slow and difficult to control optimally. However, the software pan/tilt requires duplicated sensor hardware.

Our experiments have demonstrated the basic operation of the driving pipeline and dynamic adjustment of the execution interval, vehicle speed, and aim of the sensor. We have shown that the speed of the vehicle must be reduced and the driving unit shortened in situations involving uncertainty in the map or large vehicle control error (e.g., driving in intersections). Likewise, we have shown that the vehicle can drive quickly using large driving units on well-mapped straightaways. At both extremes and across the range we have demonstrated how the scanning distance can be adjusted to maximize parallelism.

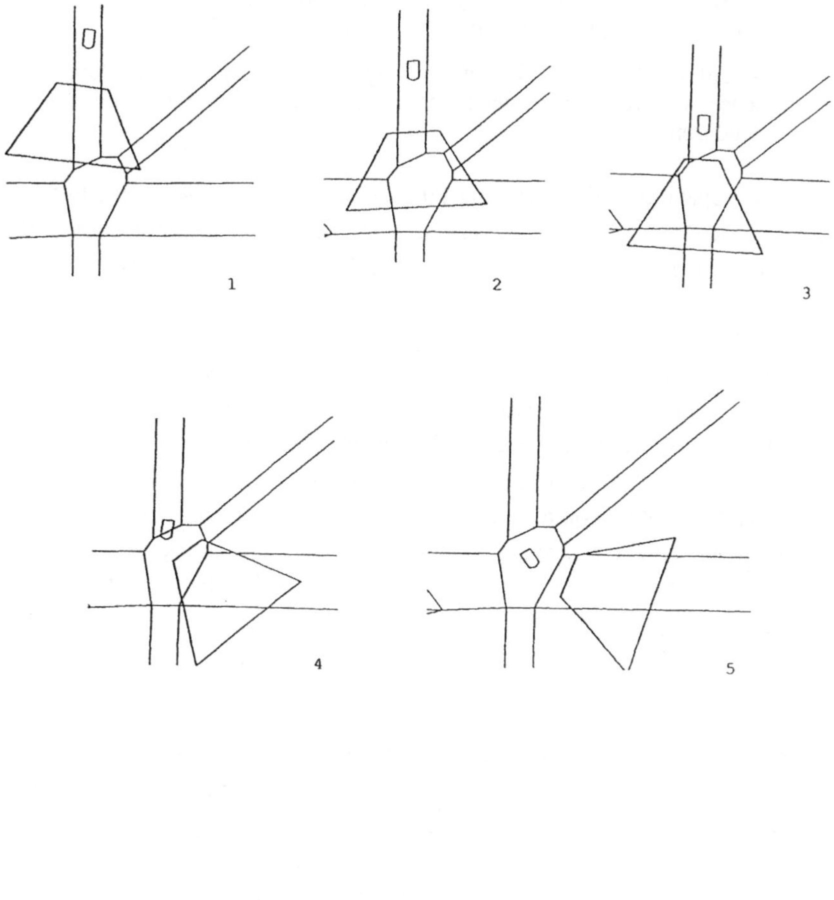

Figure 10.14: Sensor View Frames

10.5 Conclusion

This paper has described the Driving Pipeline, a driving control scheme to control a robot vehicle maneuvering in the physical world. By organizing and managing the primitive processing steps, the Driving Pipeline provides the following capabilities:

- **Continuous Vehicle Motion:** The Driving Pipeline drives the vehicle continuously by adjusting the vehicle speed and executing the Vehicle Control step in parallel with other processing steps.
- **Parallel Execution:** The Driving Pipeline executes the primitive processing steps in parallel and maintains a high degree of parallelism. Thanks to the pipelined execution, the Driving Pipeline achieves the highest possible vehicle speed.
- **Adaptive Control:** The Driving Pipeline is capable of adapting sensor aiming, vehicle speed, and execution intervals to the driving conditions.

These capabilities of the Driving Pipeline are made possible by the two key ideas of the Driving Pipeline, the driving unit and the pipelined execution of the processing steps. By using driving units, the data to be processed is divided into a sequence of driving units that can be processed separately by the processing steps. The steps themselves are designed to work in a fixed order on each driving unit. Because of the pipelined execution, the computation for these processing steps can be overlapped on successive driving units. These pipelines in both the processing steps and the data enable the pipelined execution, giving rise to parallel computation and continuous vehicle motion. The driving units also enable adaptive control. By adjusting the location, size, and interval of each driving unit, the Driving Pipeline adapts the processing to the driving situation. The pipeline execution thus enables the adaptive control in the continuous vehicle motion.

The Driving Pipeline clearly describes the driving control scheme in four aspects: primitive processing steps, organization of these processing steps, execution scheduling, and control parameters. In the case of stop-and-go motion, the last three aspects of the driving control scheme are implicit and do not need to be well defined. However, to achieve our goals -- continuous motion, parallel execution, and adaptive control -- we have developed the Driving Pipeline based on an explicit understanding of all of these aspects. This is why the Driving Pipeline is capable of controlling both geometry, such as the sensor view frames, and time, such as execution timing. Adjusting the vehicle speeds demonstrates these capabilities of the Driving Pipeline.

Although the Driving Pipeline supports continuous vehicle motion, the primitive processing steps involved in the Driving Pipeline employ only *static* algorithms. The Perception step, for example, analyzes the sensor data without taking into account the vehicle motion. Similarly, the Local Path Planning step determines the trajectory path plan as if the vehicle were not moving while the Local Path Planning step is processing. By introducing the driving units, the Driving Pipeline converts dynamic problems into a set of static problems for each driving unit. By employing the pipelined execution, the Driving Pipeline overlaps the static processing steps to perform dynamic vehicle motion. This feature of the Driving Pipeline gives two advantages. First, the Driving Pipeline makes it easier to build mobile robot systems by integrating relatively well developed processing algorithms for perception and path planning. Second, the Driving Pipeline provides a test bed for studying these primitive algorithms using real mobile robot systems.

Future research will center on expanding the concept of the driving unit and pipelined execution to accommodate multiple sensors, uncertainty in the map database, and off-road travel. Multiple sensors with different view frame sizes introduce additional synchronization points into the pipeline, thus affecting the execution flow. Uncertainty about the positions of objects in the map affects the aiming of the sensors and vehicle speed. For example, in the presence of little uncertainty, the vehicle can look far ahead and drive quickly. Off-road travel provides a new set of Prediction, Perception, and Planning steps to be incorporated with on-road travel in a single pipeline to permit multiple modes of navigation. Algorithms are needed to dynamically determine the parameters of the pipeline in these scenarios while maximizing parallelism.

10.6 References

[1] R. A. Brooks.
A Robot Layered Control System For A Mobile Robot.
IEEE J. Robotics and Automation RA-2:14-23, March, 1986.

[2] J. L. Crowley.
Navigation for an Intelligent Mobile Robot.
IEEE J. Robotics and Automation RA-1:31-41, March, 1985.

[3] E.D. Dickmanns, A. Zapp.
A Curvature-based Scheme for Improving Road Vehicle Guidance by Computer Vision.
In *SPIE Symposium on Optical and Optoelectronic Engineering*. October, 1986.

[4] Georges Giralt, Raja Chatila, and Marc Vaisset.
An Integrated Navigation and Motion Control System for Autonomous Multisensory Mobile Robots.
In *The First International Symposium on Robotics Research*, pages 191-214.

[5] Y.Goto, etc. .
CMU Sidewalk Navigation System .
In *Proc. of Fall Joint Computer Conference*, pages 105-113. Nov, 1986.

[6] Y. Goto, A. Stentz.
The CMU System for Mobile Robot Navigation.
In *Proc. of IEEE International Conference on Robotics and Automation*, pages 99-105. March, 1987.

[7] H. P. Moravec.
The Stanford Cart and the CMU Rover.
Proc. of the IEEE 71:872-884, July, 1983.

[8] S. Shafer, A.Stentz, C. Thorpe .
An Architecture for Sensor Fusion in a Mobile Robot .
In *Proc. of IEEE International Conference on Robotics and Automation*, pages 2002-2011. April, 1986.

[9] C. Thorpe, S.Shafer, T.Kanade .
Vision and Navigation for the Carnegie Mellon Navlab .
In *Proc. of Image Understanding Workshop*, pages 143-152. Defense Advanced Research Project Agency Information Science and Technology Office, February, 1987 .

[10] S. Tsuji.
Monitoring of a building environment by a mobile robot.
In *Robotics Research 2*. 1985 .

[11] A. M. Waxman, J.J. LeMoigne, L.S. Davis, T.Siddalingalah.
A Visual Navigation System for Autonomous Land Vehicle.
IEEE J. Robotics and Automation RA-3:124-141, April, 1987.

Chapter 11: Multi-Resolution Constraint Modeling for Mobile Robot Planning

Anthony Stentz

Abstract

All navigation systems for a mobile robot include a basic sense-plan-drive cycle for moving the robot about. The navigator must choose sensing points, plan paths between them, and oversee the execution of the robot's trajectory. Various constraints must be met in selecting potential sensing points, feasible trajectories, and safe configurations, all while taking positional uncertainty into account. We have developed a local planner that finds trajectories for a robot by modeling all of the constraints uniformly. At each step in the planning process, the system identifies and uses the most severe constraint to guide the search. A multi-resolution approximation to the constraint solution space is employed to reduce the number of search states, and thus the planning time. The system has been implemented and tested on real data. The results are presented.

11.1 Introduction

Mobile robots are useful for a broad range of tasks including factory automation, hazardous waste removal, planetary exploration, defense, and construction. A typical system consists of a platform equipped with wheels, treads, or legs for moving about. Sensors such as a camera, laser rangefinder, sonar proximity sensors, and inertial navigation system are employed to recognize landmarks, avoid obstacles, and guide the robot along a passageway. The software that controls the robot typically has two components: the global and local navigators. The global navigator uses pre-stored map data to plan coarse paths for the robot and oversees their execution while the local navigator coordinates the robot's sensors and low-level control functions to realize these coarse paths [21]. This paper focuses on the issues involved in building a complete local navigator.

The local navigator must plan trajectories for the robot that meet goal requirements (such as positioning to see a landmark or moving through a doorway) without endangering the robot, and must make sure that the robot is able to kinematically execute these trajectories, all in the presence of uncertainty in the robot's control and environment. A number of systems have been built that address a subset of these issues. Early planners were targeted for indoor mobile robots and assumed the environment could be modeled as a flat surface with polygonal or polyhedral objects [5, 13, 16, 17, 23]. Furthermore, the robot was assumed to be circular in shape and omnidirectional. Later, Laumond [15] and Jacobs [11] relaxed the omnidirectional requirement by modeling a car-like robot with a minimum turning radius. In a system developed at Hughes [6], the indoor environment constraint was relaxed and a planner was developed to plan paths in off-road environments. None of the above systems were able to reason about sophisticated goal requirements or uncertainty in control. The theoretical groundwork for representing and computing the propagation of uncertainty was spelled out [7, 22], but it was not shown how to plan with it. Planning with uncertainty has been explored in manipulation [8, 18], but it is unclear how to apply the domain-specific nature these of techniques to the field of planning for mobile robots.

While the contributions of the above work are very important, in many cases it is difficult to see how to extend them to address the remaining issues or how to generalize the work to other robots and environments. This paper describes a new planner that models and plans with all of the above constraints in an efficient manner, and even more importantly, develops a framework for building efficient planners to model many different types of robots, environments, goal considerations, and uncertainty effects.

11.2 The Local Navigation Problem

The local navigator generates a series of trajectories to incrementally realize the coarse path provided by the global navigator. The local navigator operates in a cycle in which it first uses the sensors to build a model of the robot's environment, then plans a trajectory to the next sensing point, and finally oversees the execution of the trajectory. Simply put, the local navigation problem is one of moving the robot between sensing points. Many mobile robots can be parameterized by two variables in position and one in heading, (x, y, θ). These parameters form a three-dimensional space, or *configuration space*. Thus, the local navigation problem becomes one of planning a trajectory from some initial configuration point to a destination configuration point, without leaving the confines of the three-dimensional space (bounded by the size of the environment).

A number of problems exist. First, the local navigator must determine the destination point. The selection depends on the particular scenario. If the robot knows its entire environment before navigating (i.e., it is mapped), then the destination point is some sensing point at which the robot can register its position relative to the environment (e.g., landmark recognition). If the environment is unknown a priori, then the destination point is some sensing point at which the robot can extend (sense) its local environment, while heading toward a goal handed down from the global navigator. Second, the navigator must address environmental admissibility. The robot cannot pass through certain configurations. In indoor environments, some configurations can place the robot in contact with other objects. In outdoor environments the situation is more complicated. Some configurations could cause the robot to topple, bottom out, or run up against rocks or other objects too large to climb over. Third, kinematic constraints come into play. Car-like robots, for example, are not omnidirectional; they have a minimum turning radius. They cannot move along a straight line between two arbitrary points in configuration space. Fourth, robots don't execute commanded trajectories perfectly. The causes can range from small error in the control vector to gross errors such as wheel slippage.

In order to be able to reason about these constraints in planning, we must be able to model them. We model the above constraints uniformly as functional inequalities f_i defined across the robot's configuration parameters p_i, such that a configuration point satisfies the constraint (admissible) if its functional value is less than some threshold, K_i, and violates it otherwise (inadmissible):

$$f_i(p_1, p_2, \ldots, p_N) \le K_i \quad (admissible) \tag{11.1}$$

$$f_i(p_1, p_2, \ldots, p_N) > K_i \quad (inadmissible)$$

The details of this model for each type of constraint are described in the following sections.

11.2.1 Goal satisfaction

As previously described, a goal for the local navigator can be a final destination point (e.g, a doorway), a point at which a landmark can be seen, or any other geometric constraint defined across the robot's configuration space. Thus, the constraint equations for goal satisfaction are merely test functions that classify a configuration point as either in or out of the goal space. The landmark visibility constraint is illustrated below as an example.

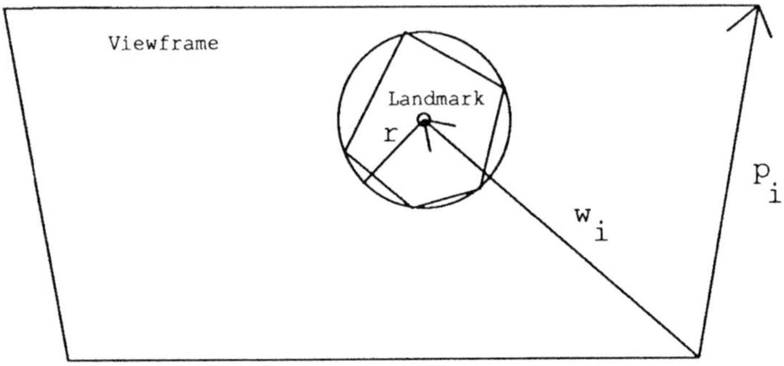

Figure 11.1: Landmark Visibility Constraint

Given a robot with a trapezoidal sensor viewframe (see Figure 11.1), we would like to determine if a landmark at position (x_l, y_l) bounded by a circle of radius r is visible to the sensor when the robot is at configuration point (x_c, y_c, θ_c). A value of zero for θ corresponds to the vehicle directed along the positive y-axis, and θ increases in the counterclockwise direction. Let p_i be a vector aligned with the i-th side of the viewframe (moving around in a counterclockwise direction), and let w_i be a vector formed by connecting the base of p_i to the center point of the landmark. Let (x'_i, y'_i) be the coordinates of the base point of the i-th viewframe vector and let θ'_i be the orientation of that vector. These three parameters are expressed in a coordinate frame affixed to the robot, with its origin at the robot's reference point. The equations for p_i and w_i are given below:

$$p_i = \langle -\sin(\theta_c + \theta'_i), \cos(\theta_c + \theta'_i) \rangle \tag{11.2}$$

$$\begin{aligned} w_i = \langle & x_l - x'_i \cos(\theta_c) + y'_i \sin(\theta_c) - x_c, \\ & y_l - x'_i \sin(\theta_c) - y'_i \cos(\theta_c) - y_c \rangle \end{aligned}$$

The landmark is entirely visible to the sensor if it lies to the left of each viewframe vector. This condition is given below:

$$\forall(i)\frac{p_i \times w_i}{\|p_i\|} \geq r \tag{11.3}$$

Note that for the above selection of p_i, $\|p_i\| = 1$. Note also that the constraint has been formulated as a functional inequality compared to a threshold constant defined solely across the robot's configuration parameters. (Negating both sides puts it in the form of Equation (11.1).) The other parameters in the equations are constants.

11.2.2 Environmental admissibility

We define environmental admissibility as follows. A configuration is considered to be inadmissible if it incapacitates the robot, rendering it unable to proceed or backtrack, or if it physically alters the robot or environment in an undesirable way. An example of the the former is a steep hill that causes the robot to tip over. An example of the latter is contact with a physical object in the environment, as is the case in indoor mobile robot planning. The treatment in the literature of trajectory planning for mobile robots on flat surfaces with polygonal obstacles is extensive [5, 13, 14, 16, 17, 19, 23, 24]. We choose to illustrate planning in less structured environments, such as those found outdoors, off-road. The constraints we describe below are similar to those used in a system developed at Hughes Corporation [6].

Consider a car-like robot equipped with a laser rangefinder capable of returning an elevation map of the terrain in front of the vehicle. For such a vehicle, three constraints determine its admissibility (see Figure 11.2). The *tilt* constraint states that the attitude of the vehicle must not exceed a safe threshold; the *support* constraint states that the robot must be able to exert sufficient force against the ground to support locomotion; and the *body clearance* constraint states that the body of the robot must not come in contact with the terrain. In order to derive the constraint equations we need a model for the vehicle. Specifically, we need to be able to determine the pose of the robot for a given placement on the terrain. Three points are needed to determine the plane of the undercarriage of the robot, but the robot has four ground contacts (the wheels). Rather than attempting to model the complexities of a real suspension, we use a simple model: that the body is supported at each wheel by springs of equal spring constants with its center of mass at the center of the undercarriage. Let (x, y, z) be a world-based coordinate system, and let (x', y', z') be a robot-based

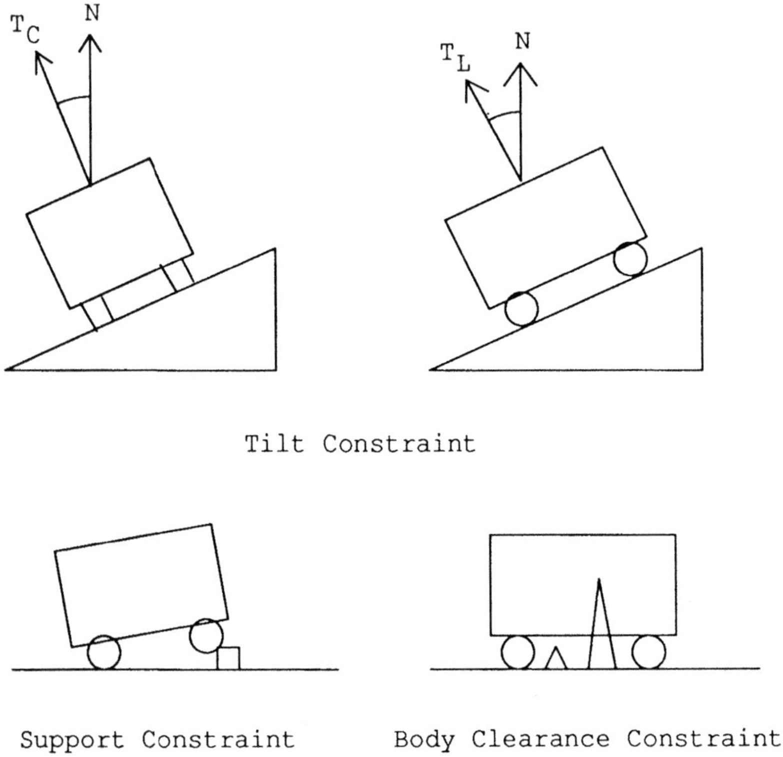

Figure 11.2: Environmental Admissibility Constraints

system in the ground plane with its origin at the center of the vehicle and the positive y-axis aligned with the heading. The plane of the undercarriage can be written as follows:

$$z' = Z_C + T_C x' + T_L y' \tag{11.4}$$

where T_C is the cross tilt, or left-to-right tilt, T_L is the long tilt, or front-to-back tilt, and Z_C is the elevation of the center of mass above the ground plane. By setting the forces and torques about the center of mass equal to zero, we get the following equations for the coefficients:

$$T_C = \frac{(z_1+z_4) - (z_2+z_3)}{2W} \tag{11.5}$$

$$T_L = \frac{(z_1+z_2) - (z_3+z_4)}{2L}$$

$$Z_C = \frac{(z_1 + z_2 + z_3 + z_4)}{4} + H$$

where z_1 through z_4 are the terrain elevation values at the four wheels (numbered counterclockwise starting with the right front wheel), and H is the rest height of the undercarriage on level ground. In order to evaluate the tilt constraint for a given configuration (x_c, y_c, θ_c), we transform the positions of the four wheels from the robot-based (x'_i, y'_i) to the world-based (x_i, y_i) coordinate frame as shown below:

$$x_i = x'_i \cos(\theta_c) - y'_i \sin(\theta_c) + x_c \qquad (11.6)$$

$$y_i = x'_i \sin(\theta_c) + y'_i \cos(\theta_c) + y_c$$

We then look up the terrain values at those four points and insert them into the above equations. We use the following equations to test for tilt:

$$|T_L| \le T_{L0} \qquad (11.7)$$

$$|T_C| \le T_{C0}$$

where T_{L0} and T_{C0} are the bounds on the safe tilt in the long and cross directions respectively.

In order to test the support constraint, we examine the *difference* in elevation at the four wheel points. The worst case for support arises when the robot must climb step-like terrain. The actual value of the maximum step is a function of the elasticity of the tires and terrain, soil characteristics, air pressure in the tires, side-wall support, tilt of the robot, and interactions between the four wheels [1, 2]. We assume the ground is approximately planar and take an empirical approach to determine the maximum height Z_w. The constraint has the following form:

$$\Delta z \leq Z_w \tag{11.8}$$

In order to test the body constraint, project the sides of the body into the plane of the undercarriage and test this extended undercarriage against the terrain beneath it. If all terrain points are below the plane, the constraint is satisfied as given below:

$$\forall(x,y) \in S, f_t(x,y) < z'(x',y') \tag{11.9}$$

where f_t is the terrain function, and S is the set of terrain points beneath the undercarriage.

Equations (11.7), (11.8), and (11.9) taken together define the functional inequalities (test functions) for environmental admissibility.

11.2.3 Kinematic constraints

Car-like robots have a non-holonomic constraint, that is, a minimum turning radius. Laumond [15] showed that a legal trajectory exists between two configuration points iff a trajectory exists consisting of turns of minimum turning radius and line segments connecting them. Assuming the planning space is smaller than twice the minimum turning radius and contains no obstacles (relaxed later), we need examine only trajectories of the form (a_1, s, a_2), where a_1 and a_2 are arcs of minimum turning radius, and s is the straight line segment that connects them. Thus, there are four possible trajectories: two left turns, two right turns, and opposing S-turns (see Figure 11.3).

Given a pair of configuration points, (x_1, y_1, θ_1) and (x_2, y_2, θ_2), the equations for the centers of curvature (centers of the turning circles) are given below:

$$x_{c1} = r_1 \cos(\theta_1) + x_1 \tag{11.10}$$

$$y_{c1} = r_1 \sin(\theta_1) + y_1$$

$$x_{c2} = r_2 \cos(\theta_2) + x_2$$

$$y_{c2} = r_2 \sin(\theta_2) + y_2$$

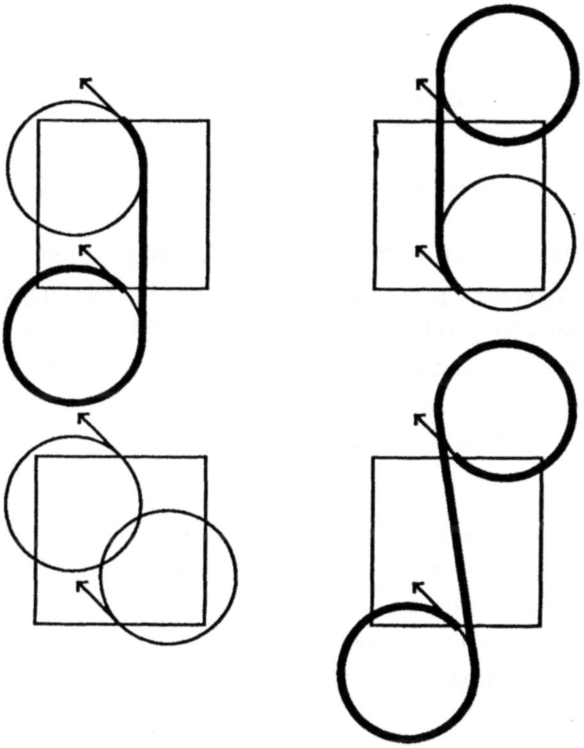

Figure 11.3: Minimum Turning Radius Constraint

where r_1 and r_2 are the minimum turning radii for the first and second turn respectively. A positive radius is a right turn.

A trajectory between the two points can fail to exist for two reasons. First, no straight line segment s can be found between the arcs. This case occurs when the centers of curvature of the two arcs are closer together than twice the minimum turning radius [15]. This constraint is formulated below:

$$(x_{c1} - x_{c2})^2 + (y_{c1} - y_{c2})^2 \geq (2r_{min})^2 \tag{11.11}$$

Second, the trajectory leaves the bounds of the planning space. Assuming the space is box-shaped with bounds of (x_{min}, x_{max}) and (y_{min}, y_{max}), the equations have the following form:

$$x_{min} \le x_{c1} - r_1 \cos(\theta) \le x_{max} \tag{11.12}$$

$$y_{min} \le y_{c1} - r_1 \sin(\theta) \le y_{max}$$

$$x_{min} \le x_{c2} - r_2 \cos(\theta) \le x_{max}$$

$$y_{min} \le y_{c2} - r_2 \sin(\theta) \le y_{max}$$

where θ specifies the heading angle along the arc. For a given pair of configuration points, we can compute the four trajectories and test their admissibility using Equations (11.11) and (11.12). Note that we must minimize and maximize the values of Equations (11.12) across the range of θ. If the above constraints are satisfied for at least one of the four trajectories, the two configurations are considered admissible.

11.2.4 Uncertainty in path execution

Uncertainty in the robot's configuration can arise from a number of sources. It can arise from systematic sources, for example, a miscalibrated robot. Other error arises from random effects in the control of the robot. Gross error due to wheel slippage or sensor malfunction can also occur. We take as an example, a gaussian random control model with bounds (covariance) large enough to subsume small, systematic effects as well.

In order to compute the propagation of uncertainty, we first need the transformation equations between configurations. Given an initial configuration, (x_i, y_i, θ_i), if we command the robot to drive a distance d with curvature k, the robot will end up at configuration (x_f, y_f, θ_f). The transformation equations are given below:

$$\theta_s = -kd \tag{11.13}$$

$$x_s = \frac{1 - \cos(\theta_s)}{k}$$

$$y_s = \frac{-\sin(\theta_s)}{k}$$

$$x_f = x_i + x_s \cos(\theta_i) - y_s \sin(\theta_i)$$

$$y_f = y_i + x_s \sin(\theta_i) + y_s \cos(\theta_i)$$

$$\theta_f = \theta_i + \theta_s$$

If we start with some gaussian uncertainty in the initial configuration and in the control vector (k, d) given by the 5×5 covariance matrix C_i, we can compute an approximate solution to the gaussian uncertainty at the final configuration given by the 3×3 covariance matrix C_f as follows:

$$C_f = J C_i J^T \tag{11.14}$$

where J is the 3×5 Jacobian mapping the initial state and control vector $(x_i, y_i, \theta_i, k, d)$ into the final state vector (x_f, y_f, θ_f). The approximation is computed by linearizing about the mean and squaring the error on both sides. The technique parallels that used for uncertainty propagation between coordinate frames [7, 22].

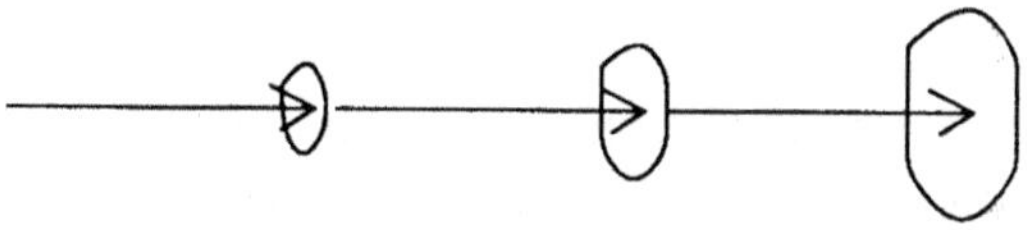

Figure 11.4: Propagation of Uncertainty along an Arc

We can construct an ellipsoid that contains roughly 99% of the probability density by bounding three standard deviations of the distribution. The propagation of uncertainty is illustrated in Figure 11.4 for a sequence of arcs. For a given configuration c, we can use these bounds to test which other configurations must also be tested for environmental admissibility and goal satisfaction to determine the admissibility of c.

11.3 Finding Trajectories

Given the means to test individual configuration points for constraint satisfaction, we now address the problem of finding trajectories. We could attempt an analytic approach in which we solve for the boundaries between environmentally admissible/inadmissible space and goal/nongoal space, determine a subspace through which the legal trajectories move and try a vertex-graph approach. However, the programming cost would be high, there would be no guarantee that an analytic solution could be found, and the solution might not be easily generalizable to more/different constraints. Furthermore, in the case of the constraints modeled in this paper, one of the functions, namely the terrain, has no analytic form.

Another approach would be to densely tesselate the space into small boxes that roughly approximate a single configuration point. Searching the resultant space would be prohibitively expensive, since the complexity is cubic in the resolution along each dimension. The problem is further complicated by the uncertainty constraints. For every configuration point c the planner examined, it would be forced to test a *set* of configuration points in the neighborhood of c for environmental and goal admissibility due to the presence of uncertainty.

Our approach is to take advantage regularities in the environment and continuity of the kinematic, uncertainty, and goal satisfaction constraints to identify large subspaces that satisfy all of the constraints, so that our tesselation is coarse where possible, and fine where need be. The principal tool we employ is the octree. Octrees were first used in CSG systems [10] for approximating the shape of solid objects. Later in planning, octrees were used for representing the admissible space for a manipulator [9] moving about in a field of obstacles. We show how this tool can be used for any constraints defining environmental admissibility and how it can be extended to incorporate goal, kinematic, and uncertainty constraints as well. The end result is that the search space is much more compact and the combinatorics are reduced in comparison to a dense tesselation approach.

11.3.1 Searching the constraint space

Heuristic search is used to find trajectories in the planning space. Heuristic search begins with the *start state* (initial configuration) and expands it to other reachable states, which are entered on the *open list*. The "best" state on the open list is selected for expansion and the algorithm recurs until a *goal state* is found or until no reachable, unexpanded states remain. Backpointers are maintained to trace the search paths. A heuristic evaluation function is applied to all states entered on the open list, and the state with the lowest value is "best". The evaluation function can be written as $f(x) = g(x) + h(x)$, where $g(x)$ is the cost incurred to reach state x and $h(x)$ is an estimate of the remaining cost to the goal. If $h(x)$ is a lower bound on the actual remaining cost, an optimal path will be found [20].

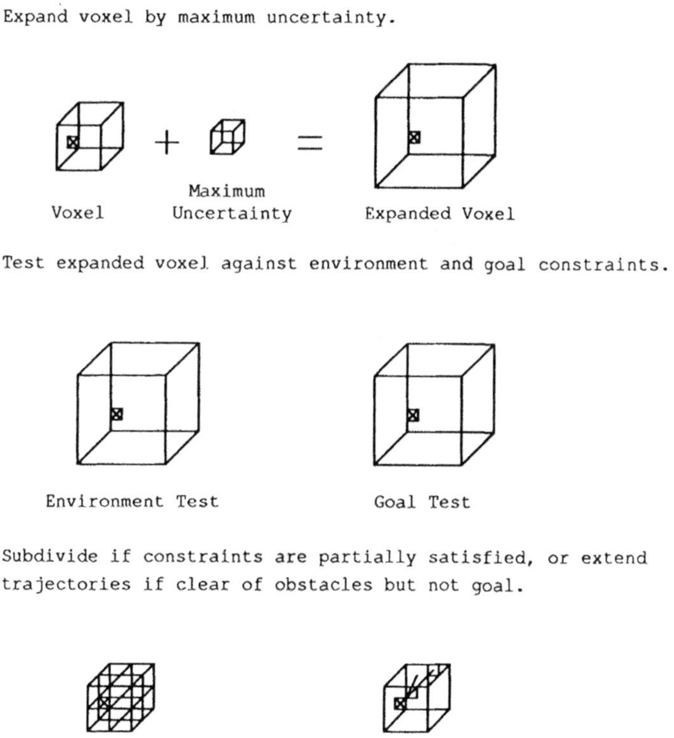

Figure 11.5: Expansion Algorithm for a State

In our implementation of heuristic search, a *state* is a subface on the boundary of a cubic subspace of the configuration space called a *voxel* (see Figure 11.5). The cost function $g(x)$ is the number of voxels through which the trajectory passes from the start state to x, and $h(x)$ is an estimate of the remaining number of voxels to the goal. The search begins with a single state, the start state, on the boundary of a voxel enclosing the entire planning space. Initially, the flag SUBDIVIDE is set to FALSE. For each state x selected for expansion through voxel v, the algorithm does the following:

1. Compute the maximum amount of uncertainty that can be generated in each dimension by moving from any configuration point in x to any other configuration point on the boundary of v without leaving v. Add this uncertainty to v along each dimension. The resultant, larger voxel is called the *expanded voxel*.
2. Test the expanded voxel for environmental admissibility. One of three cases will result:
   a. All configurations are admissible: it is safe to move through the voxel. Continue the expansion at Step 3.
   b. All configurations are inadmissible: nowhere is it safe to move. End the expansion of x.
   c. Some configurations are admissible: the search must continue at a higher resolution. If the maximum resolution has been reached, end the expansion of x; otherwise, set SUBDIVIDE to TRUE and continue at Step 3.
3. Test the expanded voxel for goal satisfaction. One of three cases will result:
   a. All configurations are admissible: the goal has been reached. End the search and return success.
   b. All configurations are inadmissible: no goal configurations in this voxel. Go to Step 4 to continue the expansion.
   c. Some configurations are admissible: the search for the goal must continue at a higher resolution. If the maximum resolution has been reached, set SUBDIVIDE to FALSE; otherwise, set SUBDIVIDE to TRUE. Continue at Step 4.
4. If SUBDIVIDE is TRUE, divide v into eight subvoxels. Attach x to the adjacent subvoxel and reenter it onto the open list.
5. If SUBDIVIDE is FALSE, compute all configurations on the boundary of v reachable from *any* configuration point in x and

enter them along with their new uncertainty parameters on the open list.

Intuitively, the above algorithm computes bounds on position of the robot as it plans and ensures that this bounded volume both steers clear of environmentally inadmissible subspaces and ends up entirely in the goal subspace. The algorithm searches at a higher resolution whenever it needs to localize the extent of the uncertainty or the boundaries of the environmentally admissible or goal subspaces. In Step 1, the expanded voxel bounds *all* of the configurations where the robot *might* be for any commanded trajectory through v. In Step 2, it's safe to move anywhere through v, safe nowhere, or the planner must subdivide v to work around the the unsafe regions. The division of v into subvoxels v_i leads to tighter bounds on the uncertainty, since the smaller size of each v_i better localizes the means of the distribution and reduces the uncertainty growth (and thus the expanded subvoxel). In Step 3, the search can terminate successfully only if the entire expanded voxel ends up in the goal space. The interpretation of this case is that even though the planner does not know exactly where the robot is, it is still guaranteed to be in the goal space (e.g., able to see a landmark). If no goal configurations exist in the expanded voxel, then the planner must extend the search paths through the voxel to neighboring voxels in search of the goal space. In the event that the expanded voxel contains a mixture of goal and nongoal configurations, v must be subdivided to find a subvoxel containing only goal configurations. If the maximum resolution has been reached, then the voxel cannot be subdivided further and the search trajectories must be extended through the voxel to find another that contains only goal configurations. In Step 4, the voxel subdivision takes place. If the state x straddles the boundaries of the subvoxels, x is subdivided into smaller states and each is added to the open list. Finally, in Step 5, the search trajectories are extended through the voxel. The reachable configurations are not added to the open list individually, but are grouped into subfaces as described in the next section.

The number of voxels traversed was chosen for the evaluation function, $f(x)$, to minimize the amount of computation needed to find a solution. For our domain, testing for environmental admissibility is an expensive operation. Since the computational cost of testing a large voxels for constraint satisfaction is only slightly higher than for small voxels (see next section), it behooves us to search for the shortest sequence of voxels to the goal. Note that the optimal-length solution to the goal in number of voxels is not necessarily the optimal-length solution in Euclidian distance. In fact, the planner will choose paths that swing wide of an obstacle through large voxels rather than expend the computation needed to explore small voxels near the obstacle's boundary. This selection has the additional benefit of finding the least complicated path, since the total number of arcs in the trajectory is three times the total number of voxels. An evaluation function that is a weighted sum of the number of voxels traversed and the Euclidian distance could be employed to strike the desired balance between search time and distance travelled.

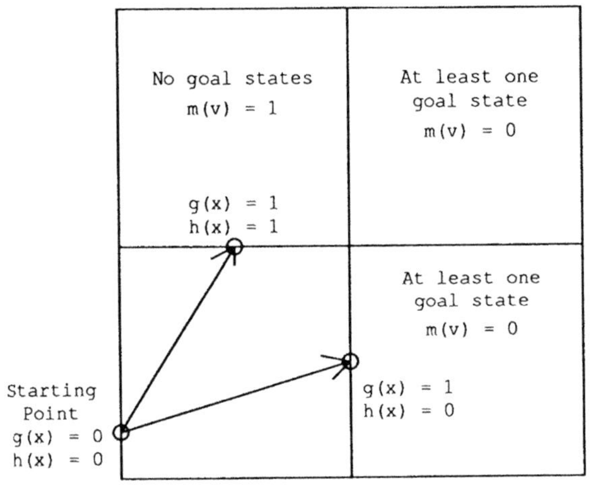

Figure 11.6: Determining the Remaining Cost

The cost function, $g(x)$, and the estimate of the remaining cost, $h(x)$, are computed in Step 5. When the state x is extended through v to state x', the cost at x' is $g(x') = g(x) + 1$. The function $h(x')$ is set equal to $m(v')$, where v' is the voxel affixed to x but opposite to v (see Figure 11.6). The function m is computed for a voxel whenever one is created through subdivision in Step 4. For a voxel v', the value of $m(v')$ is set to zero if v' contains a goal point; otherwise, $m(v')$ is set to the minimum value of m for its adjacent voxels plus one. The intuition is that if v' contains a goal point, it could reside on any of the faces, thus a lower on the remaining cost must be zero. Otherwise, the search must incur a cost of one to move through v' to another voxel. The function m maintains a lower bound on the minimum distance to the goal for every voxel as the space is explored.

11.3.2 Testing subspaces for constraint satisfaction

A number of the operations in the above algorithm could be very expensive computationally without the use of special techniques. An inefficient way of computing the maximum amount of uncertainty generated by any trajectory passing through a voxel v in Step 1 would be to individually test all trajectories between a point in state x and every reachable point on the boundary of v and record the maximum uncertainty in each dimension. Such an approach is prohibitively expensive. Likewise, testing all of the configurations in v in Steps 2 and 3 for environmental admissibility and goal satisfaction respectively would be expensive. The same is true for testing for all legal trajectories between an pair of configuration points in Step 5.

We use the SUP-INF method [3] to perform many of these operations efficiently. This technique was first developed to handle universally quantified linear constraints. It was extended in ACRONYM [4] to handle a class of nonlinear constraints as well for purposes of predicting the appearance of a three-dimensional object in a camera image. For a given constraint inequality f_i defined across a range of input parameters (e.g., a voxel v), the SUP-INF method computes an upper bound (UB) and a lower bound (LB). If $UB \leq K_i$ (see Equations (11.1)), then f_i is satisfied for all configurations in v. Likewise, if $LB > K_i$, then f_i is not satisfied by any configuration in v. Finally, if the bounds straddle K_i, then v *possibly* contains a mixture of admissible and inadmissible configurations. The bounds UB and LB are computed for f_i by treating each occurrence of the configuration parameters in f_i as independent and recursively decomposing f_i into subfunctions for which the least upper bound and greatest lower bound can be easily computed.

The SUP-INF method is used for both the environmental and goal constraints. For example, consider the tilt constraint for a voxel v. The voxel provides bounds on the configuration parameters (x_c, y_c, θ_c). These bounds are used along with the SUP-INF method on Equations (11.6) to compute bounds on the x and y coordinates of the four wheels. The terrain bounded by the four "rectangles" is examined to compute the maximum and minimum elevation (z-values) for each

Figure 11.7: The Terrain Pyramid and Rectangle Query

wheel. These bounds are subsequently inserted into Equations (11.5) to compute bounds on the planar coefficients. Finally, these bounds are inserted into Equations (11.7) to classify v as all admissible, all inadmissible, or some admissible. This classification is used in Step 2 in the above algorithm to determine which case applies. In order to efficiently search the terrain for maximum and minimum elevation values for an arbitrary rectangle, we construct a terrain pyramid (see Figure 11.7) which summarizes terrain data at multiple resolutions. The bottom level holds the iconic elevation data. Each node in the levels above it records the maximum and minimum value of the four nodes below. A typical query rectangle encloses some of the nodes high in the pyramid, thus reducing the complexity of the search.

Concerning kinematic constraints, in order to compute all configurations reachable from state x through voxel v in Step 5, we consider a subface x' of configurations at a time. By inserting the four corners of both x and x' into Equations (11.10) through (11.12), we can determine whether the subface x' lies entirely outside of the set of reachable configurations, entirely inside, or whether the boundary straddles x'. In the third case, we subdivide x' into four quadrants and reapply the test recursively. By starting with each of the six faces of v, we can construct quadtrees to efficiently represent configurations reachable from x. Furthermore, we can pre-compute a table of quadtrees of reachable configurations for all subface types. This table is indexed by voxel resolution, θ range, subface resolution, and the position of the subface on the faces of the voxel. Thus, during planning Step 5 is merely a table lookup operation.

Concerning the uncertainty constraints, in order to compute the maximum uncertainty generated by moving from state x through voxel v to any reachable point (Step 1), we must first compute the maximum propagation *coefficients*. These coefficients are the Jacobian differentials in Equation (11.14) that are multiplied by the uncertainty in x to get the uncertainty at the next state x'. We do this by considering all pairs of configuration points on the faces of v between which a legal trajectory exists. We can compute a maximum value for each element of the Jacobian across all legal trajectories. Since each trajectory consists of two arcs and a straight line segment, we must record three sets of coefficients. Note that although this operation is computationally expensive, it can be carried out entirely offline. We can construct a table of coefficients indexed by voxel resolution and θ range. While planning, we use this table of maximum coefficients in Equation (11.14) to compute the maximum uncertainty generated while moving through a voxel v. When computing the expanded voxel, we assume the pathological case, that is, that the trajectory giving rise to the maximum uncertainty touches the boundary of the voxel. Thus, we must add the maximum uncertainty generated to the voxel's boundaries in each dimension.

11.4 Experiments

The planner was tested on both simulated and real data. We expect to use it as the off-road trajectory planner for Carnegie Mellon's Navigational Laboratory (Navlab) [12]. The Navlab is a Chevy van chassis that has been modified to permit computer control of the driving/steering functions. It is equipped with a number of sensors, including a color camera and an ERIM laser rangefinder. The vehicle is controlled by specifying a sequence of arcs for it to traverse. The ERIM laser rangefinder is used to acquire three-dimensional terrain data. The ERIM scans an area about four meters to the left and right of the vehicle and six to ten meters in front of the robot. The range information is transformed from a sensor-based, polar coordinate frame to a robot-independent elevation map. Polygons are fit to approximately planar regions in the map to reduce the information size for transmission between modules in our system. Range scans from multiple viewpoints can be pieced together to construct a planning environment for the robot.

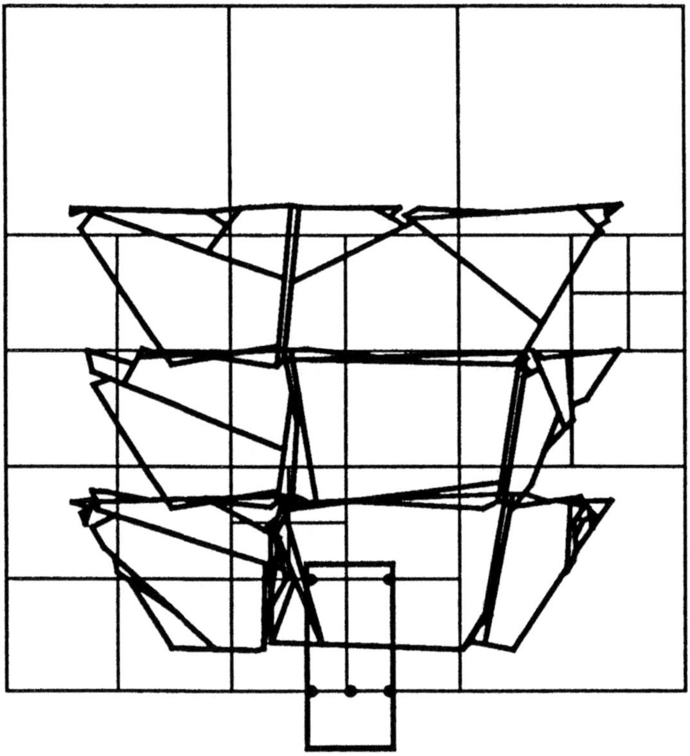

Figure 11.8: Octree Slice for Planar Environment

Figures 11.8 and 11.9 illustrate two such environments. In each figure, the three trapezoids are the ERIM sensor viewframes containing a *mesh* of polygons approximating the terrain shape. The Navlab is shown to scale. Configurations are measured relative to the center point of the rear axle, illustrated with a circle. Also shown is an xy slice of the octree representing the environment. The slice is at $\theta = 0$ corresponding to the vehicle directed straight ahead. The open squares represent admissible space, that is, configurations that satisfy all of the environmental constraints. The crossed squares represent inadmissible space,

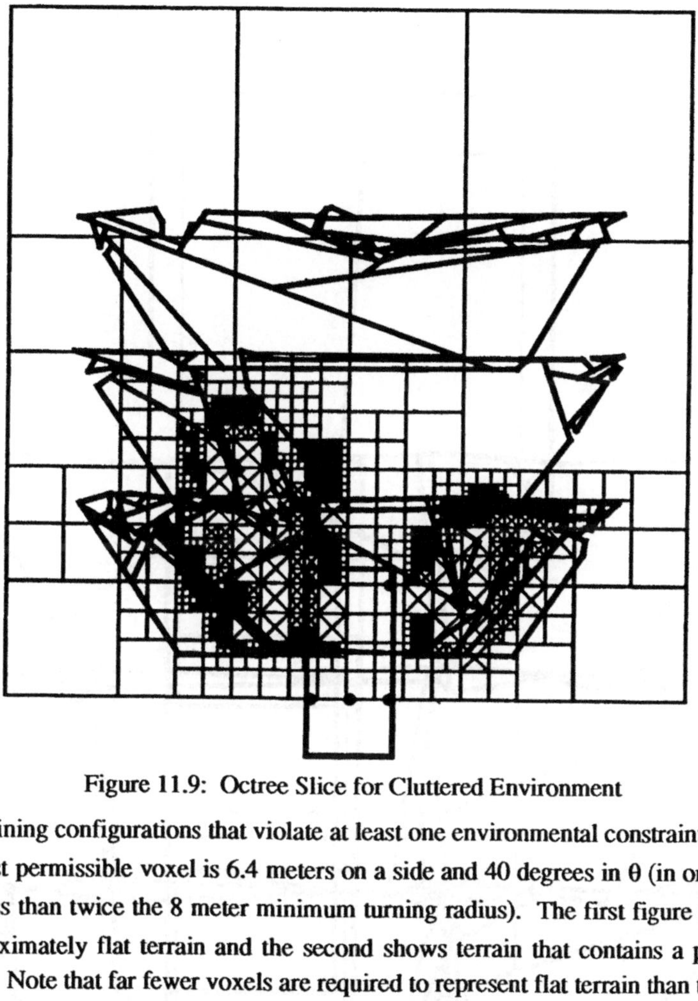

Figure 11.9: Octree Slice for Cluttered Environment

containing configurations that violate at least one environmental constraint. The largest permissible voxel is 6.4 meters on a side and 40 degrees in θ (in order to be less than twice the 8 meter minimum turning radius). The first figure shows approximately flat terrain and the second shows terrain that contains a pair of trees. Note that far fewer voxels are required to represent flat terrain than terrain with obstacles. Our tests have shown that the number of terrain accesses required to represent a typical environment with the octree is one to three orders of magnitude less than the number required to represent the environment using a dense, individual configuration point approach. Figure 11.10 shows a trajectory planned to a goal point directly in front of the vehicle off the mesh in the environment shown in Figure 11.8. The trajectory is shown by a sequence of connected line segments from the bottom of the diagram to the top. The thin-lined squares are an xy projection of the voxels expanded during the search. The

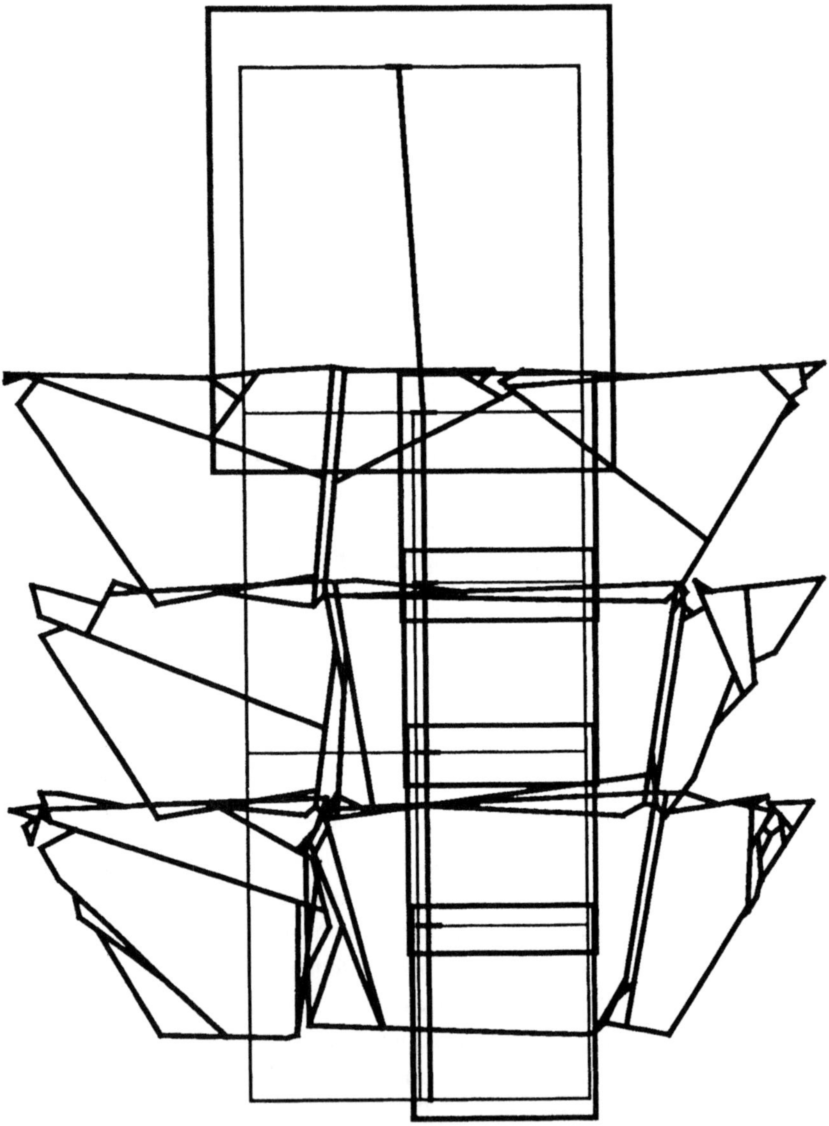

Figure 11.10: Planned Trajectory for Planar Environment

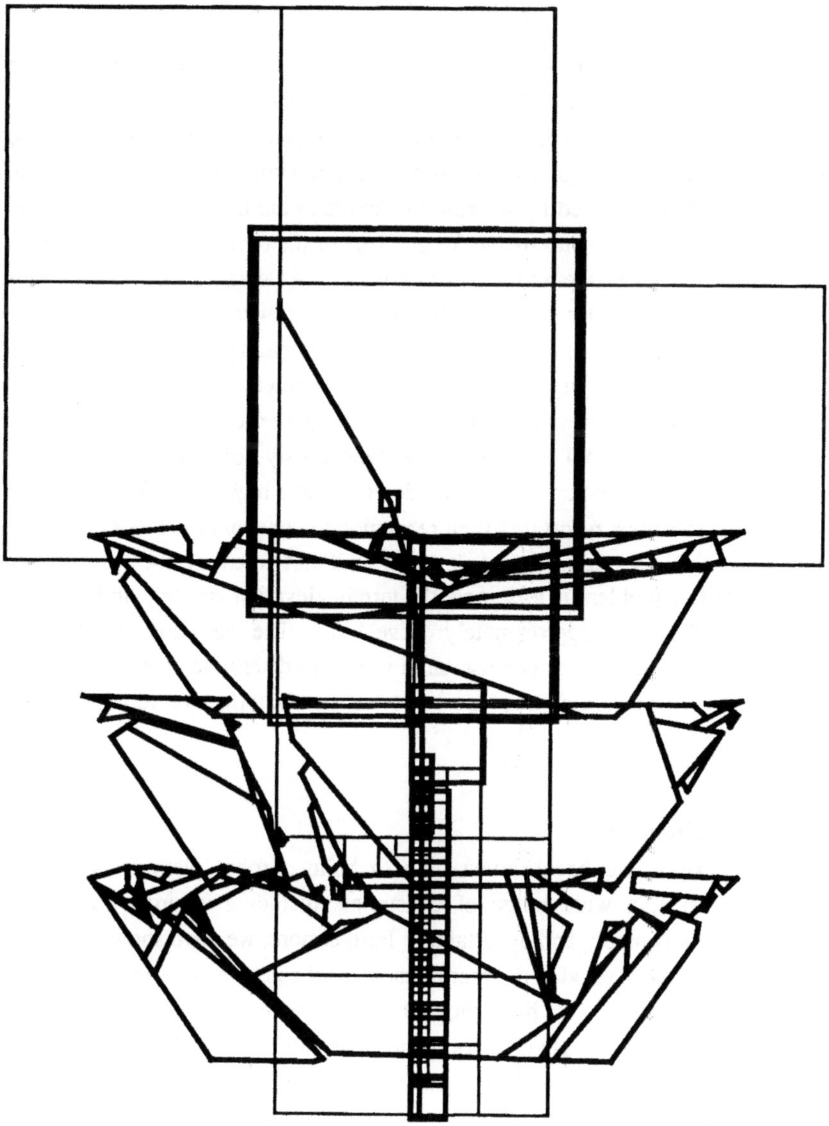

Figure 11.11: Planned Trajectory for Cluttered Environment

boldface squares are the expanded voxels (voxel plus uncertainty) along the solution trajectory. Figure 11.11 illustrates a trajectory planned for the environment in Figure 11.9 with a goal off the mesh and to the left. Note that in simple environments, the planner can find a solution quickly by planning through large voxels, while in complex environments, the planner must subdivide voxels to a high resolution to work its way around obstacles.

Heuristics and adjustments are employed to reduce the planning time. The most expensive constraint to evaluate is body clearance, since in the worst case all of the pixels beneath the undercarriage at a given configuration must be tested for intersection. For a given voxel, if the maximum disparity in elevation across the terrain spanned by the robot is less than the undercarriage height on level ground, then it is impossible for any part of the terrain to come in contact with the vehicle and the voxel satisfies the body clearance constraint. This heuristic is much faster than computing the four wheel boxes, getting the maximum and minimum terrain elevation, computing the maximum and minimum tilt, and testing the undercarriage for intersection with the terrain beneath. An adjustment to the terrain itself can also expedite the processing. If the terrain is planar but tilted (but not dangerously so), the environment is admissible but the planner must subdivide voxels to a high resolution to confirm it. The planner must resort to a high resolution to constrain the robot enough to determine that the variation in elevation will not intersect the robot. Our solution to this problem is to sample the terrain elevation coarsely and compute a rotation that will approximately "level" it. The variation in elevation disappears and the planner can use large voxels to determine admissibility. Of course, the tilt constraint must be corrected by the effective rotation of the terrain.

11.5 Conclusions

In this paper, we have shown how to build a robot planner capable of reasoning about a wide range of constraints including environmental, goal satisfaction, kinematic, and uncertainty. Furthermore, we have shown how this framework can be extended to other mathematical constraints through the use of the SUP-INF method. We have shown how this planner can reduce the number of states needed to represent the problem (and thus the complexity), by taking advantage of regularities in the environment, and continuity in the goal satisfaction, kinematic, and uncertainty constraints through the use of multi-resolution, recursive-decomposition techniques. Furthermore, we have shown how the planner can use heuristic functions to efficiently choose states to expand, thus reducing the total computation needed.

This efficiency does not come without cost, however. The preference for large voxels over small ones leads to suboptimal trajectories (in Euclidian distance space). By weighting our evaluation function, we can strike a balance between trajectory length and search time. These efficiencies lead to a loss of completeness as well. By retaining only the lowest *cost* path to a given state x, we might discard a path with lower uncertainty (but higher cost) leading to x. The higher uncertainty search path might later be terminated due to inadmissibility at a point where the lower uncertainty path could continue (i.e., "fit"). This is not much of a problem, however, since path cost and uncertainty are generally positively correlated. Finally, some completeness is lost when searching through large voxels, since the uncertainty is overestimated and this overestimate is passed along to subsequent states in the search.

11.6 Acknowledgements

We would like to thank Martial Hebert for the range processing code that generates polygonal meshes from iconic ERIM images. We would also like to thank Chuck Thorpe for his comments and the use of the Navlab and Steve Shafer for his guidance in this work. Finally, we would like to thank Takeo Kanade for all he has done to make CMU a leading university in outdoor mobile robot research.

11.7 References

[1] Bekker, M. G.
The Theory of Land Locomotion.
The University of Michigan Press, 1956.

[2] Bekker, M. G.
Off-the-Road Locomotion.
The University of Michigan Press, 1960.

[3] Bledsoe, W. W.
The Sup-Inf Method in Presburger Arithmetic.
Technical Report, University of Texas at Austin, 1974.
Memo ATP 18.

[4] Brooks, R. A.
Symbolic Reasoning Among 3-D Models and 2-D Images.
PhD thesis, Stanford University, June, 1981.

[5] Brooks, R. A.
Solving the Find-Path Problem by Representing Free Space as Generalized Cones.
Technical Report, Massachusetts Institute of Technology, May, 1982.
A. I. Memo No. 674.

[6] Daily, M., Harris, J., Keirsey, D., Olin, K., Payton, D., Reiser K., Rosenblatt, J., Tseng, D., Wong, V.
Autonomous Cross-Country Navigation with the ALV.
In *Proc. IEEE International Conference on Robotics and Automation.* 1988.

[7] Durrant-Whyte, H.
Integration, Coordination and Control of Mult-Sensor Robot Systems.
PhD thesis, University of Pennsylvania, 1986.

[8] Erdmann, M.
Using Backprojections for Fine Motion Planning with Uncertainty.
In *Proc. IEEE International Conference on Robotics and Automation.* March, 1985.

[9] Faverjon, B.
Obstacle Avoidance Using an Octree in the Configuration Space of a Manipulator.
In *Proc. IEEE International Conference on Robotics.* March, 1984.

[10] Jackins, C. L., Tanimoto, S. L.
Oct-Trees and Their Use in Representing Three-Dimensional Objects.
Computer Vision, Graphics, and Image Processing 14(3), 1980.

[11] Jacobs, P., Canny, J.
Planning Smooth Paths for Mobile Robots.
In *Proc. IEEE International Conference on Robotics and Automation.* May, 1989.

[12] Kanade, T., Thorpe, C., and Whittaker, W.
Autonomous Land Vehicle Project at CMU.
In *Proc. 1986 ACM Computer Conference.* Cincinnati, February, 1986.

[13] Khatib, O.
Real-Time Obstacle Avoidance for Manipulators and Mobile Robots.
The International Journal of Robotics Research 5(1), Spring, 1986.

[14] Krogh, B.
A Generalized Potential Approach to Obstacle Avoidance Control.
In *Proc. Robotics Research Conference.* August, 1984.

[15] Laumond, J-P.
Finding Collision-Free Smooth Trajectories for a Non-Holonomic Mobile Robot.
In *Proc. IJCAI-87.* August, 1987.

[16] Lozano-Perez, T., Wesley, M. A.
An Algorithm for Planning Collison-Free Paths Among Polyhedral Obstacles.
Communications of the ACM 22(10), October, 1979.

[17] Lozano-Perez, T.
Spatial Planning: A Configuration Space Approach.
IEEE Transactions on Computers C-32(2), February, 1983.

[18] Lozano-Perez, T., Mason, M., Taylor, R.
Automatic Synthesis of Fine-Motion Strategies for Robots.
International Journal of Robotics Research 3(1), 1984.

[19] Moravec, H. P.
The Stanford Cart and the CMU Rover.
Proc. IEEE 71(7), July, 1983.

[20] Nilsson, N. J.
Principles of Artificial Intelligence.
Tioga Publishing Company, 1980.

[21] Shafer, S., Stentz, A., Thorpe, C.
An Architecture for Sensor Fusion in a Mobile Robot.
In *Proc. IEEE International Conference on Robotics and Automation.* April, 1986.

[22] Smith, R., Self, M., Cheeseman, P.
Estimating Uncertain Spatial Relationships in Robotics.
In *AAAI Workshop on Uncertainty.* 1986.

[23] Thompson, A. M.
The Navigation System of the JPL Robot.
In *Proc. IJCAI-77.* 1977.

[24] Thorpe, C. E.
FIDO: Vision and Navigation for a Mobile Robot.
PhD thesis, CMU Computer Science Dept., December, 1984.

Chapter 12: NAVLAB
An Autonomous Navigation Testbed

Kevin Dowling, Robert Guzikowski, Jim Ladd, Henning Pangels, Sanjiv Singh, and William Whittaker

12.1 Introduction

The Navlab is a testbed for research in outdoor navigation, image understanding, and the role of human interaction with intelligent systems. A mobile navigation habitat, it accommodates researchers and significant onboard computing. Applications for navigation vehicles include mapping of hazardous waste sites, off-road haulage, material handling at construction worksites, and exploration of planetary surfaces.[1]

The Navlab is a roadworthy truck modified so that humans or computers can control as occasion demands. Because it is self-contained, it is not subject to telemetry bottlenecks, communication faults or dependence on stationary infrastructure, and can travel to confront navigation problems of interest at any site.

The Navlab controller manages all locomotion. It interacts with a computer host and human operator to implement varying degrees of autonomy. The Navlab controller queues and executes commands originating from a computer or human host, providing a "Virtual Vehicle" interface that hides details of the actual hardware.

The Navlab configuration consists of a chassis, drivetrain and shell. The chassis is a modified, cut-away van with a computer-controllable, hydraulic drivetrain. Driver's controls allow a human monitor to override automatic control for overland travel, setup and recovery from experimental errors. The shell houses all onboard equipment including computers, controllers, telemetry, and internal sensors. In addition, it provides a working area for operators, allowing research within the confines of the vehicle. Equipment racks, monitors, lighting, air-conditioning, seating and desk space create a mobile environment for research.

---

[1]An earlier version of this paper appeared as CMU Robotics Institute report CMU-RI-TR-87-24. Reprinted by permission.

Humans can monitor and supervise the Navlab from the operator's console for setup, error recovery and tuning. Interface modes include Virtual Vehicle instructions, joystick motion control, and direct servo motion commands. The console also incorporates several displays to show the current states of both the vehicle and control computer.

The Navlab supports a choice of sensing to accommodate many types of navigation research. Video cameras provide color and intensity images for scene interpretation. Navlab vision experiments use a single camera to analyze road edges through intensity, texture, and color segmentation. A scanning rangefinder sweeps the surroundings with a distance-measuring laser that provides useful three-dimensional information about the geometry and reflectivity of the environment. Laser experiments navigate through geometric features like trees and buildings. Taken together, data of color, intensity, range and reflectance provide a rich basis for building natural scene descriptions. Sensor information from several sources can be fused to achieve more robust perception. A blackboard architecture integrates the distributed processes that sense, map, plan and drive.

The Navlab represents continuing evolution in the design of navigation vehicles. Fully self-contained, it is a milestone in mobile robotics.

12.2 Controller

12.2.1 System Architecture

The Navlab Controller provides a Virtual Vehicle interface for commands and queries. Using the virtual system, many of the high-level modeling and planning functions can port across a number of different physical systems that can be controlled with the same command primitives. Only the lowest level control computing must deal with the physical differences of the system.

Complex systems usually defy any attempts at mathematical modeling techniques, which makes control parameters impossible to even estimate. A set of pseudo constants, tunable from an operator's console, adjust the parameters and gains. The Navlab maintains these constants in file structures that remain on disk for power-up initialization. The system always starts up with default values established from the most recent tuning of the system.

The control computing accepts commands from a host or human operator who can intervene at various levels of control to insure safe operation during experiments. The assumption during development is that the higher levels of computing will not succeed in all situations. The control computing thus provides a graceful transition between computer and human control when failures occur. The hooks for human inputs are also useful for setup and recovery during experiments.

The sensors monitored by control computing reflect the state and ability of the system to respond to commands issued by the cognitive planning layers. The values of the observed parameters have fixed maximum limits that are characterized by the physical system. These limits, however, are not static and can move inward during certain operating conditions. Physical parameters such as heat and pressure can diminish Navlab's mechanical ability to respond to commands. The parameter limits are dynamically adjustable by the controller to protect the Navlab. When controlling with a powerful physical plant like the Navlab, erroneous plans and commands have significant impact. The control computer should never execute commands blindly, so mechanisms are needed for validating and rejecting commands, with advisories communicated to the source of commands.

Control computing, the lowest of the levels within the autonomous mobile system architecture, interacts with the physical system. The design criteria set forth for the Navlab low-level controller include an open-ended architecture, a virtual system interface and multiple command streams.

12.2.2 Virtual Vehicle

A Virtual Vehicle is a man-machine interface that accepts conceptual commands and provides a clean separation between the navigation host and vehicle control. This interface masks implementation details of the physical vehicle, facilitating adaptability to future navigation testbeds.

Higher-level planners (generically called the Host) communicate with the controller through three kinds of messages:

- *Commands* issued by the Host
- *Queries* by the Host about the status of devices
- *Reports* initiated by the Virtual Vehicle on completion of commands and in case of emergency.

In the current implementation, the vehicle is directed along circular arcs because arcs are quickly computed and absolute position is not critical. More recently the Fast Navigation group has upgraded these arcs to quintics of constantly changing radii to traverse more complex paths. Because it is not possible for a vehicle to switch between an arc of one curvature and another instantly, path transitions are inexact. Errors are compensated for by dynamically planning arcs to reach subgoal points along the path.

12.2.3 Motion Control

Both drive and steering can be controlled both manually and automatically. Autonomous mode control uses motor controllers, with different versions of our controller using boards from Galil or Creonics. Manual mode allows the user to steer with the Navlab's steering wheel, and to control the drive system through a potentiometer controlled by a foot pedal.

Dash Panel Control

The vehicle operates manually to simplify transport to and from test sites. Manual operation doesn't require any computing or generator power. The electronic components active during manual operation are powered by the Navlab's 12 V battery system.

A human interface is incorporated for safe and easy use by drivers of standard automobiles. Figure 2-6 shows illuminated pushbutton controls mounted within reach of the driver.

- **High, Neutral, & Low**: allow the operator to choose gears. Because switching directly from one gear to another produces an unsafe lurching of the vehicle, a hardware logic function allows switching only by first selecting Neutral.
- **Forward, Reverse**: select the direction in which the vehicle moves.
- **Auto/Manual**: a pull-push switch that switches between manual and automatic control.
- **Emergency Stop**: disables autonomous locomotion and brings the vehicle to a rapid, controlled stop. Servo-lock of steering is disabled; steering is returned to manual control.
- **Brake Pedal**: as in commercial cruise control systems, a light touch of the brake pedal brings the vehicle back under manual control.
- **Speed Control Pedal**: activates a 20K ohm potentiometer to produce a voltage proportional to the angle of deflection of the pedal.
- **Throttle Control**: this dial sets the vehicle engine RPM as detailed in Section 12.4.2.
- **HE**: this switch turns on/off the heat exchanger fan for the hydraulic system.
- **HP, LP**: These lamps are lit when the dirty oil filters in the high and low pressure hydraulic systems indicate an alarm.

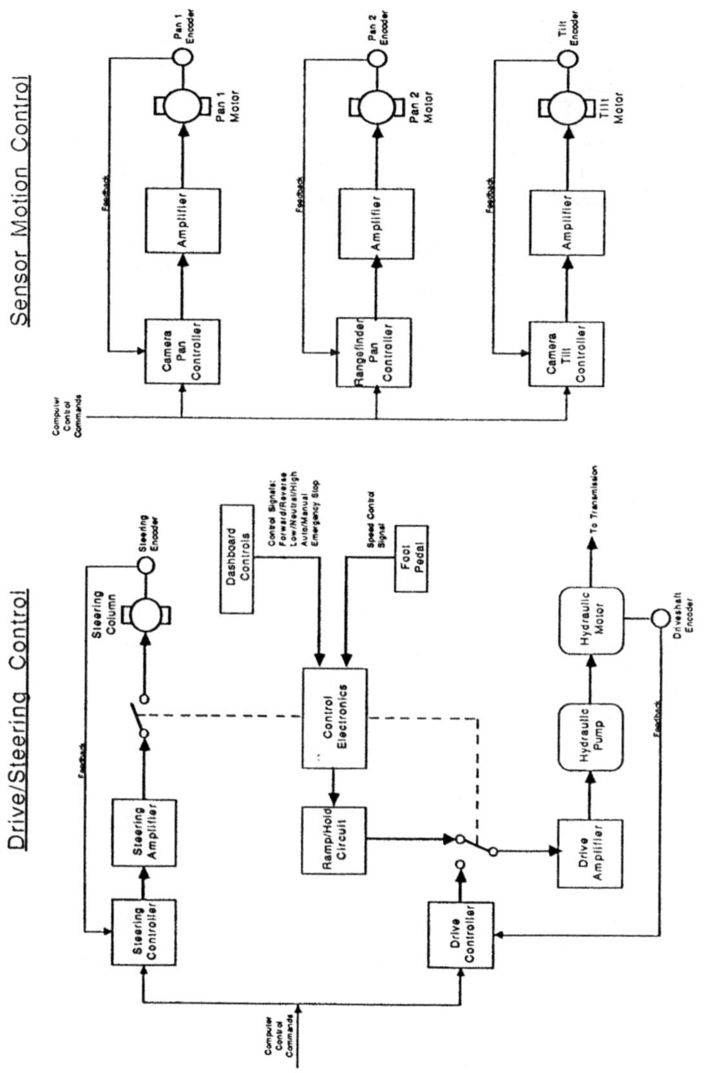

Figure 12.1: Motion Actuation

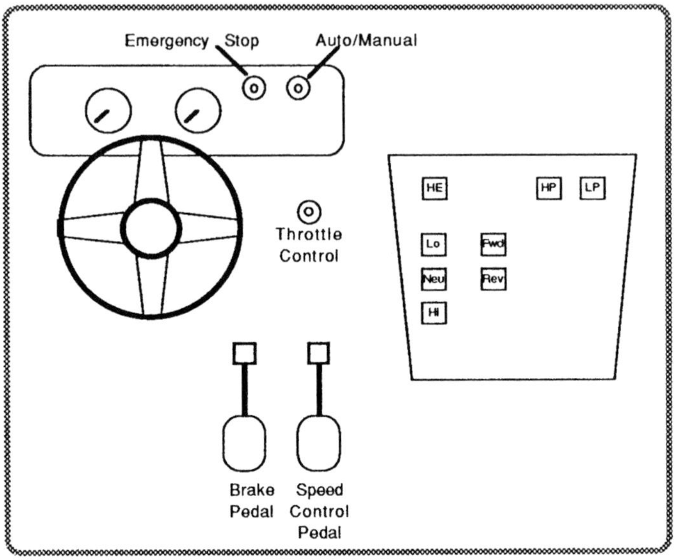

Figure 12.2: Dash Panel Layout

By default, when the vehicle is powered up, it is put into manual mode, neutral gear, and forward direction. It is necessary to provide the ability to override the autonomous mode in a fast but controlled manner if an emergency develops. To ensure reliable operation, manual override is a hardwired electronic circuit with sealed electromagnetic relays instead of sequential logic gates. This design proved to be immune to the noise and power fluctuations common to automotive electrical systems. Because this circuitry is essential to vehicle locomotion, it is powered by the vehicle 12 V system rather than the generator.

An electronic ramp/hold circuit in series with the foot pedal provides adjustable limits on acceleration and deceleration and ensures that abrupt movements of the foot pedal do not cause the vehicle to lurch. This feature was included both for safety and ease of driving. A second ramp/hold unit ensures a smooth deceleration in case of an emergency stop.

Steering Control

The steering control system consists of a computer-controlled DC servo motor linked to the steering column by a toothed belt. A single axis motor controller (DMC-201) uses feedback from an optical 1200-line incremental encoder mounted directly on the motor shaft to maintain tight position control over the steering wheel. A servo-amplifier converts the +/- 10V control signal from the motor controller to drive the DC motor with up to 11 amps of continuous current. At maximum speed, the steering mechanism can be moved between its two extreme positions in 2 seconds.

Feedback is obtained from an encoder on the motor shaft that is mated to the steering wheel, which is always turned a specified amount. Differences between intended and achieved radius occur due to linkage non-linearities and factors such as friction between the road and the wheels, grade of the road, vehicle speed, and speed with which the steering wheel is turned.

Limit switches on the steering linkage are hardwired inputs to the controller board and provide both a safety stop to protect the steering mechanism and a reference point for roughly calibrating the steering control system to a known position on power-up or system reset.

Drive

A single axis motor controller services the drive system. The voltage (-10V to 10V) produced by the motor controller is converted to a current signal (-100mA to 100mA) by an amplifier that directly operates a hydraulic servo valve to set the speed of the hydraulic motor. Acceleration of the vehicle is limited by a ramp/hold circuit, mentioned in Section 12.2.3, in effect providing a low pass filter to the input signal. An optical 300-line incremental encoder mounted on the hydraulic motor shaft provides feedback to the motor controller. Because the transmission is downstream of the hydraulic motor (i.e., between the motor and the driveshaft), the encoder pulses must be interpreted differently for high and low gears.

12.3 Vehicle Shell

Navlab's foundation is a 1985 General Motors Vandura cutaway chassis chosen as a commercial base to simplify development. As acquired, the vehicle consisted of a chassis frame, a drivetrain and a cab. A custom shell was constructed to house the onboard AC power generation, power distribution, control and computing equipment. Space for operators is provided, allowing research activity within the confines of the vehicle. The original configuration also included a 350 ci V-8, cruise control, an automatic transmission, dual rear wheels, power steering, power brakes and a 33 gal fuel tank.

The chassis was originally rated for 10,000 lbs. gross vehicle weight. The final vehicle weight was established as 12,000 lbs., thus necessitating a more robust suspension. In order to achieve the necessary load rating the front coil springs were upgraded and two extra leaves were added to the rear springs, increasing the gross vehicle weight by 2400 lbs. Heavy duty gas/oil shock absorbers were installed to minimize a slight tendency to pitch due to the extra weight. In addition, the original equipment tires were exchanged for Goodyear radials with a higher load and all-weather rating.

Modifications to the frame were minimized to preserve strength and stiffness. However, in order to mount some of the larger hydrostatic components alterations had to be made. The main forward crossbeam, transmission rear support beam and surrounding floor were completely removed. The crossbeam was replaced by a box section which bridges the hydrostatic pump. A channel section was added as support for the pump and also provides additional frame strength.

12.3.1 Exterior Design

The shell was custom-built with particular attention paid to strength requirements, anticipating needs for extensibility. The roof and cab support air-conditioning, antennas, sensors, and working personnel. The floor of the shell supports about 2000 kg. The shell is dimensioned so that researchers can stand inside; five equipment racks are housed side by side along one side of the vehicle. Figure 12.3 shows a rear and side view of the vehicle.

The shell is made entirely of steel. Heavy gauge was used on the front and back walls while lighter gauges were used along the side walls and roof. A metallic blue paint protects the entire shell. There are compartments for the generators and power-related equipment. Louvered metal doors provide outside access; there is no access to these compartments from inside the vehicle to keep fumes from entering the shell.

A wiring port in the floor behind the driver's seat allows wiring from the underside to enter the vehicle. Another access vent in the shell above the passenger compartment enables wiring from cameras and range sensors to enter the vehicle.

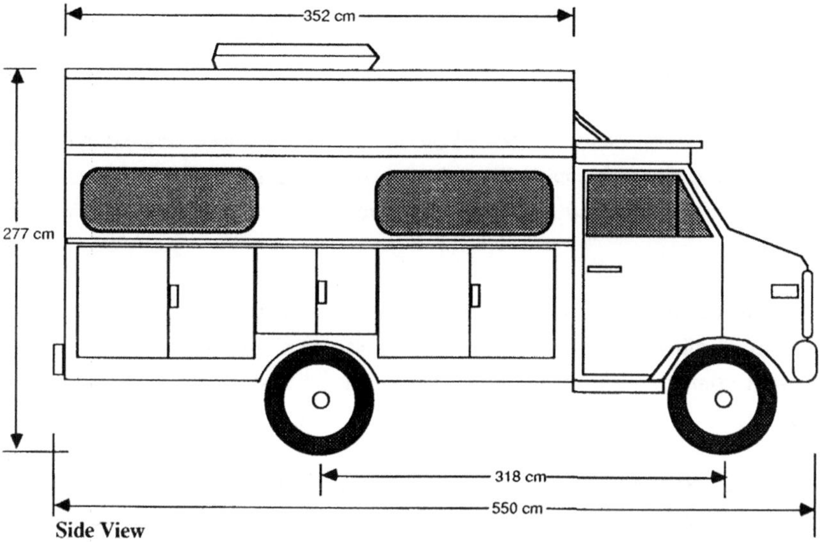

Side View

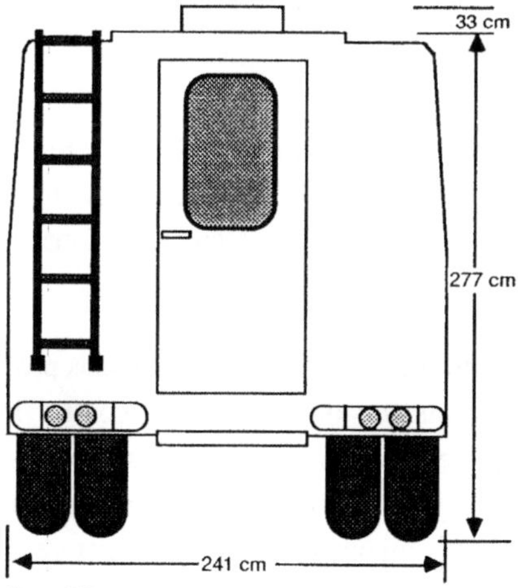

Rear View

General Navlab Specs

Total Weight: 5449kg
Minimum Turning Radius: 750 cm
Center of Gravity: (112cm, 244cm)
(x,y,) with rear corner on the driver's side as the origin.

Figure 12.3: Side and Rear View of the Vehicle

12.3.2 Interior Design

Figure 12.4 shows a topdown view of the Navlab. The cab has two seats, one for a driver and one for a passenger. A console located between the two seats allows the operator to control and monitor the transmission. A research area behind the cab contains computing, sensing, and control equipment, as well as space for two researchers.

Five equipment racks are located on the left side. A desk area extending the length of the research area is located opposite the racks across the aisle. Three video monitors mounted above the desk area can swivel to a desired viewing position.

Along the rear edge of the desktop an outlet strip provides power for the various terminals and test equipment. Bolt-down inserts with elastic straps prevent computer equipment from sliding on the desktop when the vehicle is moving.

Cabling between devices passes through cable trays mounted close to the ceiling. The tray design securely holds video and communication cables but allows for easy removal or addition. Track-mounted lights above the desk area provide independently aimed illumination. Seating is simply removable solid chairs from the offices and other labs.

Cooling

The heat generated by power conditioning, lighting, and electronics would damage some of the experimental computing. Thus in addition to the air conditioning provided on the van, a standard roof-mounted recreational vehicle air conditioner provides cooling.

Fiber insulation between the shell and interior panels also provides protection from the heat. Insulation inside and outside the shell helps control interior climate. Underfloor insulation keeps heat from the hydrostats and exhaust from entering the interior. Flat sheets of fiberglass covered with thin gauge aluminum are inserted in floor areas between frame members. High temperature silicon-based insulation covered with heavy gauge aluminum foil covers exhaust pipes.

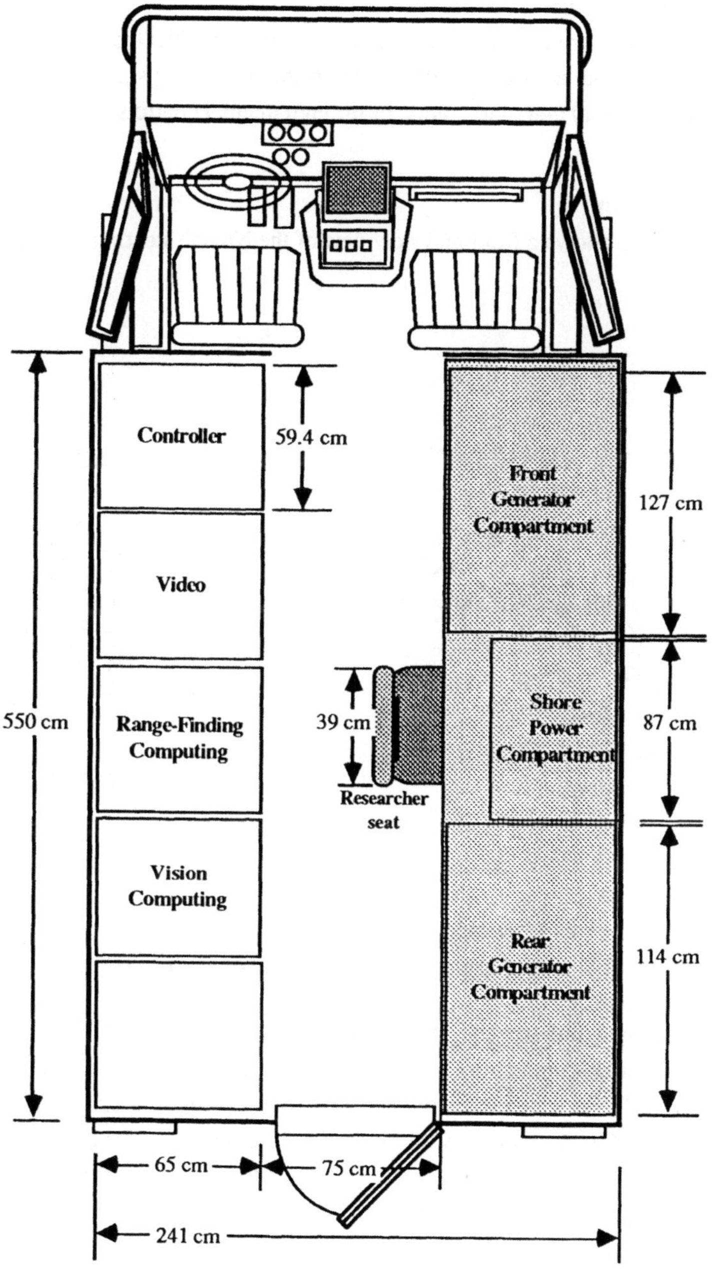

Figure 12.4: Interior Layout of Vehicle

12.4 Locomotion

Steering and drive motions coordinate to drive the Navlab through planned trajectories. Both axes of motion are controlled by analog signals issued by the controller while in automatic mode or through manual controls.

12.4.1 Steering

Figure 12.5 shows a front and side view of the Navlab steering mechanism.

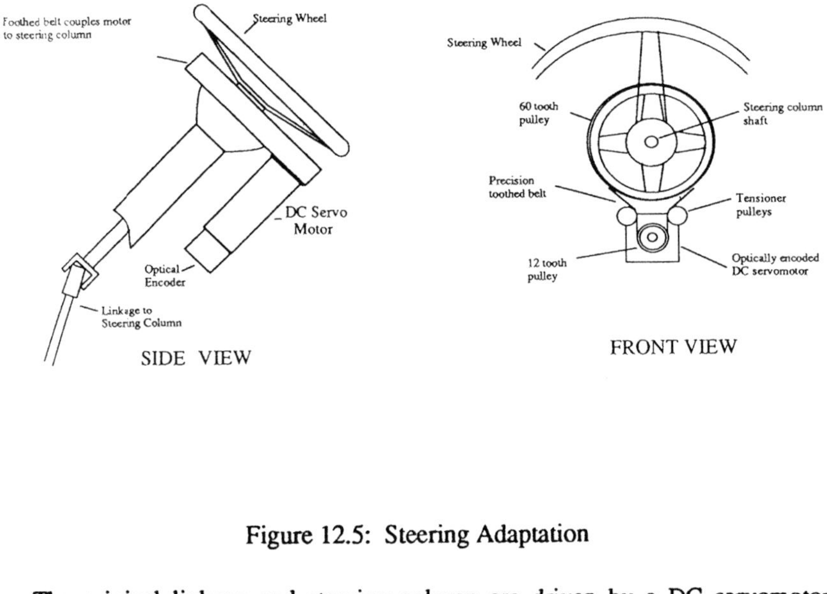

Figure 12.5: Steering Adaptation

The original linkage and steering column are driven by a DC servomotor mounted below the steering column. The motor is connected to the steering column shaft via a precision toothed belt; two toothed pulleys provide a gear reduction of 5 to 1. This configuration provides high enough torque to turn the steering shaft but low enough for the operator to overpower the steering motor in an emergency. A special hub ties all the steering elements together and a safety enclosure houses the moving parts. Limit switches at the extremes of steering travel prevent command error from damaging the system.

12.4.2 Drive

The original transmission and drive shaft were eliminated and new hydraulic components added. A hydraulic pump and motor combination comprises Navlab drive. This hydrostatic combination was selected because it provides precise control of position, speed and acceleration that a traditional driveline could not. Hydrostatic equipment also has a long history of smooth control and finely adjustable response.

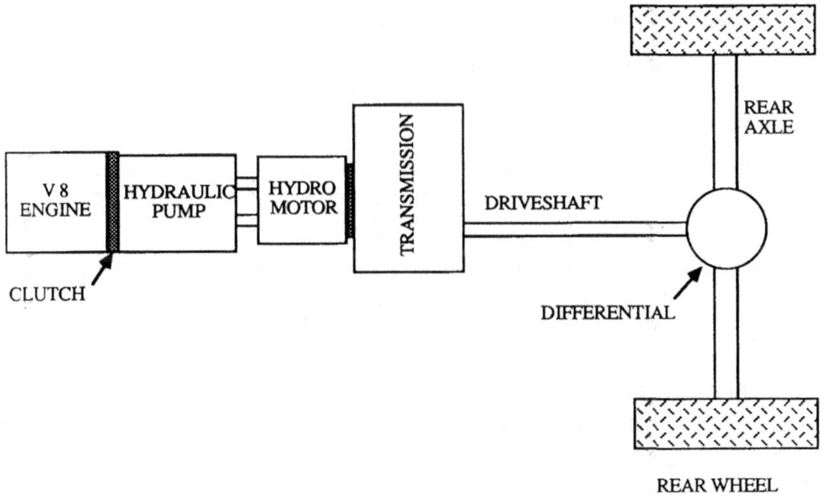

Figure 12.6: Schematic of Vehicle Drivetrain

Drive power comes from the main vehicle engine. Engine RPM is limited by a governor to prevent overdriving the attached hydraulic pump. Pump output is controlled by an analog signal. This signal originates from either a foot pedal that replaces the standard gas pedal or a drive controller, depending on whether the vehicle is in manual or automatic mode. The displacement of the pump (proportional to the signal from the footpedal) determines the speed at which the hydraulic motor moves. The motor in turn powers a two-speed transmission which operates at either a 4:1 (low gear) or 1:1 (high gear) ratio, turning the driveshaft connected to the vehicle differential. Figure 12.6 shows a schematic of the drivetrain.

The configuration described above makes control of vehicle motion simpler than if the vehicle transmission and brakes had to be controlled to produce desired velocity. The standard braking system is intact but is only used in case of emergency because the analog signal to the hydraulic pump controls both accelerations and decelerations.

Vehicle Engine

The standard 350 ci V-8 engine is the main source of driving power. The following modifications were made:

- An electrically actuated clutch was installed to couple the vehicle engine to the hydraulic pump. The clutch is disengaged to isolate it from the engine when the engine is being started.
- The alternator was upgraded to a 120 amp dual output unit to satisfy the additional requirements of the two-battery, 12 volt system onboard.
- One stock emission control air pump was substituted by a hydraulic power takeoff unit. It is driven from the crankshaft end and shares a stock V-belt with the power steering pump.
- An engine oil cooler was installed to reduce oil deterioration caused by the constant high engine temperature.

Engine RPM Control

An engine RPM control keeps the vehicle engine running at a determined range of RPM irrespective of grade and speed. A magnetic pickup on the output shaft of the engine provides feedback to a specialized controller that maintains a constant RPM by moving an actuator linked to the engine carburetor. Figure 12.6 shows a schematic of the mechanism.

Hydraulic Pump

The output of the engine drives a hydraulic pump through a flexible coupling. The pump is suspended from a frame crossmember with rubber shock mounts to allow movement with the engine.

The pump, a Sunstrand axial piston pump, is equipped with an electronic displacement control valve that alters the angle of an internal swashplate between 0 and 18 degrees, depending on an input signal that varies from -10 to 10 V. Negative voltages cause the pump to turn in the reverse direction. At 0 volts the pump has a holding torque to keep the vehicle stationary. At 10 V the displacement is maximum, corresponding to maximum driving speed.

Hydraulic fluid is supplied to the pump from the reservoir by an integral

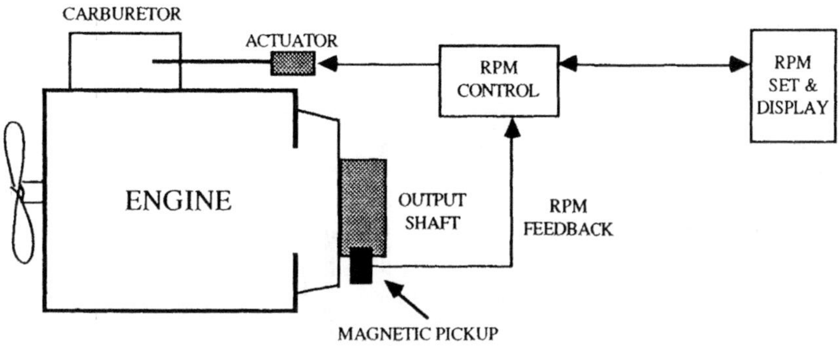

Figure 12.7: Mechanism for Engine RPM Control

charge pump to replace the fluid pumped to the motor while an equal amount of surplus hot oil is drained from the pump case and passed through the main heat exchanger.

Hydraulic Motor

Hydraulic power from the pump is transmitted to a matching fixed displacement motor attached directly to the transmission. High-pressure flexible hose couples the motor and pump. Because the motor is a fixed displacement type, it always turns the same amount for every unit volume of fluid pumped in, resulting in an RPM of the motor that is directly proportional to the input signal of the pump.

A 10 micron filter cleans the return leg of the high pressure system. An additional crossmember supports the hydrostatic motor. The motor unit has an SAE standard shaft mounted to a mating flange on the driveshaft's forward universal joint yoke.

Transmission

The Funk transmission, an electrically shifted two-speed gear box installed between the hydraulic motor and the driveshaft, is bolted to a reinforced frame member. It is mechanically coupled to the motor on the input side and to the driveshaft on the output side. The transmission provides a ratio of 3.950:1 in low gear and 1.0441:1 in high gear. Low gear supports low-speed experimentation

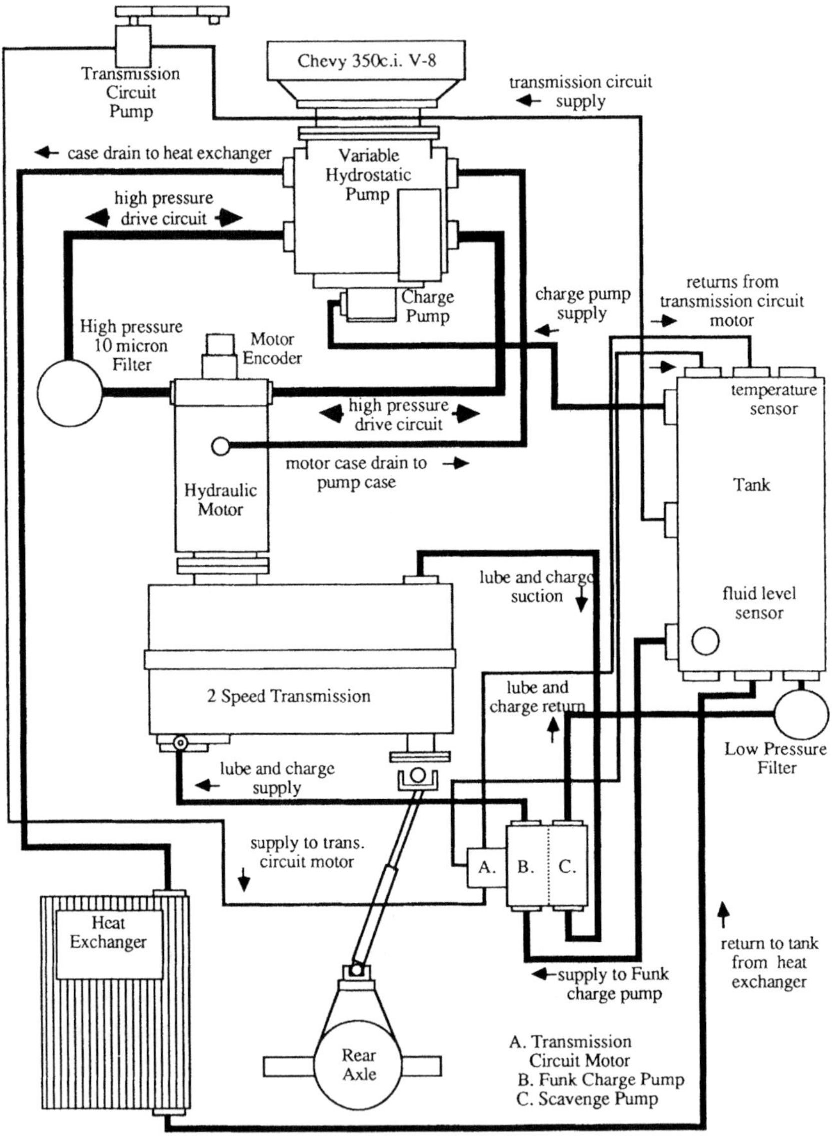

Figure 12.8: Hydro Drive System

(0-20 km/h); high gear (0-40 km/h) transports the vehicle along public roads in manual mode. The gear is selected electronically by applying a voltage to one of two solenoids on the transmission; if neither solenoid is activated, the transmission is in neutral gear. It is necessary to shift into neutral when changing from one gear to another.

A flow-through lubrication and charge system was added to the transmission to circulate hydraulic fluid. The fluid is returned to the reservoir through a low-pressure filter.

Reservoir, Heat Exchanger, and Filters

A reservoir holds about 80 liters of hydraulic fluid. Because seals and bearing surfaces are sensitive to temperature and contamination of the hydraulic fluid, oil returned to the reservoir must be allowed enough time to de-aerate and cool. Heat is removed by passing oil from the pump case drain through a heat exchanger. Cooled oil is directed back to the reservoir. Dirt in the oil is filtered at two points: in the return leg of the high-pressure system and between the transmission and the reservoir.

A series of valves assist in the cooling and circulation of working fluid. A shuttle valve and a low pressure bleed-off valve act together to allow a small portion of the working fluid to circulate through the oil cooler and reservoir. A make-up pump replenishes the fluid that is removed via a bleed-off valve.

The reservoir is equipped with a thermistor and a level gauge to relay tank status to the vehicle controller.

Hydrostat Sensor and Control System

Figure 12.9 shows the sensing and control system associated with the hydraulic drive system. All the components are located on the underside of the vehicle so all lines enter the vehicle through a wiring port in the floor behind the driver's seat.

Control lines include:

1. Hydrostatic pump displacement: This line controls the swash plate angle in the pump regulating the displacement of hydraulic fluid to the motor.
2. Gear selection: This line controls the gear (high or low) of the transmission.
3. Heat exchanger fan control: This line controls the on/off state of a fan that cools the hydraulic fluid.

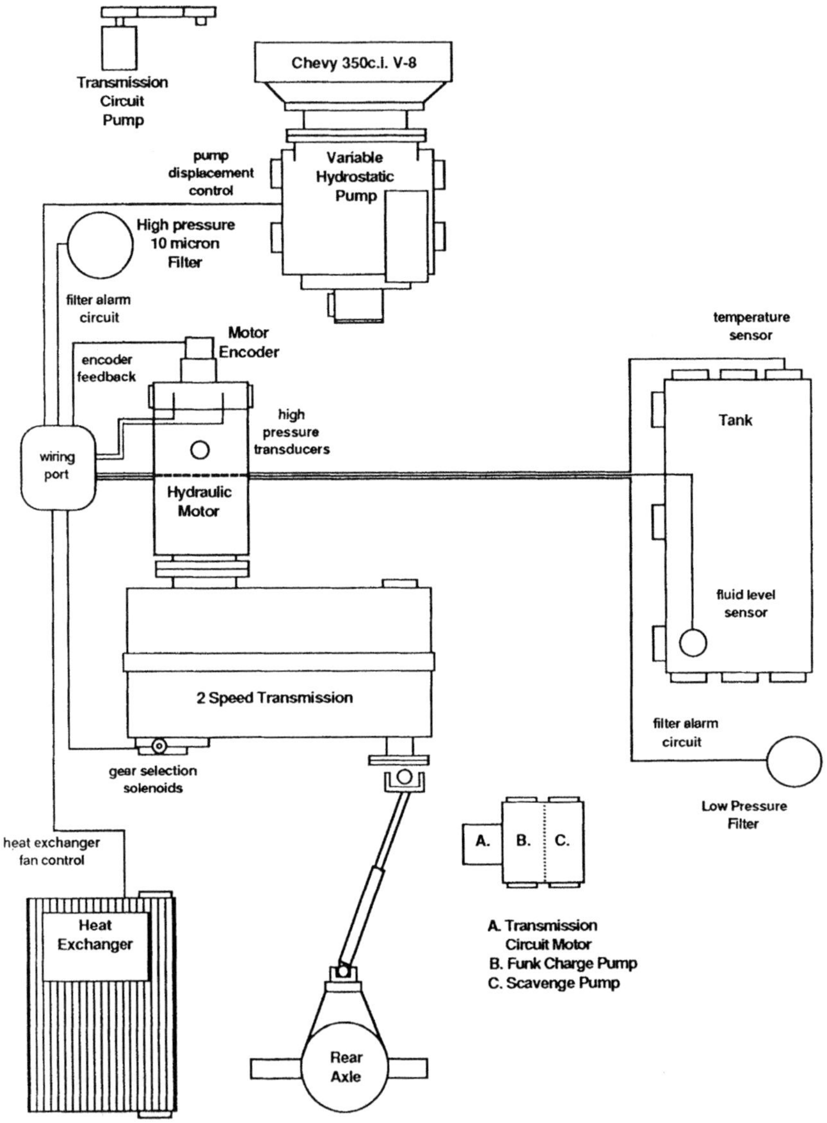

Figure 12.9: Hydrostat Sensor and Control Lines

Sensors include:

1. Dirty filter sensors: one dirty filter sensor is installed in each of the high- and low-pressure legs of the hydraulic system. These sensors trigger an alarm when they become clogged.
2. Pressure transducers: These read system pressure at input and output of the hydraulic motor.
3. Reservoir sensors: These measure fluid level and temperature of hydraulic fluid in the reservoir.
4. Motor encoder feedback: An optical encoder mounted on the shaft of the hydraulic motor provides feedback to the drive controller.
5. Steering limit switches: Limit switches are installed on the steering mechanism to signal an alarm if the wheels are cranked beyond acceptable limits.

12.5 Electrical System

All the electrical power needed by the Navlab is available onboard the vehicle. Electrical power can, however, be brought in from a shore power plug while the vehicle is in a fixed location. Power is distributed such that the generators are not needed to drive the Navlab manually.

12.5.1 AC Power

Figure 12.10 shows a schematic of the AC power system onboard the Navlab.

Generators

The generator supplies 120 VAC power to the variety of devices on the Navlab. The generator resides in a compartment accessible only from the outside of the vehicle, insuring the separation of noxious fumes from the interior.

Two compartments house an engine that is hydraulically coupled to a hydraulic generator unit in the forward compartment. This arrangement allows a single source of power up to 20 KW. Fuel to supply the engine comes from the vehicle fuel lines and the electrical power to start the generator is supplied by the vehicle 12 VDC system.

Shore Power

The Navlab can plug into power from a building when stationary, alleviating constant generator operation. An extension cord from a nearby power outlet (220 VAC 50A) mates to a 220 VAC single-phase plug mounted in the outside center compartment. This allows operation of all on-board computing when the Navlab in not in use on the road. Additionally, a link to the campus-wide

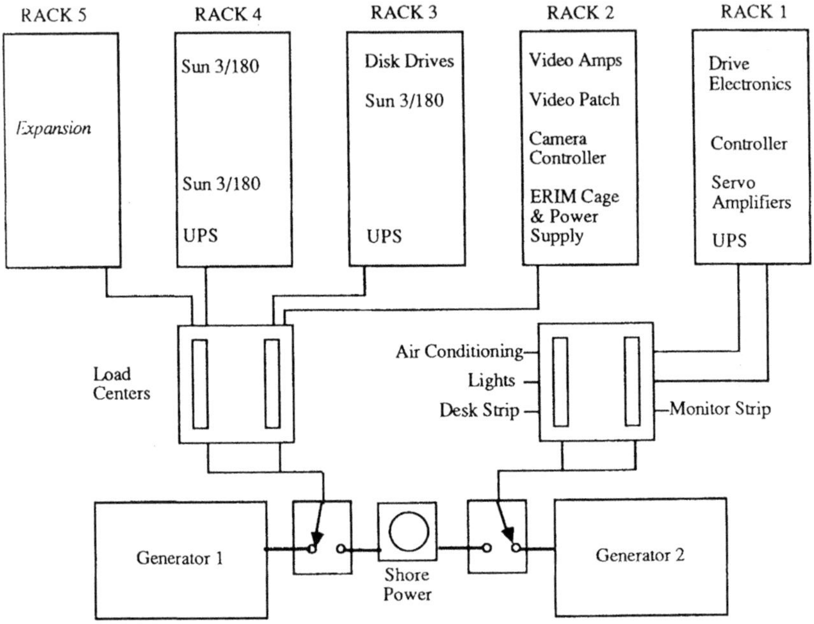

Figure 12.10: Wiring Schematic for AC Power

Ethernet network are connected at that time to allow access over the Internet to the Navlab computers for development purposes.

Power Conditioning

Because variations in load and temperature affect generator power output, the power from the generators must be be conditioned to protect sensitive machinery from spikes and brown-outs. This is done by passing power through Uninterruptable Power Supplies (UPS). These devices not only condition the power from the generators but also provide full-load backup for up to 15 minutes, even if the generators or shore power are shut down, allowing a graceful system shutdown if power fails. Three UPS devices provide a total of 6KW of conditioned power, which suffice for a complete configuration of computing equipment. The lights, air conditioner, video monitors, and servo-amplifiers do not receive conditioned power because they are much less prone to fluctuations in generator output.

12.6 Telemetry

Telemetry to the vehicle was thought to be useful where environmental sensitivity, location, or size of computing equipment precludes installation onboard the Navlab. This feature has not yet found use in practice. Navlab telemetry provides control and monitoring from a remote site, allowing stationary computers to be used in navigation experiments.

Current capabilities include video over radio or microwave,, high and low-speed data, as well as cellular phone technology.

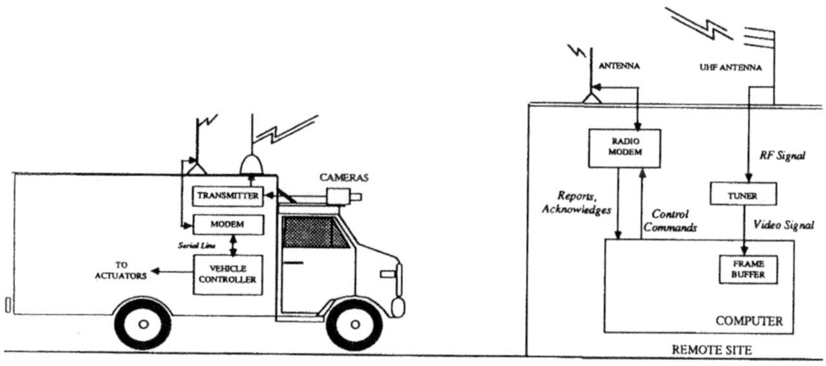

Figure 12.11: Telemetry Configuration of Navlab

The scenario in Figure 12.11 shows the closed loop set-up of a vehicle experiment where computing might be distributed offboard. The camera outputs a video signal that is broadcast over a UHF frequency and picked up by an antenna and receiver located on the Carnegie Mellon campus. The receiver provides the video signal to a frame buffer within the computer that processes the image. The signal is digitized and then analyzed. Commands to the vehicle are sent over a serial line to a wireless modem. A radio modem on the vehicle picks up this signal and feeds it to the controller.

An experimental radio license obtained from the FCC covers several broadcast frequencies. The license covers 2 UHF television channels, a full duplex radio link, and a 2 MHz microwave link.

The video signal is transmitted on the video transmitter while range data are transmitted over an aural sideband of one UHF channel. Because transmission rates can be as high as 56 K baud, the other aural sidebands not currently in use could serve several other data transmission needs.

Two sets of 1200 baud radio modems are used for simple, low-bandwidth digital communication. These devices provide a transparent RS-232 connection between computers and facilitate sending commands to and from an offboard machine and the Navlab controller.

Separation of vehicle from stationary base facilities involves not only communication over distance for the machines but also for humans. For this reason, a cellular phone using existing mobile communications networks within the city has been installed.

12.7 Perceptive Sensing and Computing

A variety of sensors can be mounted on the Navlab depending on the type of research being conducted. Vision, laser ranging, and sonar ranging are the most popular sensing modes.

12.7.1 Video

Cameras provide a standard RS-170 video image to the frame buffer. Vision processing transforms this image into a scene description to support navigation.

Typically, a single camera mounted in the front of the vehicle provides a wide-angle view of the scene. Some vision algorithms, however, call for a stereo pair of cameras. Broadcast-quality cameras that provide red-green-blue color signals are used. Remote control units allow control over camera functions like gain, color balance, and iris size. Presently, camera focus and zoom must be controlled manually.

12.7.2 Laser Ranging

Laser ranging is useful in areas where vision algorithms fail -- in detecting depth discontinuities in scenes where the edges are not obvious and in those scenes that have uneven lighting because of shadows. Whereas the camera is a passive instrument, the laser rangefinder is an active device that emits a beam in a raster fashion and captures the reflection to provide two types of information -- distance and reflectance. The data are then analyzed to provide a scene description. Laser ranging provides a direct 3-D description of the scene while vision requires more expensive computation to extract this information. Range readings are particularly useful because they are not affected by ambient light.

The current laser ranging device, manufactured by ERIM, provides a 256x64x8 bit depth map. The scanner output is processed by a Motorola M68000 processor and sent to a Sun computer dedicated to ranging.

12.7.3 Computing Configuration for Sensing

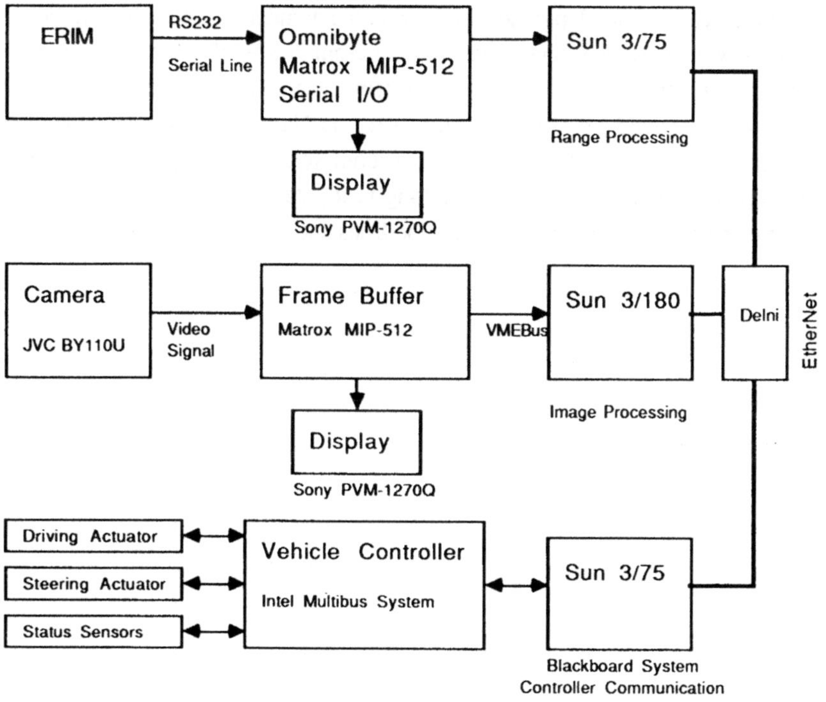

Figure 12.12: Typical Architecture

Figure 12.12 shows configuration of computing for simple, perceptive sensing. Much more complex setups are common. Each sensor commonly requires its own workstation or specialized processor. Another computer runs the blackboard system that integrates perception, modeling, and planning.

12.8 Conclusion

The Navlab has been and continues to be a significant development in mobile robots. The Navlab provides a stable infrastructure for developing new methods and techniques in sensor processing, robot architectures, data fusion, map building and navigation, high speed computing and code development. What made this possible is the dependability of the mechanism and infrastructure, the elimination of fragile and costly communication links, and the incorporation of researchers within the robot environment. The Navlab is, literally, a mobile robot laboratory. Several years of operation and the flow of research results from projects have demonstrated the long term capabilities of the Navlab.

What would we do differently? Many upgrades over the years have increased capabilities of Navlab. Upgrades range from the mundane to the sublime. Power systems were increased three times to the present 20KW of power. Cooling has been added incrementally to the rack systems, access panels, a new generation of controller, sound proofing etc. However, the Navlab has not fundamentally changed from our initial conception of a robot vehicle to support our research. This reflects our concern to provide long term, stable fertile environments for robotics research in our people, in our facilities and in our machines.

Chapter 13: Vehicle and Path Models for Autonomous Navigation

Dong Hun Shin and Sanjiv Singh

13.1 Introduction

Most implementations on the Navlab have used a simple Virtual Vehicle specification to control the vehicle. The Virtual Vehicle is a low level instruction set that accepts conceptual commands and provides a clean separation between a navigation host and vehicle control, masking implementation details of the physical vehicle. While such a scheme is modular and allows various groups of researchers to easily implement their algorithms on the vehicle, the price is paid in terms of limited performance. This cost has been acceptable for most applications so far since they require speeds on the order of 1 mile per hour.

We have analyzed the problem of navigating at speeds an order of magnitude higher than currently pursued by other researchers on the Navlab and propose a paradigm of path tracking which is based on the use of explicit position information. Such a strategy is particularly suited to navigation in featureless terrains where the autonomous vehicle cannot rely on distinguishable road edges or guard rails, as is required by current road following systems.

In this paper, we describe vehicle and path models relevant to navigation of a conventionally steered vehicle. As is the case for most robotics systems, it is essential to develop kinematic models that provide geometric relationships between the various state variables. We show how adopting particular kinematic conventions simplifies control. The importance of modeling vehicle dynamics is discussed, and a distinction is made between two separate phenomena—actuator dynamics and vehicle-ground interaction. While composite models of dynamics involving many different considerations (hysteresis, limited steering velocity and acceleration, tire deformation, etc) are necessary for simulating the vehicle, these models are too cumbersome for use in the actual control of the vehicle. Instead, simpler, dominant first order models are used in the compensation of vehicle dynamics.

Control of the vehicle along a desired path is framed in terms of a classical tracking problem in which performance can be improved if a priori knowledge of future reference states exists. A partitioned control scheme is presented that accomplishes feedback to minimize position error from the path, as well as a feedforward scheme that uses a first order model of the actuator dynamics and future path information to compensate in an anticipatory fashion. Results are shown that illustrate the effect of the feedback and feedforward schemes separately and in combination.

13.2 Vehicle Representation

We identify three areas in which it is necessary to model an autonomous vehicle: *kinematics*, *dynamics*, and *systemic effects.*

13.2.1 Vehicle Kinematics

In this section we present a simplified kinematic model of the Navlab. Kinematic relationships for control of speed and steering are discussed along with practical issues that arise in application of the model to the real vehicle.

A simplified view of the steering and drive systems is shown in Figure 13.1. The vehicle is rear wheel driven through a hydrostatic transmission (Hydro), and, front wheel steered through a power steering mechanism (PS). Control for such a vehicle is more complex than the simple case of the front wheel steered and front wheel driven tricycle configurations that are common to many factory robots for the following reasons:

- All four wheels travel on different arcs and therefore at different speeds. The reference velocity of the vehicle is specified in terms of a particular point, usually along the longitudinal axis of the vehicle. Since this *guide point* is not located at one of the wheels, it is not simple to measure and control its true velocity.
- The steering angle is not easily measured, either, because the front wheels turn unequally to make up for the fact that they travel on arcs of varying radii.

A bicycle model is chosen to represent Navlab as shown in Figure 13.2. Both wheels are hypothetically placed along the longitudinal axis of the vehicle. The advantage of such a model is that the steering angle ϕ, corresponding to a path with an instantaneous radius of curvature r, can be computed independent of speed if the *guide point* of the vehicle is chosen to be the center of the rear axle (the center of the hypothetical rear wheel).

There are several advantages for choosing the center of the rear axle as the

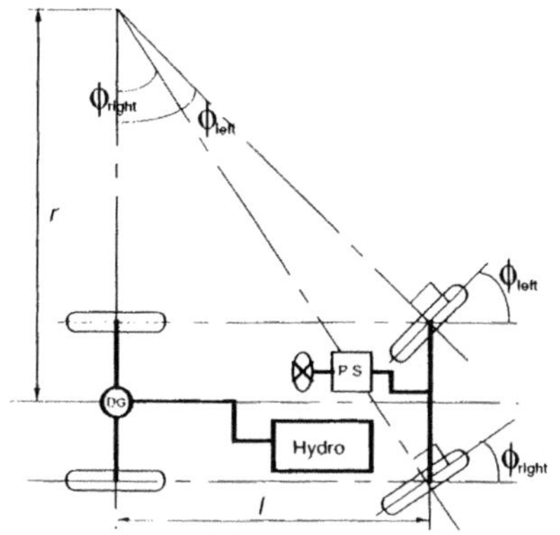

Figure 13.1: Physical configuration of Navlab

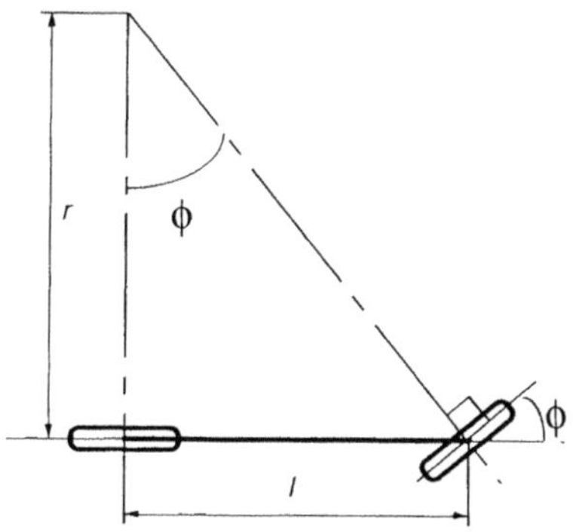

Figure 13.2: Bicycle model of the vehicle

guide point:

- Steering and velocity control of the vehicle are decoupled; if the guide point is chosen elsewhere, the expressions for ϕ and ω are more complex than those given in (13.1), (13.2).
- For a given steering angle, points on the rear axle follows a smaller turning radius than do points in front of or behind the axle.
- The heading of the vehicle is aligned with the tangent direction of the path. This provides a more reasonable vantage point for a camera or laser range finder mounted on the vehicle.

Typically, a vehicle is commanded in terms of the instantaneous radius of curvature r, and, a velocity v. Given these inputs, the two degrees of freedom in the control of the vehicle, steering angle, and the angular velocity of the drive shaft, are obtained as follows. For a vehicle with wheelbase l, the steering angle ϕ is

$$\phi = \text{atan}\left(\frac{l}{r}\right) \tag{13.1}$$

Desired vehicle velocity must be transformed into angular velocity of the drive shaft because there is no way of directly regulating the speed of the guide point. The angular velocity of the drive shaft, ω_d, given the radius of the rear wheel r_w, the rear differential gear ratio, k_d, and a desired vehicle speed v, is:

$$\omega_h = k_d \frac{v}{r_w} \tag{13.2}$$

Given the instantaneous curvature of a reference path, it is trivial to determine the hypothetical steering angle as described above, but somewhat more complex to determine the amount that the steering wheel must be turned to produce the hypothetical steering angle. Figure 13.3 shows experimental data that relates the amount that the motor on the steering column is moved to the radius of curvature of the resultant path. It should be noted that even though all steering column movements were made from the zero position, marked non-linearities were found in the response of the vehicle, i.e. for the same amount of steering motion, varied response was noted in the resultant radius of curvature.

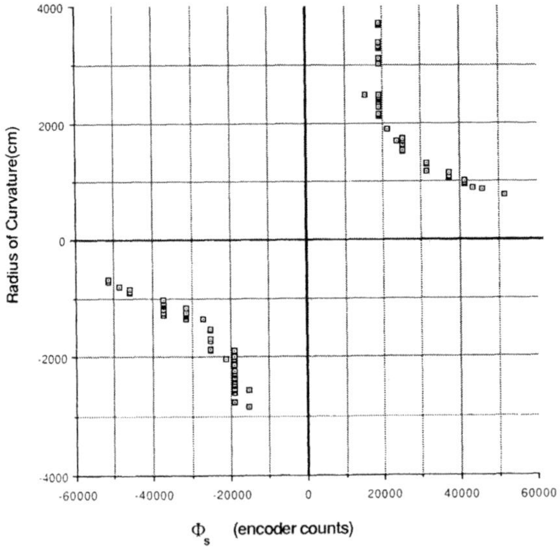

Figure 13.3: Radius of curvature obtained for various steering positions

The shape of the curve in Figure 13.3 corresponds to the expected *atan* relationship, and, assuming zero steering angle (straight ahead) as a nominal operating point, a linear gain k_s is used to relate the steering motor angle ϕ_s, with the hypothetical steering angle, ϕ_h:

$$\phi_s = k_s \cdot \phi_h \tag{13.3}$$

k_s is obtained by taking the ratio of the covariance and variance of (ϕ_h, ϕ_s) obtained from experimental data:

$$k_s = \frac{\Sigma\phi_h \cdot \phi_s}{\Sigma\phi_h} \tag{13.4}$$

One problem with the use of a bicycle model to approximate the kinematics of a four wheeled vehicle like Navlab is that it is not possible to measure the actual steering angle. ϕ_h must be estimated either by a reverse transformation of (13.3), or in the following manner. c, the reciprocal of the radius of curvature, r, is given by:

$$c = \frac{d\theta}{ds} \tag{13.5}$$

Since $\dot{\theta}$ is the rate of change of heading with respect to time and v is the forward velocity of the vehicle:

$$\dot{\theta} = \frac{d\theta}{dt}, \quad v = \frac{ds}{dt} \tag{13.6}$$

c can be written as:

$$c = \frac{\dot{\theta}}{v} \tag{13.7}$$

ϕ_h can then be estimated:

$$\phi_h = \operatorname{atan}\left(\frac{l \cdot \dot{\theta}}{v}\right) \tag{13.8}$$

Typically (13.8) is used for feedback purposes, i.e to compute an error vector, while the reverse transformation of (13.3) is used to make a forward calculation.

It is worth noting that the steering on the Navlab shows significant non-linearities even in tests where the transient response is not an issue. We chose not to develop higher order models that would incorporate considerations such as initial steering position, speed, and surface friction, into the lowest level of control (the Virtual Vehicle convention described in [3]). Instead, the simple linear model described above is used; deficiencies in repeatability and accuracy are left to be compensated by higher levels of control.

13.2.2 Vehicle Dynamics

It is common in the control of mobile robots to establish the necessary kinematic relationships, but to ignore an explicit representation of the dynamics or, alternatively to compensate for them using "fudge factors". We have found it important to model the dynamics of the vehicle for two reasons:

- Vehicle actuators have less than perfect response. The better they can be modeled, the better the chance of compensating for these effects.
- A good model of the vehicle is required for simulation so that algorithms can be tested without having to experiment directly on the vehicle.

The dynamics of the vehicle can be separated into two phenomena: *actuator dynamics*, and *vehicle-ground interaction (VGI)*. The former is due to finite response time of the actuators while the latter is due to the flexing and slipping in the contact between the vehicle and the ground.

Actuator Dynamics

Figure 13.4 shows a block diagram of the vehicle steering and drive systems. The steering system consists of a DC servo motor mounted on the steering column, whose motion is transferred to wheel articulation through a power steering system and then a four bar linkage common to conventional vehicles. The drive system consists of a V8 engine that supplies power to a hydraulic pump and motor, a two speed transmission, drive shaft, and differential gear mechanism.

Since this system was designed more for simplicity of control than performance, it is not surprising that the finite response time of the actuators is significant—to the order of 0.5 s for steering angle and 2 s for drive velocity. Our experience is that these responses can be modeled sufficiently as first order systems with a time constant. This is especially the case for the steering system, which is an inherently overdamped mechanism. A simple prediction of ϕ_h, given a time constant α of the steering system, can be made as follows:

$$\phi_h(t) = \frac{\phi_s}{k_s} \cdot [1 - e^{-t/\alpha}] \tag{13.9}$$

This model turns out to be quite accurate for small steering angles. For larger signals, hard limits must be imposed on steering velocity and accelerations

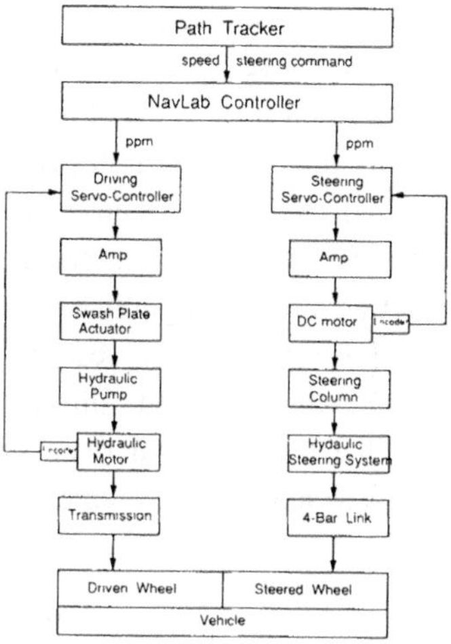

Figure 13.4: Block diagram of vehicle and steering and drive train

Vehicle-Ground Interaction

The relationships that try to capture vehicle ground interaction describe how the vehicle moves for given steered wheel angles, wheel angular speeds, vehicle mass, friction between the road and the wheels, and tire stiffness. In general, VGI is a very complex phenomenon that cannot be modeled easily with analytical expressions. Graettinger and Krogh [5] proposed a method for avoiding the explicit representation of wheel slip for trajectory evaluation. They postulated that ellipsoidal boundaries on the wheel friction vectors can characterize operating conditions for which wheel-slip is negligible. That is, they assumed there is a bounded region of operation within which the slip is negligible. It must then be guaranteed that the system is operated within the assumed constraints by "time scaling" (slowing down by the amount necessary to stay within the no-slip bounds). In this manner, for every trajectory, an upper limit of speed can be obtained for which the vehicle would not slip, reducing the problem to a kinematic one. The assumption that Graettinger and Krogh made is that the tires act as rigid bodies. However, previous researchers [4] have shown that the principal VGI effects are caused due to the distortion of the tires, making the rigid body assumptions intractable. It is possible to avoid tire deformation altogether by "time-scaling" the trajectories but this would only be feasible at very low speeds and curvatures.

Tire distortion is characterized by *slip angle*, and *under/oversteer*. Slip angle is the angular difference between the centerline of the tire carcass and the velocity direction of the wheel. The slip angle does not result from actual slippage of the tread, but instead from distortion of the tire sidewalls, allowing the wheel rim (and vehicle) to move in a different direction from the tread contact patch. The difference in motions is the slip angle. The under/oversteer phenomenon characterizes the deviation of a straight running vehicle from a straight course when subjected to a steady side load applied at the center of gravity [9]. If an applied side force produces equal front and rear side-slip angles, the vehicle moves off on a new straight line path. A front side slip angle greater than rear slip angle will result in a curved path away from the applied force, a rear slip angle greater than the front slip angle will result in a path toward the applied force. The terms neutral steer, understeer, and oversteer apply to these three behaviors, as indicated in Figure 13.5.

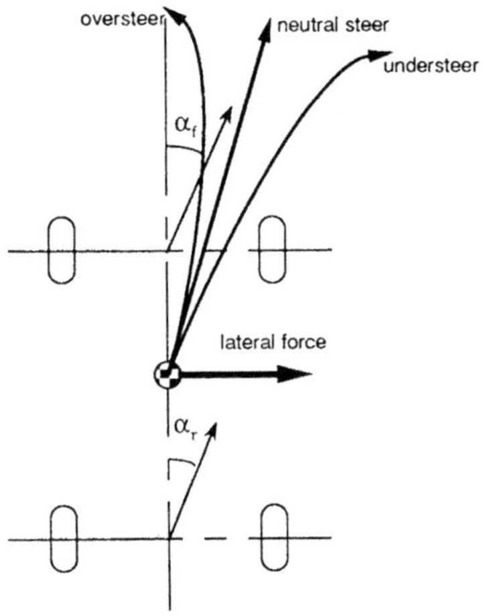

Figure 13.5: The under/oversteer effect

Slip angle relationships are obtained from tire specifications as amount of slip (in terms of steering angle) for various amounts of lateral force on the tires. Since the VGI phenomena are directly attributed to lateral force on the tires, the effect is almost negligible at low speeds. We have chosen to model the VGI effects in our simulations, only for speeds higher than $3m/s$.

13.2.3 Systemic Effects

Often, it is easy to ignore systemic effects like communication delays in the modeling of a vehicle, especially because they are not visible at low tracking speeds. There are several delays and timing considerations in the Navlab control scheme that are only due to the I/O and computing conventions used.

Our approach has been to ignore them at low speeds. At high speeds we have had to violate the black box convention of the *Virtual Vehicle*, by providing for explicit control of the steering mechanism, thus reducing delays. In either case, these delays can be modeled as "dead bands" for simulation purposes.

Timing of control commands is another issue of concern at high speeds. Using an operating system like Unix, the procedure is to poll a clock to ensure that commands are sent at regular intervals. However, such operating systems do not guarantee real-time response and the interval between commands can vary up to 0.5 s on the computers resident on the Navlab.

13.3 Path Representation

In this section we present criteria for paths that are suited to path tracking. We compare performance of a vehicle tracking paths that were designed without concern for ease of tracking, with a path that is intrinsically easier to track.

As with any control system, the response of an autonomous vehicle in tracking a reference path depends partly on the nature of the path. Early work in the development of path representation sought to develop schemes that would produce paths that were intrinsically easier to track. In particular, it was postulated that "good" paths

- are continuous in position, heading and curvature
- have curvature that varies linearly with the distance along the path

In the simplest form of path specification, where a path is represented as a sequence of straight lines and arcs, discontinuities of curvature exist at the point where the two arcs of differing radii meet as shown in Figure 13.5. For a robot whose guide point is at the center of its rear axle, such discontinuities are troublesome since an infinite acceleration of the steering wheel is required to track the reference path. A real robot (with finite acceleration) traveling through such transition points will experience an offset error along the desired path. In addition, the extent to which steering motions are likely to keep a vehicle on a desired path can be correlated to the linearity of curvature of the path, since linear curvature along a path requires linear steering velocity (assuming that less control effort is required for an actuator to provide a linear velocity profile than an arbitrary nonlinear one).

Certain spline curves [8] are candidates for path representation because they guarantee continuity in position, heading, and curvature. However, these spline

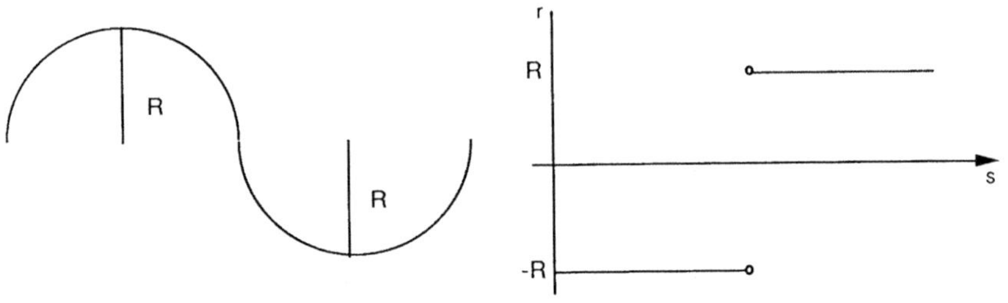

Figure 13.6: Discontinuity in curvature at the junction of arcs of differing radii

curves do not guarantee linear gradients of curvature. Clothoid curves, by contrast, do have the property that their curvature varies linearly with the distance along the curve.

Kanayama proposed pairs of clothoids segments as a method to interpolate between via points [6]. The problem with clothoids is that though they guarantee smoothness and linearity as described above, it is non-trivial to generate clothoid segments for arbitrary starting and ending vehicle configurations; the expressions for these curves are underconstrained, and further, no closed form expression is available—the expressions must be evaluated using numerical methods. Kanayama's scheme is tenable under the simplification made by the requirement that the starting and ending curvatures at the via points are always zero. This model is adequate for simple paths for which symmetric pairs of clothoid curves can be found, or for paths with mild curvatures that can be easily broken down into separate clothoid segments, as were used by Dickmanns [2].

Singh and Shin [11] compared the response of a robot vehicle to different reference paths. Simulation results show that in the open loop case (position error is not compensated for), a path that has discontinuities in curvature results in larger steady state tracking errors. This is particularly the case if the actuators are slow and the communication delays are significant. Open loop response of a vehicle following a path composed of arcs and straight lines, and another composed of clothoid segments, is shown in Figure 13.7 and Figure 13.8. In

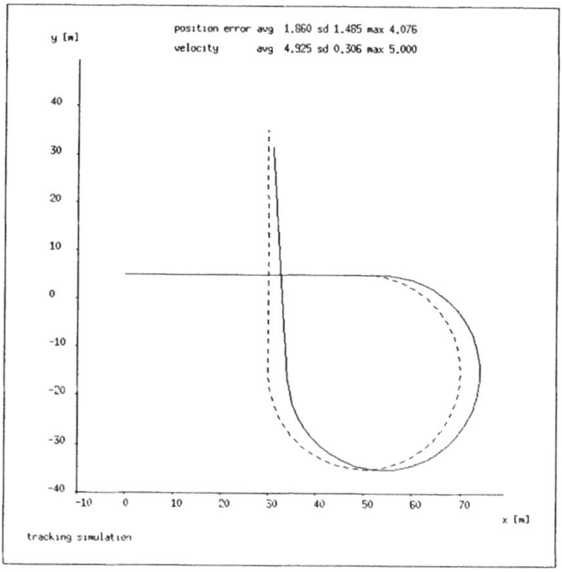

Figure 13.7: Open loop performance on a path composed of arcs and straight lines

each case, the reference path is indicated by dashed lines while the actual path is indicated by solid lines.

While the results shown above support the claim that "good" paths are intrinsically easier to track, it turns out that using an optimized closed loop scheme and fast actuators, tracking performance is only marginally better for smooth, linear-curvature paths than for paths composed of arcs and straight lines, especially at low speeds. Intuitively, at low speeds, there is enough "slop" in the vehicle actuators that the compensating controller makes up for errors due to curvature discontinuities and non-linearities along with errors due to imperfect actuators. For high speed path tracking of paths that have significant curvature, it makes sense however, to ensure at least curvature continuity.

13.4 Path Tracking

In this section, we analyze the path tracking problem and motivate a general control structure for path tracking by mobile robots. Work by other researchers is compared and a scheme is presented to autonomously track paths.

The Navlab has non-holonomic steering constraints. A non-holonomic system is one where a geometric constraint cannot be expressed analytically as

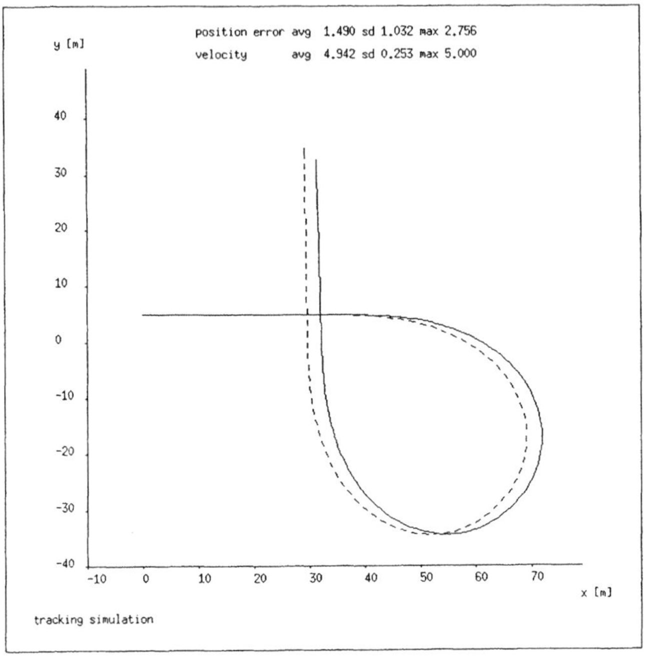

Figure 13.8: Open loop performance on a path composed of clothoid segments

an equation relating position and time. The most important type of non-holonomic constraint is one which can only be expressed as an equation relating infinitesimal variations to coordinates. In systems that have such constraints, the number of degrees of freedom is less than the number of degrees of independent generalized variables. For our system we have three variables (x,y,θ), but only two degrees of freedom: drive and steering. That is, it is not possible to move sideways.

Given the non-holonomic nature of robots that share the physical configuration of Navlab, it is not sufficient to servo the steering to a premade list of commands. Imagine that a step input is given to the steering system. After the transient response has subsided, the steering mechanism will have reached the desired steering angle. However, in failing to track the step input exactly, offset errors have been introduced in the resultant path. The resultant path will turn out to have the required curvature but will be parallel to the reference path.

It is not surprising, then, that path tracking is better thought of as a classical "tracking" than a "regulation" problem. The controller can be transformed into a regulator if the error is taken to be in terms of a particular desired state (in our case, vehicle position) rather than in terms of the reference input (in our case, steering angle and the resultant path curvature). This can be accomplished by using two feedback loops. The inner loop provides steering angle position feedback control, while the outer loop feeds back current position to the steering planner to generate the reference steering angle for the inner loop. The control structure is described in Figure 13.9.

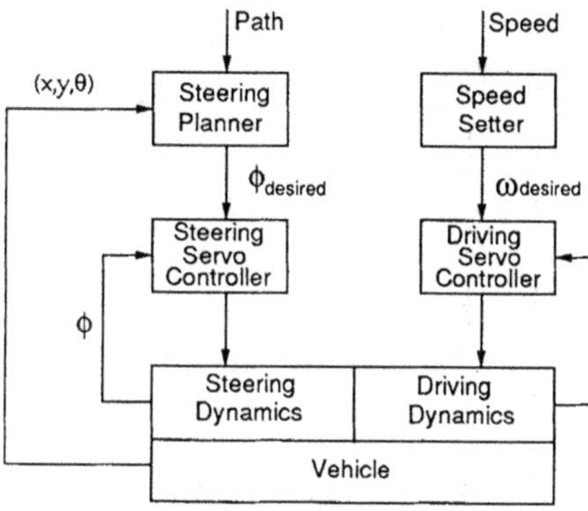

Figure 13.9: Nested feedback control loops

One characteristic of a tracking control problem is that a priori knowledge of the future inputs can be used to improve performance. Intuitively, it is possible that some of the system errors are fortuitous in the light of future commands, and thus can be exploited. Another way of looking at this is that feedback control only compensates for errors after the fact, whereas knowledge of future commands can be used to anticipate appropriate inputs to the controller. This sort of anticipatory control (also called feedforward control) is a well known part of human performance in tasks like driving [10]. Figure 13.10 contrasts a conventional feedback scheme with a scheme that uses a "preview" of future inputs. In Figure 13.10 (b), at time t, the controller uses knowledge of the desired states in the interval $[t \rightarrow t + l_a]$ where l_a is the "preview" distance, to plan the next control input.

A viable controller for high performance tracking should have both feedback and feedforward capabilities as described above. The controller can be partitioned into two parts:

- *feedback* compensator that provides compensation for errors between the vehicle's actual path and its desired path.
- *feedforward* compensator that uses current inputs but also incorporates the future course of the path.

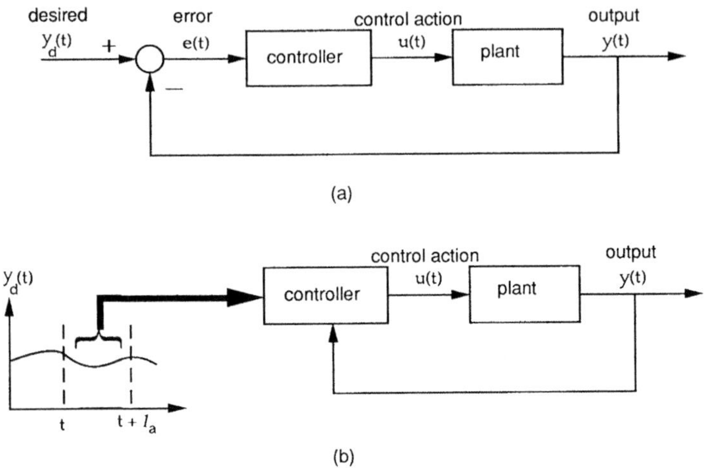

Figure 13.10: (a) conventional feedback scheme,
(b) "preview" control scheme

A control structure that has the above components is shown in Figure 13.11.

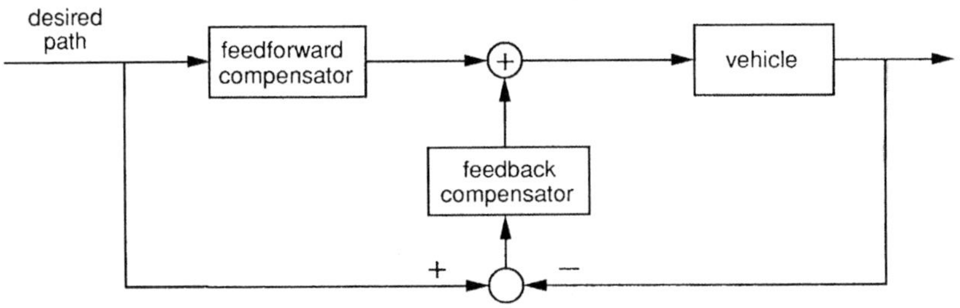

Figure 13.11: Partitioned control scheme using both feedforward and feedback schemes

In general, the problem of generating the control inputs U_i might be thought of as:

$$U_i = R_i + K \cdot e_i \tag{13.10}$$

where R is the reference input, e is the error between the reference input and the actual state of the system, and K is a gain. It can be seen that $K \cdot e_i$ term corresponds to the feedback control whereas the use of the R term corresponds to feedforward control.

As early as 1967, Bernstein [1] had identified the necessary components of a scheme that would autonomously guide a vehicle along a path. He identified three *receptors* (sensing elements), one for each of:

- lateral error from the path
- heading error from the path at some fixed distance ahead
- the effective curvature error of the path.

Wallace *et al* [12] proposed a scheme for a vision-guided, road following vehicle that attempted to keep the road image centered in its field of view. They expressed the control input (in this case, $\dot{\theta}$ the rate of heading change with respect to time) in terms of the distance between a ray cast straight ahead through the center of the field of view and the road centerline at a fixed distance in front of the vehicle. Lateral position and heading errors as identified by Bernstein were used to produce a path that, in the ideal case, converges exponentially with the road ahead. This approach was first used on a skid steered vehicle and then updated for road following experiments on the Navlab. A typical sequence of actions might look like the following:

- determine lateral and heading errors from road image,
- compute the desired $\dot{\theta}$, corresponding to these errors, given the constant reference velocity v,
- compute the instantaneous road curvature corresponding to $\dot{\theta}$ (= $\dot{\theta}/v$),
- compute the steering angle corresponding to this curvature and issue it to the controller.

The above is repeated as soon as the next road image has been analyzed and the current arc being executed is interrupted with a steering command. This scheme made simplifications that were justifiable at low speeds:

- the path ahead of the vehicle is approximated by a straight line segment
- curvature error is not incorporated into the planning of the next steering angle

This scheme is purely geometric in that it does not compensate for any vehicle dynamics, and corresponds to the $K \cdot e$ term in (13.10). The only feedback to the navigation system is at a high level and occurs at the frequency that the images are processed.

Nelson [8] specified control inputs as shown in (13.11). ϕ_{ref} is the precomputed steering angle for the position on the path corresponding to the current vehicle position. e_n, e_t are errors in the lateral and tangential directions respectively. Gains (K_1, K_2) are decided by trial and error.

$$\phi = \phi_{ref} + K_1 e_n + K_2 e_h \tag{13.11}$$

In some sense, the ϕ_{ref} corresponds to the R part of (13.10), though since it doesn't include future path information, it is not strictly doing any feedforward control. The rest of the expression is equivalent to the feedback part of (13.10). Since K_1, K_2 are chosen purely by trial and error until satisfactory results are produced, they effectively compensate for significant dynamical effects if any exist. In a similar way, Kanayama [7] applied a PID control law to the feedback compensation for the path tracking of his differential wheeled indoor mobile robot. In both cases, the robots traveled indoors at the rate of a few inches per second; it is conceivable that for such scenarios, controllers can be tuned to ensure good tracking behavior.

Dickmanns [2] developed a visual feedback control scheme to guide a vehicle on well structured roads. Road edge curvature, position offset from the road centerline and heading error relative to the centerline, are estimated by analysis of a visual image from cameras mounted on the front of the vehicle. Anticipatory R_i (as in (13.10)) are computed by estimating the road edge curvature a large distance ahead of the vehicle. Feedback error gains are chosen to stabilize a state space model of vehicle dynamics relative to the road edge at a fixed look ahead distance close to the vehicle. Nonlinear vehicle dynamics are approximated by a time varying linear model whose coefficients are stated as a function of vehicle speed. Dickmanns was able to sufficiently capture the dynamics of the vehicle using system identification methods to enable road following at speeds up to 60 miles/hr. His state space scheme includes a model of the slip angle produced due to lateral forces on the tires as well as an integrator model of the steering system.

We propose a path tracking scheme of the form shown in Figure 13.11, that doesn't require explicit modeling of VGI in the control scheme. Since the steering and velocity control are decoupled, path tracking can be reduced to steering planning. The path tracking problem, then, can be formulated as one of obtaining a list of inputs ϕ_i to be supplied as inputs to the steering mechanism, given a list of reference positions P_i that the guide point of the vehicle must attain, such that the position error between the actual path of the vehicle (P_{ai}) and (P_i) is minimized.

13.4.1 Feedback Control

At any given moment, the vehicle will have errors in all three of lateral position, heading and curvature. The proposed feedback control scheme minimizes these errors. Consider a path as a continuous function x(s), y(s) and a vehicle that is currently at P_a. An error vector can be calculated that represents error in the distance transverse to the path (ε_0) relative to P_0 as in Figure 13.12, in heading (β_0), and in curvature (γ_0).

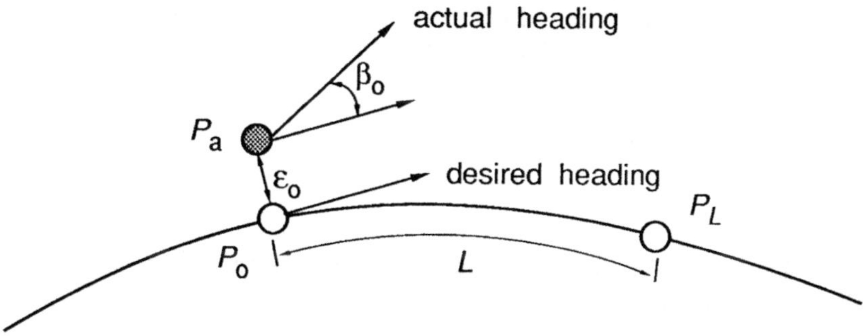

Figure 13.12: Computing an Error Vector

If the robot is to be brought back onto the specified path within distance L (measured along the reference path), six boundary conditions can be stated corresponding to the initial errors and to zero errors at P_L.

$$\varepsilon(P_0) = \varepsilon_0 \; ; \; \varepsilon(P_L) = 0 \tag{13.12}$$

$$\beta(P_0) = \beta_0 \; ; \; \beta(P_L) = 0$$

$$\gamma(P_0) = \gamma_0 \; ; \; \gamma(P_L) = 0$$

These boundary conditions can be sufficiently satisfied by a polynomial of fifth degree. A quintic polynomial can be constructed to describe the replanned path (in error space) as follows:

$$\varepsilon(s) = a_0 + a_1 s + a_2 s^2 + a_3 s^3 + a_4 s^4 + a_5 s^5 \tag{13.13}$$

where $s \rightarrow [0, L]$

Solving for $a_0, a_1, a_2, a_3, a_4, a_5$:

$$
\begin{aligned}
a_0 &= \varepsilon_P \\
a_1 &= \beta_P \\
a_2 &= 2\,\gamma_P \\
a_3 &= \frac{-(13\,\gamma_P\,L^2 \;+\; 12\,\beta_P\,L \;+\; 20\,\varepsilon_P)}{2\,L^3} \\
a_4 &= \frac{9\,\gamma_P\,L^2 \;+\; 8\,\beta_P\,L \;+\; 15\,\varepsilon_P)}{L^4} \\
a_5 &= \frac{-(7\,\gamma_P\,L^2 \;+\; 6\,\beta_P\,L \;+\; 12\,\varepsilon_P)}{2\,L^5}
\end{aligned}
\tag{13.14}
$$

The expression for $\varepsilon(s)$ gives the error along the replanned path from P_0 to P_L. Variation in the steering angle from the preplanned path (or in error space) can be written as:

$$
\frac{d^2\varepsilon}{ds^2}\Big|_s = 2a_2 + 6a_3\,s + 12a_4\,s^2 + 20a_5\,s^3 \tag{13.15}
$$

Steering angle variation obtained from the above method can be computed as follows:

$$
\phi_{feedback}(s) = \mathrm{atan}\left(\frac{d^2\varepsilon(s)}{ds^2}\Big|_s\right) \tag{13.16}
$$

Computing the steering angle $\phi_{feedback}$ over the entire interval $[0 \rightarrow L]$ is not necessary. Instead, during every control loop cycle, the coefficients $a_1..a_5$ are computed based on the most recent estimate of position heading and curvature errors, and $\phi_{feedback}$ is computed only for a small path distance immediately ahead of the vehicle. The sensitivity of the scheme described above can be modulated by the adjustment of the parameter L which represents the distance at which all three of position, heading, and curvature errors are forced to be zero. Figure 13.13 shows the difference in variation produced by the quintic polynomial for three different values of L. The solid lines denote the variation in position error while the dashed lines indicate the variation in the curvature.

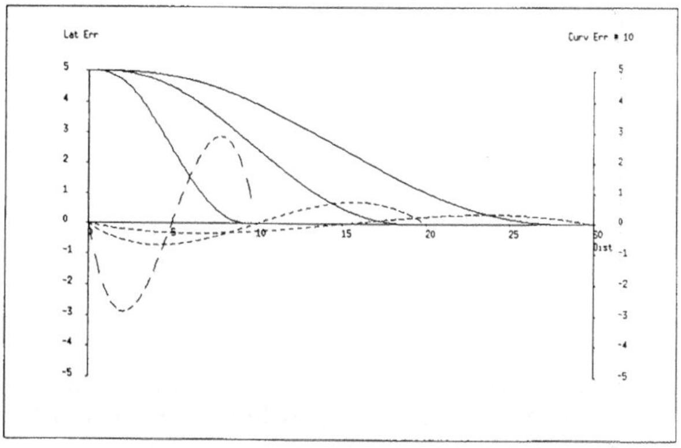

Figure 13.13: Modulation of the parameter L

L is chosen based on vehicle speed, much as a human driver would choose to look out further along the path, the higher the vehicle speed. A larger value of L corresponds to a milder contribution from the feedback scheme than for smaller values. Once again, a simple linear model is use to calculate L, the gain for which is determined experimentally.

13.4.2 Feedforward Control

In this section we describe the feedforward part of the proposed control scheme. Results with and without such compensation are compared.

As mentioned before, feedforward control can be used to improve performance if there exists a priori knowledge of the reference inputs along with an approximate dynamical model of the system. For purposes of a dynamical model to be used in the feedforward path, a first order model as in (13.9) is used such that both actuator dynamics and vehicle-ground interaction are described by a single time-constant.

The effect of the feedforward compensation can be seen in Figures 13.14 - 13.17. Figure 13.14 shows the response (solid lines) of the steering system to given reference commands (dashed line). The reference commands correspond to a path composed of a straight line, a circular arc, and a straight line. The reference and resultant paths in this case are shown in Figure 13.15.

Feedforward compensation is accomplished by sending reference commands in advance by the time constant of the first order model. In this manner, the steering starts moving before the arc is reached. If the vehicle dynamics are

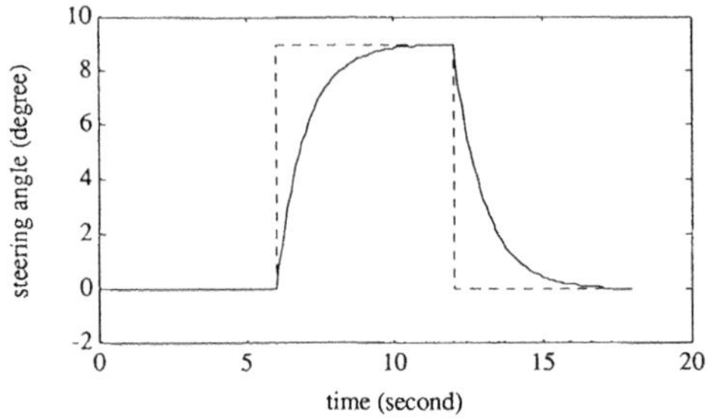

Figure 13.14: Performance of first order model of the the steering system, in response to path consisting straight lines and an arc

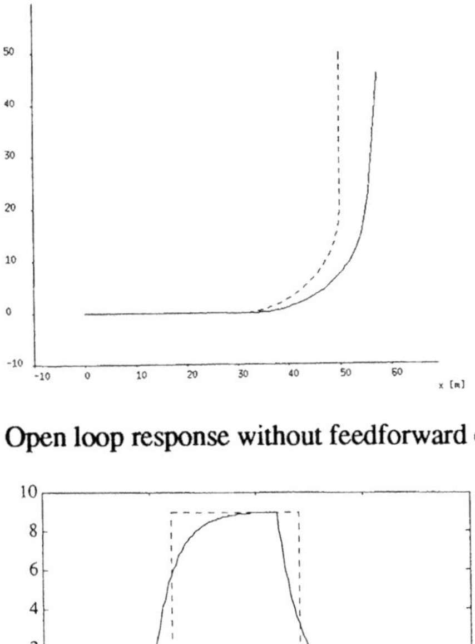

Figure 13.15: Open loop response without feedforward compensation

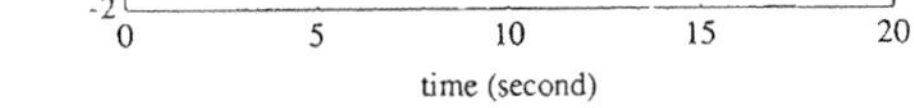

Figure 13.16: Compensated steering performance modeled as a first order lag

modeled exactly by a first order lag, then the steering would have reached 63% of the desired value by the time the arc is reached. An intuitive argument can be made that such a scheme produces an area under the actual curve, (solid line in Figure 13.16) that approximates the area under the reference curve (dashed line in Figure 13.16), and at the required time. Figure 13.17 shows the improvement produced by anticipating the response of the vehicle dynamics. Real vehicle

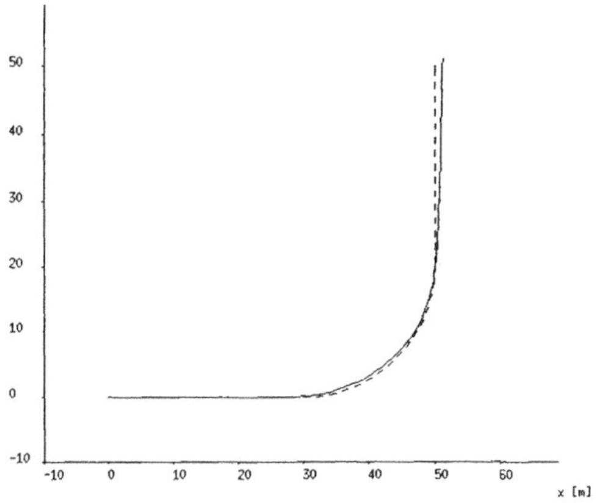

Figure 13.17: Open loop response with feedforward compensation

responses are not as nicely predictable as shown but the above scheme works well especially since reference steering commands have much smaller step changes (due to discrete time control) than those shown in the example.

13.4.3 Speed Control

One of the benefits of choosing the guide point to be the center of the rear axle is that speed and steering can be controlled separately. This is convenient because typically the speed for a robot maybe based on several constraints:

- maximum speed limit in a particular area
- distance to possible obstacles
- path curvature, i.e for constant lateral acceleration, curvature of the section of path immediately in front of the vehicle can be used to constrain the vehicle speed.

Speed can be chosen to be the minimum of the above constraints and this speed is sent as input to the drive control. Another advantage of the decoupled steering and velocity control is that the dynamics of the drive train do not have to be compensated for. This is significant because the drive train with its many components is highly nonlinear in responding to speed commands.

13.5 Results

Simulation results are considered for four different cases:

- The vehicle is made to follow a path in open-loop fashion. In this case only the steering angle is regulated, while position errors are not compensated for (Figure 13.18).

- Position errors are compensated for using the feedback control scheme discussed in Section 13.4.1 (Figure 13.19).

- The only compensation made is a feedforward one, using the control scheme discussed in Section 13.4.2 (Figure 13.20).
- Both feedback and feedforward schemes are used (Figure 13.21).

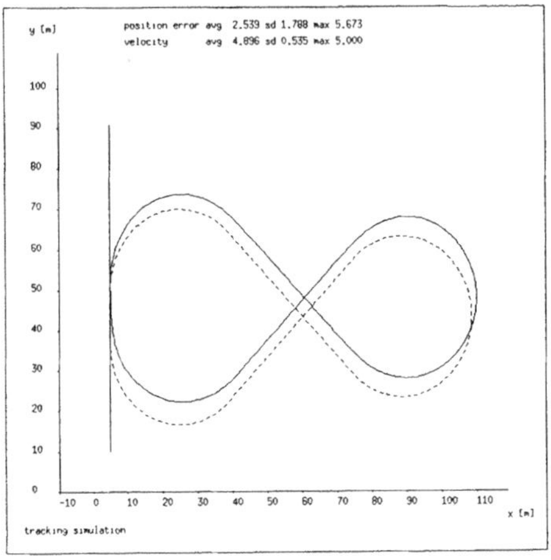

Figure 13.18: Tracking without compensation of position errors

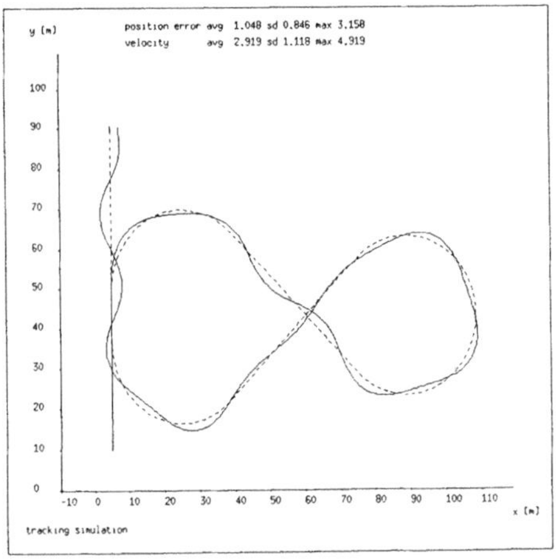

Figure 13.19: Tracking with feedback compensation of position errors

In the open-loop case, the resultant path is often parallel to the desired path. This is because, after the desired curvature is obtained, no further correction is made unless the input is changed. In the feedback only case, the vehicle oscillates around the nominal path because the feedback gain makes the vehicle sensitive to position errors. For lower value of feedback gain, much less

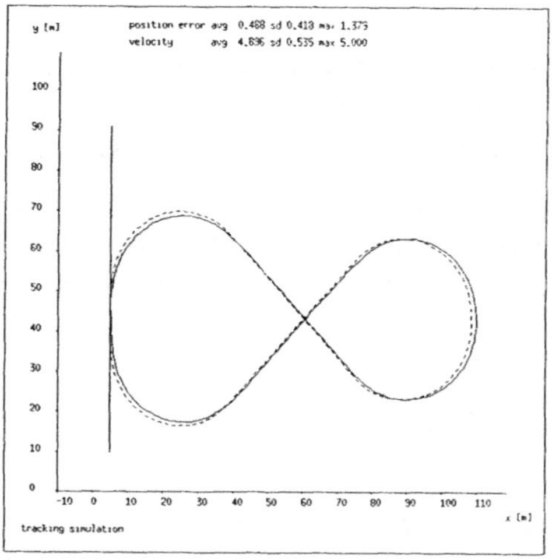

Figure 13.20: Tracking with feedforward compensation

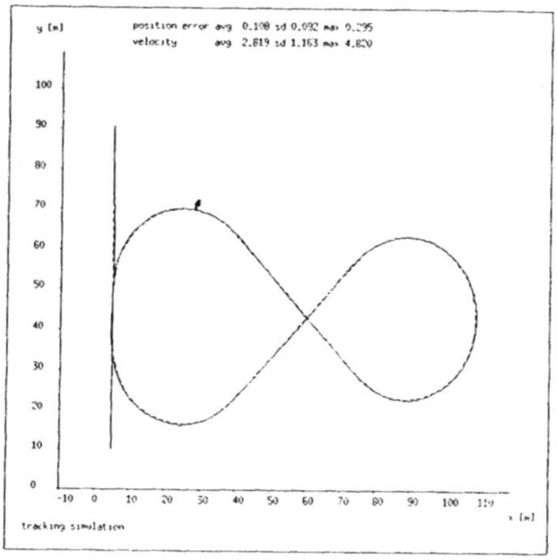

Figure 13.21: Tracking with feedback and feedforward compensation

overshoot is noticed but the feedback response is very sluggish. In the feedforward only case, the algorithm was able to compensate for the vehicle in a predictive fashion. However, such errors accumulate without feedback in the general case. In the last case, it can be seen that the combined feedback and feedforward schemes complement each other and the resultant path is qualitatively superior to those obtained by application of the other methods.

13.6 Conclusions

While the expressions for control have been developed specifically for vehicles that can be kinematically approximated by a bicycle model, extrapolation to the case of skid steered vehicles is straightforward. In that case, the control should be expressed in terms of rate of heading change, $\dot{\theta}$. For a given forward velocity, v, and width between tracks w, the velocities for the left and right drive trains, v_l, and v_r respectively, can be expressed by:

$$v_l = v - \frac{w}{2} \cdot \dot{\theta} \qquad (13.17)$$

$$v_r = v + \frac{w}{2} \cdot \dot{\theta}$$

In this paper we have described a partitioned control scheme that can be used to guide a robot along a desired path. In particular, the feedback scheme uses geometry of the errors from the desired path to keep the vehicle on track. A simplified first order characterization of vehicle dynamics is used in a feedforward path to anticipate future reference commands. Better performance can be obtained by refining the vehicle model used in the feedforward path. Given the non-linear and time-varying nature of the system, it would be of further benefit for the controller to learn the nature of the vehicle through actual experimentation.

13.7 References

[1] N. Bernstein.
Some Emergent Problems of the Regulation of Motor Acts.
The Coordination and Regulation of movements.
Pergamon Press, 1967.

[2] E. D. Dickmanns.
Vehicle Guidance by Computer Vision.
In *Proc. Intl. Workshop on High Precision Navigation*, pages 3.2-1 to 3.2-11. May, 1988.

[3] K. Dowling, R. Guzikowski, J. Ladd, H. Pangels, J. Singh, and W. Whittaker.
Navlab: An Autonomous Navigation Testbed.
Technical Report, Carnegie Mellon University Robotics Institute, November, 1987.

[4] H. Dugoff, P.S. Fancher, and L. Segel.
An Analysis of Tire Traction Properties and Their Influence on Vehicle Dynamic Performance.
SAE paper 700377, 1970.

[5] T. J. Graettinger and B. H. Krogh.
Evaluation and Time-Scaling of Trajectories for Wheeled Mobile Robots.
Proc. American Control Conference , June, 1988.

[6] Y. Kanayama and N. Miyake.
Trajectory Generation for Mobile Robots.
In *Robotics Research: The Third International Symposium, ISIR*, pages 333-340. Gouvieux, France, 1986.

[7] Y. Kanayama, A.Milipour and C. A. Nelm.
A Locomotion Control Method for Autonomous Vehicles.
In *IEEE Intl. Conference on Robotics and Automation*, pages 1315-1317. April, 1988.

[8] Winston L. Nelson.
Continuous Steering Function Control of Robot Cart.
IEEE Transactions on Industrial Electronics , Submitted for publication, 1988.

[9] M. Olley.
Independent Wheel Suspension-its whys and wherefores.
Trans. Soc. Auto. Engrs. :41-49, February, 1935.

[10] T. B. Sheridan.
Three Models of Preview Control.
IEEE Trans. on Human Factors in Electronics :91-102, June, 1966.

[11] Dong Hun Shin and Sanjiv Singh.
Path Generation for Robot Vehicles Using Composite Clothoid Segments.
Technical Report, Carnegie Mellon University Robotics Institute, December, 1989.

[12] R. Wallace, A. Stentz, H. Moravec, W. Whittaker, C. Thorpe, T. Kanade.
First Results in Robot Road-following.
In *Proceedings of IJCAI.* Los Angles, August, 1985.

Chapter 14: The Warp Machine on Navlab

Jill D. Crisman and Jon A. Webb

14.1 Introduction

The Carnegie Mellon Warp[1] machine is a systolic array computer developed by H. T. Kung's group, and used for many applications incuding image processing and mobile robot control [1]. We relate the history of the use of the Warp machine on Navlab and evaluate the Warp machine in light of this experience. As we will demonstrate, the Warp and Navlab projects influenced each other in several ways; this influence led to increased capabilities in the Warp machine and useful applications experience, as well as increased capabilities for Navlab.

We begin with a short history of the Warp machine on Navlab. Next we describe the major Navlab systems that were implemented using the Warp machine. Then we evaluate the Warp machine using experience from these systems.

14.2 History of the Warp Machine on Navlab

This section traces the history of the development of the machine, its software, and its application on Navlab. The earliest systolic array designs that led to the Warp machine were two-level pipelined arrays by Kung, et al. [6, 9], described in 1981, 1982, and 1983. The systolic array formed one pipeline, because the linear array of cells could pipeline data from one cell to the next, and within each cell the floating-point pipeline formed another. These designs were shown to be capable of convolution.

With the introduction of the Weitek 1032 floating point chips in 1983, it became possible to implement a machine based on these ideas using ordinary engineering effort. Professor Tai-Ming Parng, a visitor to Carnegie Mellon from National Taiwan University, created a systolic array design in the fall of 1983 which was shown by Kung [7] to be capable of performing one- and two-dimensional convolution as well as the Fast Fourier Transform (FFT). It was this design that was the basis of the Warp cell developed by Kung and

---

[1] Warp is a service mark of Carnegie Mellon

Menzilcioglu [8].

At this point the Warp cell included the Weitek floating point chips, which were fed data from a pipelined register file, two input and output queues connecting each cell, and some onboard cell memory. Addresses were supplied externally via a third queue. No data-dependent branching or address generation was possible.

In early 1984 a group of researchers, including hardware, software, and applications designers began planning the design of the Warp computer. The design of the machine changed rapidly, and became much more general. Data dependent control flow and program memory was added. A crossbar, originally with limited interconnection and later with full generality, was added to connect the various functional units on the cell. The pipelined register files were replaced by Weitek register files.

The cell at this point had several features that were eliminated later. The address queue was still the sole source of addresses for cells. It was thought that address generation for complex addressing operations such as FFT could be factored out from the array and performed on a special board called the Interface Unit. The address and data paths between cells fed into RAMs with read and write counters that could be incremented or held – not queues. This made it possible for the queue to be used as an auxiliary scratchpad register file. However, it was not possible to switch back and forth between using the RAM as a register file and as a queue, since the counters could not be saved or restored. Thus, using the RAM as a scratchpad register file eliminated one data path to the cell. There was also a loopback feature where the output of a cell could be fed back into its input queue in order to allow a cell to simulate multiple cells.

The Warp cell was built in prototype form, as a two-cell array with an interface unit, called the "demonstration" machine. Assembler development proceeded in parallel through two stages. The first stage was W0 (originally, Warp0). W0 controlled the cell at the 16-bit, 100 ns level (hardware datapaths were 16 bits wide; the 32 bit-wide functional units were supplied with data in two 100 ns steps). W0 was defined and an assembler was implemented in the summer of 1984. Applications development was also already underway. Even before assemblers or simulators were available, programs such as affine transformation, clipping, histogram, median filtering, and binary image processing were being written. This work influenced the design of the Warp cell.

The second stage of assembler development was W1, which controlled the cell at the 32-bit, 200 ns level. (Each W1 instruction corresponded to two instructions at the W0 level). W1 was defined and an assembler was implemented in the fall. As soon as W1 was available, it replaced W0 for early applications programming and generation of test programs.

In order to build more than a few machines, Carnegie Mellon needed an industrial partner. An RFP was prepared and sent out in December 1984. General Electric and Honeywell Corporations were selected to build the first full-scale prototypes. They participated in the design and construction of the first two-cell demonstration machine and in the design of the external host software.

The first demonstration Warp cell array was completed at the end of May, 1985. The array consisted of two Warp cells and an interface unit. The array was controlled and fed data by a Sun 2, which also ran applications code not running on the Warp array. (The external host had not yet been completed). This array was used for demonstrations of Warp programs written in W1 on June 21, 1986.

FIDO, a stereo vision system used to drive a robot vehicle, was a key application of the Warp machine in the early part of the project. It was proposed to speed up FIDO by a factor of ten, from about 30 seconds/step to about 3 seconds/step. Implementation of FIDO algorithms on the Warp machine started in the summer of 1984.

The start of the Parallel Vision and Road Following projects in January of 1985 led to early use of the Warp machine in real situations. In most hardware projects, applications of the hardware to real problems occurs only after much of the software and hardware is already developed. But these projects helped provide focus and direction for the Warp project even as the hardware and software were being defined. A simple color-based road-following program based on an algorithm due to Terry Dunlay of Martin Marietta was demonstrated on pre-recorded images July 2. The algorithm took the difference of the blue and green images, subjected them to edge-preserving filtering (twice), and thresholded them using an adaptive histogram method. Later, this algorithm was refined by Richard Wallace of Carnegie Mellon; this work led to the SCARF algorithm discussed later. On September 9, the demonstration machine was used in several live runs of the Terregator (images were transmitted from the Terregator to the Warp machine using a low-power television antenna, or at times via coaxial cable). To our knowledge, this is the first application of a supercomputer to actual control of a robot vehicle. These runs set records for speed and distance of the Terregator (this reflects more the successful integration of the Warp machine into the control of Terregator than any advance in image processing). On September 9 the Terregator navigated for 105 meters at 0.5 km/hr, using a simple color-based road following algorithm. On September 16, the Terregator made several runs of over one hundred meters at 1 km/hr, on a broken asphalt path and on a concrete sidewalk.

In parallel with the applications of the demonstration Warp array, the development of the W2 compiler proceeded. Early in the compiler's design, it was realized that a design error in the cell made it difficult to generate code efficiently. The problem was that on transfer of a word of data from one cell to another, the receiving cell had to explicitly increment its queue counter when the word arrived [1, Section IVa]. In order to generate such code, the sending and receiving loops had to be unwound three times in general. This led to very large code bodies. A hardware change was necessary to overcome this problem. The hardware change was proposed in the summer of 1985, approval from DARPA was requested and received at the beginning of August, and by the end of September W2 programs were running on the new hardware. The old machine (and W1 programming) was retired at this point and the new machine (with W2) was used exclusively.

In December of 1985 we began serious plans for installation of a Warp machine on Navlab. Since Navlab was to be a self-contained machine, it was not practical to do the image processing remotely using the Warp machine, as we had with Terregator. But installing a new computer like the Warp machine in a moving environment required careful planning, both to ensure that it was useful to Navlab and to guarantee that this almost-unique machine was not damaged. Issues like cooling and vibration of the Warp cell array were considered in particular. In fact, as we later learned, the critical issues were cooling of the external host MC68020 processor and memory boards and connector damage as the Warp cells were removed and replaced in the backplane. Cooling of the Warp cells was not difficult because they dissipated much less heat per area than the commercial external host boards, which were tightly packed with chips. Vibration was easily dealt with by ordinary measures like mounting a plate to hold the Warp cells in place, and mounting the rack holding the Warp machine with shock-absorbing mounts.

As the first full-scale prototypes were being built, we began to look forward to the production Warp machines, which would be built using printed-circuit boards. The change from wirewrap to printed-circuit boards allowed some redesign. The first serious discussion of the technical issues within the Warp group as a whole took place during a retreat in October 1985. This discussion was strongly influenced by the experience we had already had with the first prototype: specifically, no use was made of the RAMs with read and write counters between cells other than as queues, so that they could be replaced by much smaller special-purpose chips; this freed up board area for onboard address generation as well as a larger (eight times) per-cell data and code memory. We also saved board space by eliminating the "loopback" feature; explicit multiple cell simulation was never done, since data partitioning could more easily achieve the same goals. In addition we seriously discussed replacing the Weitek floating point chips with newly-available AMD chips that would give us the capability of 64-bit floating point arithmetic; but this idea was later abandoned because it would have meant too many changes to the cell data path. An unused port in the Weitek register files was used to add a second scratchpad register memory, and a special address crossbar was used to generate addresses for this and the onboard cell memory. The final change (and probably the most expensive in terms of design time) was to allow cells to block when attempting to write to a full queue on an adjacent cell, or to read from an empty queue. The result of all these changes was to create a much more powerful cell, with flexibility comparable to a standard computer. Notification of the award of the commercial Warp machines was made to General Electric on December 9, 1985. With all the changes, the new PC Warp machine required considerable redesign time, which was largely done at General Electric; the first full-scale PC Warp machine (called the "production" machine) was accepted at Carnegie Mellon April 15, 1987.

Towards the end of 1986 Hamey began developing the Apply compiler. Apply had been previously developed as a C subroutine package for writing image processing functions; the programmer would write a simple subroutine that processed a window of an image and the Apply subroutine would "apply" the subroutine all across an image. This was done to speed up image processing using a subroutine package for accessing images in different formats; the C Apply subroutine buffered the image especially efficiently. Hamey adapted the C subroutine idea to a code generator for the Warp machine that took W2-like Apply programs and generated W2 programs. In the summer of 1987, Wu developed the first "full" Apply compiler, that took Ada-like Apply programs and generated W2 code [5]. This compiler took advantage of the Warp cell's capabilities and generated efficient W2 programs for local image processing functions. At the same time, Ribas developed a library of approximately one hundred Apply programs.

The new Warp prototype was used extensively with Terregator. FIDO, the stereo vision system, ran successfully on April 25, 1986, although not quickly; the step time was approximately the same as on the Vax, i.e., 35 seconds/step. The prototype machine was demonstrated on August 29, 1986. The demonstration included obstacle detection using pre-recorded data from our ERIM scanning laser rangefinder. The Warp machine drove the Terregator from the lab with an algorithm that included color classification and a geometry module. FIDO was demonstrated in continuous motion with a speed of ten seconds/step.

The Navlab group work on the Warp machine began in June 1986. The work included development of a geometry module for color road classification, which was tested from the beginning on the Warp machine. One of the wirewrap prototype machines was mounted in Navlab January 1987.

At this point the Warp software environment used a complex linking procedure to combine Warp code with the C program that ran on the Warp host. This procedure was fine as long at the Warp and C programs were rarely changed, but quite difficult to use in a research environment. As a result, in early 1987 we constructed a runtime code downloading procedure together with an interface that allowed calling Warp routines remotely with an Ethernet interface. This interface greatly aided development of Warp code.

In May 6, 1987 we demonstrated a five-to-one (50 cm/sec instead of 10 cm/sec in November) speedup of Navlab over the Navlab November speed. This was color-based road following and ERIM-based collision avoidance, both on the Warp machine. Both the color and the ERIM code were run on the same Warp machine; we thought we could get better performance with two Warp machines, and tried this idea later in the year.

A small four-cell PC Warp system was mounted on Navlab in November 1987. With two separate Warp arrays we could do the ERIM processing in parallel with the color-based road following. This two-Warp machine system was used in a demonstration December 18, 1987; ERIM obstacle avoidance happened in parallel with color-based road following. After the demo, however, the second external host, which was on loan from General Electric, was removed and returned.

There were serious problems with mounting a second Warp machine on Navlab. The Warp cells were not a problem; a ten cell array could easily be split into two arrays. But the external host boards had to be duplicated, something which we were loath to do because of the expense. Moreover, the external host was one of the least reliable components; duplicating it reduced its reliability correspondingly.

In this period we seriously addressed the issue of cooling for the Warp machine, particularly its external host. We installed a special air conditioner for Warp, and added temperature sensors that would automatically turn off the Warp machine when the temperature went too high. This allowed us to run the Warp machine continuously on Navlab, giving us a three Warp system; two in the laboratory, and one on Navlab, which could be used remotely when the Navlab was at home and connected to Ethernet.

The major application of the Warp machine from here on was color-based road following. In June of 1988 an adaptive color classification algorithm, using one or two cameras (one with the iris wide open and one with the iris nearly closed, to increase the dynamic range) and a simple geometric model was implemented. The SCARF road following algorithm was re-implemented on the Warp machine in October with a speed of four seconds per image. This was sped up to two seconds per image in November. The resulting system was demonstrated in December; Navlab was driven at one meter/second, with a processing speed of ten to one hundred over the same algorithm running on the Sun. SCARF speed was further improved in February 1989.

At this point development of ALVINN, for neural net-based road following, began. A three-layer neural net was trained to recognize driving direction in graphics-generated road images. The training was done off-line, in an eight-hour run on the Warp machine in the laboratory. The resulting trained network was then used to drive Navlab. Runs began approximately in February of 1989. Images could be processed as quickly at 0.75 s/image; on March 16 a new Navlab record of 1.3 m/s was set. Later, in June, we found that we could train the network "on the fly" by feeding it live road images and driver steering angle while Navlab was under human control. This training was done using the Warp machine on the Navlab. The resulting technique was very powerful; we could for example train the network by driving the Navlab halfway along a test course under driver control, and then allow the network to take over vehicle control.

In order to see further speed improvement both in the color-based and the neural net-based road following work it was thought that the Sun 3 Warp host should be replaced with a Sun 4. To do this for the Warp host would involve extensive changes to the Warp software. Moreover, the Sun 4 was several times more powerful than the Sun 3 and its integrated, general-purpose nature made this power more usable than the Sun 3/Warp machine combination. With the addition of a commercial (Androx) array processor accelerator board many of the things done on the Sun 3 and the Warp machine could be done almost as quickly on the Sun 4, with an increase in reliability and programmability, as well as a large decrease in power, space, and cooling requirements. Accordingly, the Warp machine was taken off Navlab September 6, 1989, and replaced by a Sun 4 with an Androx accelerator board.

14.3 FIDO

FIDO (Find Instead of Destroy Objects) was a stereo vision navigation system used for the control of robot vehicles; it included a stereo vision module, a path planner, and a motion generator. This system descended from work done by Moravec at Stanford [12]. After Moravec came to Carnegie Mellon in 1980, work was done by Thorpe and Matthies [11, 15], who gave the system its name. More recently, work was continued by Klinker, Crisman, and Clune [2] as well as others. This vision system was unusual in its longevity and in the range of speed over its span of development: Moravec's original algorithm, which was heavily optimized (though different in many important ways from the FIDO algorithm), took fifteen minutes to make a single step while running on an unloaded DEC KL10; the Vax 780 implementation ran at thirty-five seconds per step; the Sun 3 implementation took 8.5 seconds per step; and the implementation on the Warp machine took 4.8 seconds per step.

14.3.1 FIDO Algorithm

FIDO was a feature-based algorithm. A *feature* was a point that was detected with FIDO's interest operator operator and located in three-dimensional space by correlation between the left and right images. An *obstacle* was a feature that the vehicle could not drive over – i.e., a feature sufficiently above ground level. It was assumed that all actual obstacles to the vehicle would have enough image features to be detected by FIDO as obstacles.

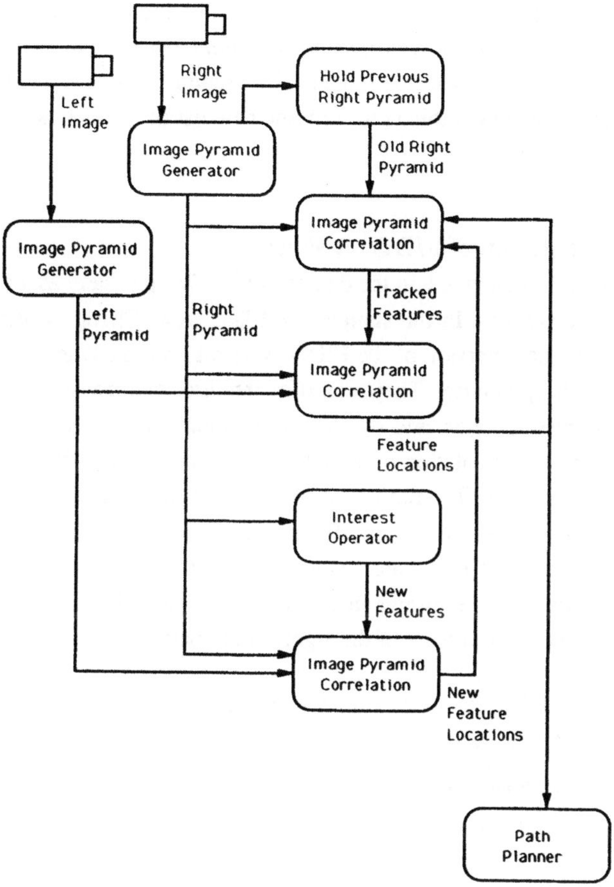

Figure 14.1: FIDO Block Diagram

FIDO performed the following steps, as shown in figure 14.1. First, it took two 512×512 images of its environment, a left image and a right image. These two input images were reduced by the *Image Pyramid Generator* by successive factors of two, creating images of size 256×256, 128×128, and so on. Then *Image Pyramid Correlation* was used to locate all of the previously known features in the new right image. If the feature could be seen in the new scene, it became a tracked feature. *Image Pyramid Correlation* was then used to identify the tracked feature from the new right pyramid in the new left pyramid. Once the corresponding features were located, the three-dimensional position of the obstacle creating that feature was identified and the vehicle was directed one step towards its goal by the *Path Planner*. The *Interest Operator* picked new features in the right pyramid so that new obstacles moving into the scene could be detected. Again *Image Pyramid Correlation* was used to find the corresponding point features in the left image pyramid. The new features and the tracked features were combined to form the new list of "previously known" features for the next image.

14.3.2 Implementation of FIDO on Warp

FIDO has been implemented on the demonstration Warp system as well as the prototype Warp machine. In the summer of 1984, Dew, Chang, Matthies and Thorpe designed a new version of the FIDO system to run on Warp, which was then in its initial design phase. They identified the three major vision algorithms (correlation, interest operator, and pyramid generation), which were considered to be suitable for implementation on a systolic array such as the Warp machine. Then they redesigned FIDO to run on the Warp machine. Next the three vision modules were implemented using Warp microcode by Klinker on the demonstration Warp system. Later, when the prototype Warp machine was available, the modules were reimplemented by Clune using W2 and the external host and the Warp array ran parts of the algorithm in parallel.

Each of the modules that were implemented on the Warp machine will be now described and their performance will be given.

Image Pyramid Generation

The image pyramid consisted of seven levels, starting with a 512×512 image and ending with an 8×8 image. Areas of 2×2 pixels were replaced by one pixel in the next level of the pyramid. The new pixel value in a lower resolution image was computed by averaging over a window in the higher resolution image. The simplest averaging was to take a 2×2 pixel area and average it to one pixel. The initial implementation on the Warp array used overlapping 4×4 windows, which gave slightly better results than 2×2 windows.

The pyramid generation algorithm was implemented in W1 in a systolic scheme, as suggested by Kung for convolution-type algorithms [7]. The algorithm accumulated sixteen pixels in a 4×4 window and then normalized to produce one reduced pixel value. This was mapped onto the Warp array as nine modules, with the first eight each adding two new pixel values to the accumulated partial sum, and the ninth module normalizing the result. The second, fourth and sixth module also stored the partial results until the necessary pixels from the next row underlying the 4×4 window had arrived at the module.

The new data and the partial results were then sent together to the next module.

A simpler sequential algorithm (with non-overlapping reduction windows) took about one second on a Vax/780. Nine Warp cells provided a speed-up of 14, which was relatively small. The implementation of the pyramid generation algorithm on the Warp machine was communication intensive: it used the adder effectively only half of the time (in every other row). It did not use the multipliers at all (except for a normalization). Each Warp cell was used as a 2.5 MFLOPS machine, for 25 MFLOPS from the array. This explains the relatively small speed-up of the pyramid generation algorithm. Adding more cells would not increase the speed since this would not reduce the communication requirement.

This module was later reimplemented as a C program to run on the cluster processors, since very little computation was done here. This made it possible to do the two pyramid generations in parallel using the two cluster processors. In this implementation non-overlapping 2×2 windows were used instead of the overlapping 4×4 windows in the implementation on the Warp machine, to simplify the computation.

Interest Operator

FIDO detected features with an interest operator, which was designed to detect points that could be localized well in different images (for example corners). Such points had image intensities that changed rapidly in all directions. The interest operator took squared pixel differences in the 3×3 neighborhood around the point [3, 15]. The output of the operator was the minimum of the squared differences in the vertical, horizontal, and both diagonal directions. The interest values were locally maximized in one hundred subimages that were arranged in a 10×10 grid. The maximum of all subimages were stored in a list ordered by decreasing interest values. This gave a set of point features, distributed across the image, which could be localized in other images.

Only the first part of this algorithm, accumulating squared pixel differences in all four directions for every pixel, was implemented in W1 on the Warp array. In the demonstration system, the processing stopped here. Later, we implemented the minimization, maximization, and list formation on the cluster output processor.

The interest operator did not offer a good partitioning into modules with similar timings. We thus did not try to implement it in a systolic scheme, as was described for the pyramid generation algorithm in the previous section. Instead, we used the *input partitioning* model [10] where data was divided into equally sized parts. In this scheme, each cell performed the complete algorithm on a portion of the data. An $m \times n$ image was divided into c vertical stripes to be processed on c different cells. For the interest operator, the stripes had to overlap by four pixels, due to the width of the operator window. Thus, every cell ran on $m \cdot (\lceil \frac{n}{c} \rceil + 4)$ pixels. The systolic communication facilities were then used like a "bus": each cell received data from the previous cell and sent it to the next cell. The host sent the data interleaved such that each cell could use every c^{th} pixel for itself. At the beginning of every new iteration, c new pixels were sent over the "bus." The offset between programs that ran on neighboring cells was two cycles such that each cell started a new iteration exactly when a new pixel arrived.

The sequential algorithm ran in about 2.65 seconds on a Vax/780. Ten Warp cells provided a speed-up of 26.5. The adder was the most used resource of the interest operator. It was used in forty out of sixty-five cycles of the innermost loop. The multiplier was barely used (four multiplications in sixty-five cycles). The algorithm thus used each cell as a 3.4 MFLOP machine. The addition of more cells would greatly improve the speed. In the described implementation, each cell needed a new pixel every sixty-five cycles. Thus, maximally sixty-five cells could have been used in parallel before the interest operator had become I/O limited.

Image Pyramid Correlation

For a given pair of images and a given list of point features in one image, the correlation algorithm found the corresponding point features in the other image. The search for the most likely correspondence was performed on the image pyramids, starting at the lowest resolution (8×8) image. At each level, a 4×4 template around the interesting point was correlated with an 8×8 search area in the other image at the same resolution. The best matching position of the template in the search area determined the position of the search area in the next higher resolution image in the pyramid [12].

A pseudo-normalized correlation was used, as given by this formula [3]:

$$CORR_{lm} = \frac{S_t - t_{mean} S_1}{t_{var} + (S_2 - S_1^2)/16}$$

$$\text{with } S_1 = \sum_{i,j=0}^{3} I_{i+l,j+m}, \quad S_2 = \sum_{i,j=0}^{3} (I_{i+l,j+m})^2,$$

$$S_t = \sum_{i,j=0}^{3} t_{ij} I_{i+l,j+m}$$

where t_{ij} denotes the template element at position (i,j), and $I_{i+l,j+m}$ denotes pixel at position $(i+l,j+m)$ in the image.

In the W1 version of the algorithm, the Warp machine found the positions of all features for one given pyramid level at a time. First, templates for all pyramid levels were sent. The cells stored the templates and computed their means and variances. Then the search areas of each level were given to the Warp array in the same sequence as the templates. The cells correlated the current template with the current search area and sent the correlation results for every template position to the output cluster. The cluster processor then found the best position of each template within its search window and determined the search areas for the next higher resolution. The process was repeated for all of the images in the pyramid [3].

The correlation algorithm was implemented in a systolic programming scheme, just as in the pyramid generation algorithm. It was designed as nine modules. Each of the first eight modules covered two template elements. The algorithm was designed so that initially, each module received the template elements and stored the respective template elements of each template. The mean and the variance of all templates were computed and stored in the ninth module. Then, in the correlation phase, each module got the pixels of the search areas and the partial sums S_1, S_2, and S_t from its left neighbor and updated the partial sums before it sent them to its right neighbor with the next pair of pixels. As in the case of data pyramid generation, the second, fourth and sixth module stored the derived partial results until the pixels of the next row, underlying the current window position, arrived. The ninth module combined the partial sums and the mean and variance of the current template into a correlation value that denoted how well the template fitted the data in the search area at the current position.

The sequential algorithm took about 2.3 seconds on a Vax 780. Nine cells provided a speed-up factor of seventy-eight. This was a much higher speed-up than that achieved by the pyramid generation algorithm and the interest operator because the multiplier was used in every cycle and the adder was used in every other cycle. Each cell thus ran here as 7.5 MFLOP a machine. The communication facilities were also used in every other cycle. Therefore, the correlation algorithm was a fairly well balanced algorithm. The maximum speed-up would have been reached if eighteen cells had been used (due to communication requirements).

This module was originally written as a systolic program, but could not be reimplemented in W2 in this way because the prototype W2 compiler allowed only homogeneous code. Instead, it was implemented using input partitioning, like the interest operator.

14.3.3 Performance of the Vision Modules

The reimplementation of FIDO led to a total system time for one step of 4.8 seconds, which was a large speedup over the original time, but still relatively small compared to the time that had been achieved by that time on a Sun 3 alone (8.5 seconds). In this section we will analyze the performance of the FIDO system on the Warp machine.

Most interesting was the pyramid generation module on the Warp machine. It actually took longer to run on the Warp machine than on the Sun alone. This was because the data flow between the clusters and the Warp machine was unbalanced. Time consuming manipulations were required to order the data correctly for the Warp machine in this implementation, but the actual pyramid generation on the Warp array was not computationally intensive. The array was virtually starved for data. This was a case where the ordering of data was too complex for the Warp machine (specifically the clusters). A more efficient implementation would be for the cluster processors to send the pixels in the order that they were stored in memory so that data could flow rapidly into the array, and the Warp array could reorder the data.

The interest operator and correlation functions did not perform at the predicted speeds on the prototype machine, although they were faster than the comparable Sun functions. If the startup times on the Warp machine were subtracted (the startup time was much lower on the production Warp machine), then the actual times were close to the predicted times.

The interest operator required about 0.1 seconds of Warp array processing time for the ten cell implementation compared with a one second Sun 3 time. Additional time was spent starting the Warp array (about 25 milliseconds). However, most of the time was spent in post processing. After the interest operator was run, the cluster processors sorted and selected the resulting data. This was about 28% slower than the Sun 3 processor, because of a slower clock rate.

The correlation function had less than a factor of three speedup, compared to a Sun 3 alone. As with the interest operator, the time required for the correlation function on the Warp array was small. However the time spent processing data for the Warp array on the cluster processors dominated the total execution time. This time included the following:

- Startup overhead of 25 ms. In one step, correlation was called seven times, for a total overhead of approximately 0.2 seconds.
- Rearranging data for the Warp machine. Complex addressing was needed to send the image patches from the different pyramid levels to the Warp machine.
- Fixed loop function of the W2 compiler. A fixed number of features must be processed in every correlation, in our case fifty, although the average number of features in a correlation was approximately twenty-five.

Work on FIDO stopped in 1987. The move from Terregator to Navlab, with its ERIM laser range scanner, ended it. FIDO's stereo vision was not as reliable, and could not be made significantly faster, than the ERIM scanner. While FIDO could locate a small number of "feature points" in a few seconds of Warp machine time, the ERIM scanner provides a dense three-dimensional array of points in one-half second scanning time and a few seconds of processing. Moreover, ERIM worked much more reliably than FIDO – it could even be used at night, and FIDO's interest operator, designed to look for object corners in indoor images, never performed very well outdoors; it was confused by image clutter, such as leaves, in outdoor images.

14.4 SCARF

SCARF (Supervised Classification Applied to Road-Following) is a road-following navigation system used to drive the Navlab. The system labels every pixel in an image as road or off-road depending on how well the color of the pixel matches road and off-road colors from previous images. The road location is determined by matching an ideal road shape model with the labeled image data. This location is then used to update the stored road and off-road colors and to steer the robot vehicle. This system has evolved over a period of four years and is still being used as a research tool today.

The work began with Wallace in 1986 [16]. His system stored one road and one off-road color from the previously processed image. After an image was digitized, each pixel was labeled as road or off-road. Then the labeled pixels were grouped into blobs using a connected components algorithm and polygons were fit to the blobs. Candidate road edges were selected and the best fit of these edges to the ideal road model decided the road location. This location was then used to sample road and off-road colors from the current scene. The new road colors were then used to re-label each pixel and the road model was again fit.

In 1987, Crisman and Thorpe continued work on the classification system. Rather than storing only one road and one off-road color, the system stored several road and several off-road colors. This allowed different colors, for example for shadowed and sunny road. They also replaced the polygon-based road fitting technique by a more reliable method based on Hough space.

In the following years, this basic system has been the basis for much experimentation. A new texture feature was added so that pixels could be labeled by how well they match both the color and texture of the sampled road or off-road. The system was adjusted to include two camera inputs at fixed iris locations to extend the dynamic range of the input. At first the two images were combined into one image and SCARF processed the combined image. Later both images were used for classification, so that pixels were labeled by their match to both "open" iris stored colors and "closed" iris colors. Also, the SCARF system was implemented using different sized input images and using different sized images for classification.

SCARF has had several implementations on the Warp machine. The first of these implementations was written in W2 and Apply on the prototype Warp machine. This implementation showed only a factor of two speedup over the Sun 3 version of the code. This speed improvement would have been larger, but SCARF had already been fine tuned on the Sun 3. Later versions of SCARF were implemented on the production machine. We used the Warp machine to process two larger images rather than the one smaller image of the prototype implementation. In this case we saw a speedup of six over the Sun 3 implementation. The fastest SCARF system had a total one second Warp machine time and a total time of three seconds counting all the overhead including vehicle control. Compared to a Sun 3 implementation, the speedup was thirty for the Warp machine time or ten counting all overheads.

In the next section, we will describe the SCARF algorithm in more detail. Then we will show equations for each SCARF module that was implemented on the Warp machine and describe their implementation. Finally, we will discuss how the later SCARF implementations were derived from the first and discuss in general terms the timing of the systems.

14.4.1 SCARF Algorithm

The program flow and data transfer between the different SCARF modules is shown in Figure 14.2. SCARF starts with (480×512) RGB images from the color camera. The *Image Pyramid Generator* creates an image pyramid for each of the RGB input images. The *Texture Operator* takes the blue image pyramid and creates an image corresponding to the texture seen in the scene. The smallest level of the RGB pyramid, the *RGB Images* and the *Texture Image* is then sent to the *Classifier*. The *Classifier* compares the color of each pixel in the image with remembered road and off-road colors described by the *Color Model*. Each color pixel is assigned a value in the *Probability Image* representing the likelihood that the pixel is a road pixel. This image is used as voting weights by the *Road Hough* module. Each pixel votes, using its assigned probability, for all of the possible roads that contain that pixel. The *Road Location* with the largest accumulated vote is selected as the best road. The resulting *Road Location* is used by the *Color Model Generator* to label pixels in the image as road and off-road. The labeled pixels are then used to formulate new road and off-road colors models. The *Road Location* is also used to generate motion commands for the vehicle.

14.4.2 Implementation of SCARF on the Warp Machine

SCARF has been implemented on the prototype Warp system as well as the production machine. In the initial system, four modules were picked for implementation on the Warp machine: the *Texture Operator*, the *Classifier*, the *Road Hough*, and the *Color Model Generator*. The modules were initially implemented by Crisman and Webb for the prototype machine. Later, Chen and Crisman implemented a different version of SCARF on the production machine. This version was more computationally expensive than the original system. A final SCARF system was implemented by Crisman. It had only one module to process the entire SCARF algorithm.

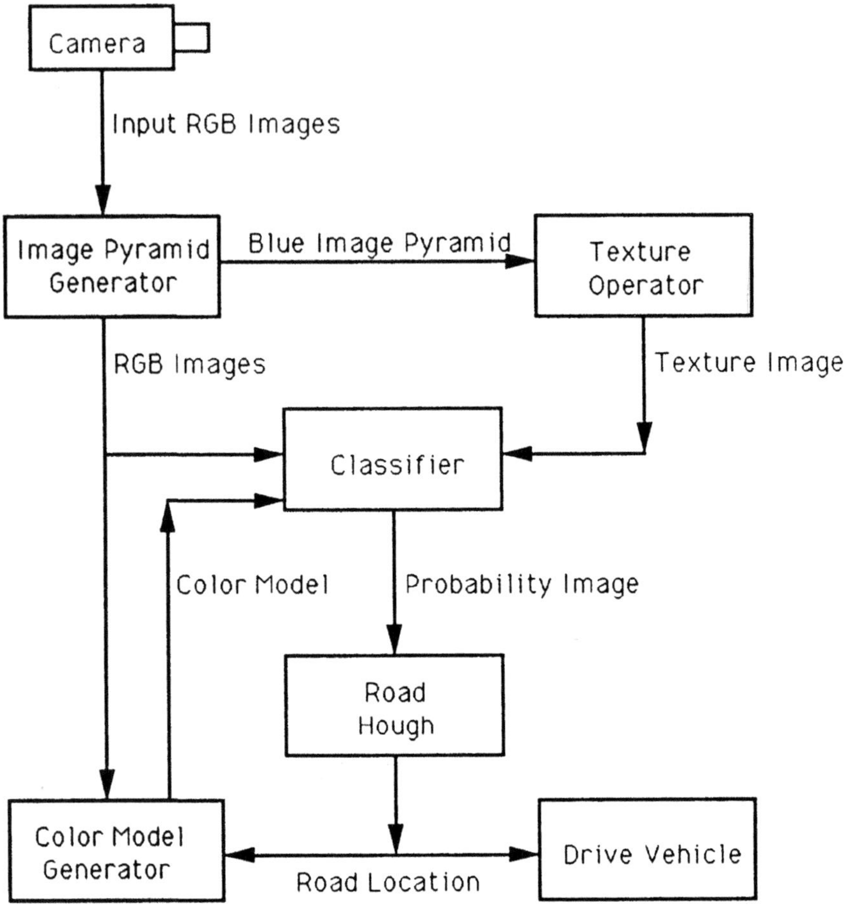

Figure 14.2: SCARF Block Diagram

The next sections will describe each module that was implemented in more detail. The general equations will be given and a brief overview of how the algorithms were divided among the Warp cells. Finally we will discuss how these modules were combined to for the SCARF systems and give overviews of their performance.

Texture Operator

The *Texture Operator* consists of four separate phases that derive a *Texture Image* from the blue pyramid of images as shown in Figure 14.3.

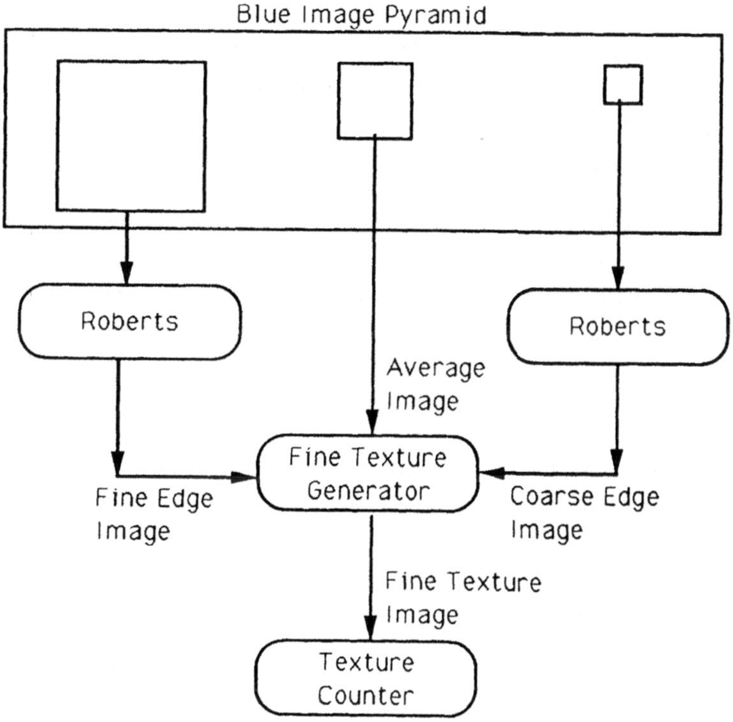

Figure 14.3: Texture Operator

The first and second phases are *Roberts* edge operators [14]. While Roberts performs somewhat poorly as an edge detector, it works quite well to distinguish highly-textured areas such as grass from smooth areas such as road. The first operator is run over the 120×128 blue image to form a 120×128 *Fine Edge Image*. The second edge operator is run over the 30×32 image to form the 30×32 *Coarse Edge Image*. A Roberts operator computes a edge value by looking at the input image, *in*, values around the edge pixel location. Therefore edge value at row *i* and column *j* is calculated by

$$edge[i][j] = |in[i][j] - in[i+1][j+1]| + |in[i+1][j] - in[i][j+1]|$$

The third phase of the *Texture Operator* is the *Fine Texture Generator* and it creates a *Fine Texture Image* from the *Fine Edge Image*, the *Coarse Edge Image*, and the *Average Image*. The *Average Image* is the 60×64 image in the input blue image pyramid. The a pixel located at row i and column j of *Fine Texture Image* is calculated by

$$fine\_texture[i][j] = THRESHOLD\,(fine\_edge[i][j]\ /\ (\alpha\ coarse\_edge[i/4][j/4] + (1-\alpha)\ average[i/2][j/2]\ )).$$

THRESHOLD is a thresholding function that outputs a 1 if its argument is greater than a particular threshold value and 0 otherwise. The constant α=0.2 is a weighting value; it was set heuristically.

The *Texture Counter* is the final phase of the *Texture Operator*. It forms the *Texture Image* from the *Fine Edge Image*. The output 30×32 *Texture Image* is computed by summing the *Fine Texture Image* values over a corresponding 4×4 window. In particular, the pixel value located at row i and column j in the *Texture Image* sums the value of the pixels located between rows $4i$ and $4i+3$ and between columns $4j$ and $4j+3$ of the *Fine Texture Image*

The implementation on the Warp machine used three different modules, the first two of which were Roberts edge operators while the third was a combination of the *Fine Texture Generator* and the *Texture Counter*. Although the the algorithm was the same for the edge operators, they were implemented separately since the input images were different sizes.

The edge operators were written in Apply and the last module was implemented in W2. The input images were divided column-wise among the cells. To speed up the processing time, the loops were unwound and each pixel of the output *Texture Image* was calculated immediately after the calculation its corresponding 4×4 block of the *Fine Texture Image*. Therefore there were 16 explicit equations in the W2 code for each of the fine texture pixels in the block, each of which was followed by a counter keeping track of the sum.

Classifier

The classification module of SCARF uses a Bayesian classification technique to determine the likelihood that each pixel is a road pixel by matching pixel colors with remembered road class colors. A Bayesian classifier takes a d-dimensional measurement vector, $\mathbf{x}$, and chooses the best class label, w_j, from a set of K classes, using a previously computed, *class conditional* probability, $P(\mathbf{x} \mid w_j)$, for each class [4, Section 2.8]. For our case $\mathbf{x} = [Red\ Green\ Blue\ Texture]^T$. We assume that the *class conditional* probability can be modeled by a Gaussian distribution and therefore, is totally specified by $\{\mathbf{m}_j, \mathbf{C}_j, N_j\}$, the mean color and texture, the covariance matrix describing the relationship between the colors and texture, and the number of samples in class w_j. This classifier can be shown to be equivalent to picking the class that maximizes the following likelihood:

$$\lambda_j = \ln(N_j/N) - d/2\ln(2\pi) - 1/2\ln(|\mathbf{C}_j|) - 1/2(\mathbf{x}-\mathbf{m}_j)^T\mathbf{C}_j^{-1}(\mathbf{x}-\mathbf{m}_j)$$

where each pixel provides a four dimensional measurement vector (d = 4). To get the *Probability Image* value the exponential function is applied to the maximum likelihood value. The sign is negated if the maximum class is an off-road class.

This module was implemented in W2 by once again dividing the input 30×32 *RGB Images* and *Texture Image* into column stripes. The input statistical color models were duplicated on each cell. Notice that the first three terms of the likelihood calculation can be computed only once for each class rather than once for each pixel and was passed as input into the Warp array. To get the desired probability measure, an exponential function is needed. An approximation to an exponential was implemented on the Warp cells.

Road Hough

This SCARF module searches through all possible road interpretations for the road having the greatest accumulated probability based on the *Probability Image* from the *Classifier*. We assume the road is locally nearly straight, and can be parameterized using (v, θ) where v is the column where the center of the road intercepts with the vanishing row in the image and where θ is the angle difference from perpendicular where the center line lies (see Figure 14.4.) These two parameters are the axes of an accumulator space used for collecting votes. Each pixel in the probability image votes for all the roads that contain that pixel by adding its probability to the proper positions in the accumulator. For each angle θ_i, a given pixel location (r, c) will vote for a set of vanishing points lying between v_s and v_e given from the equations below:

$$v_s = c + (r - horiz)\tan\theta - (w/l)(r - horiz)$$

$$v_e = c + (r - horiz)\tan\theta + (w/l)(r - horiz).$$

where *horiz* is the horizon row in the image, w is the road width at the bottom of the image, and l is the length from the horizon row to the bottom of the image. The maximum value of the accumulator is chosen to be the road.

This module was implemented by distributing the input *Probability Image* column-wise among the cells. There was no overlap of the input between cells.

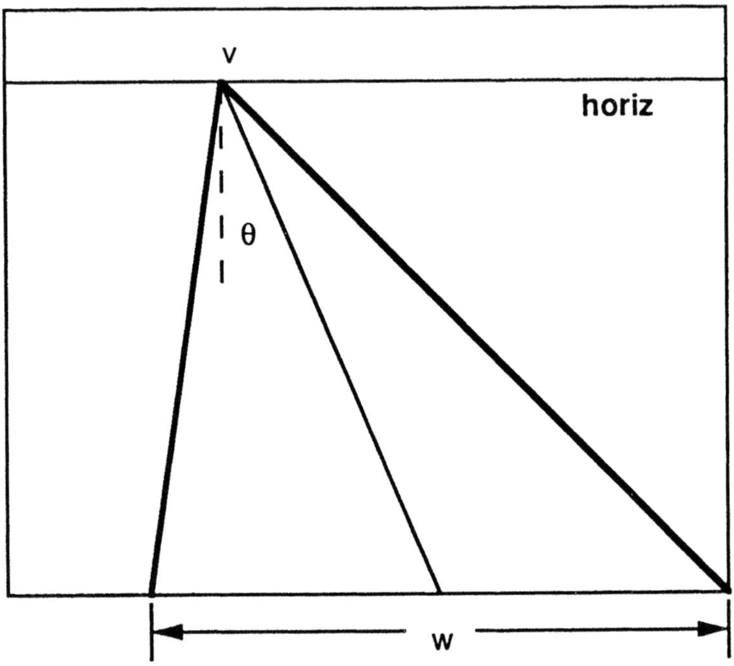

Figure 14.4: Road Hough

However, a complete Hough space was calculated on each cell. To get the output Hough, the individual cell's Hough spaces were added as the output Hough Space was passed through of the array.

Color Model Generator

This function calculates the road and off-road color models after each image is processed so that the system can adapt to changing illumination conditions. The texture model is not adapted and therefore is not computed after each image like the color model. The road and off-road color models are modified in three steps. First a set of pixels in the current image is chosen as road and off-road training sets. Next the training sets are subdivided into classes using an ISODATA clustering algorithm. Finally, the new statistical color models are calculated for each class.

The ISODATA clustering algorithm was implemented by a pair W2 modules; the *Sample* module and the *Adjust* module as shown in Figure 14.5. It started

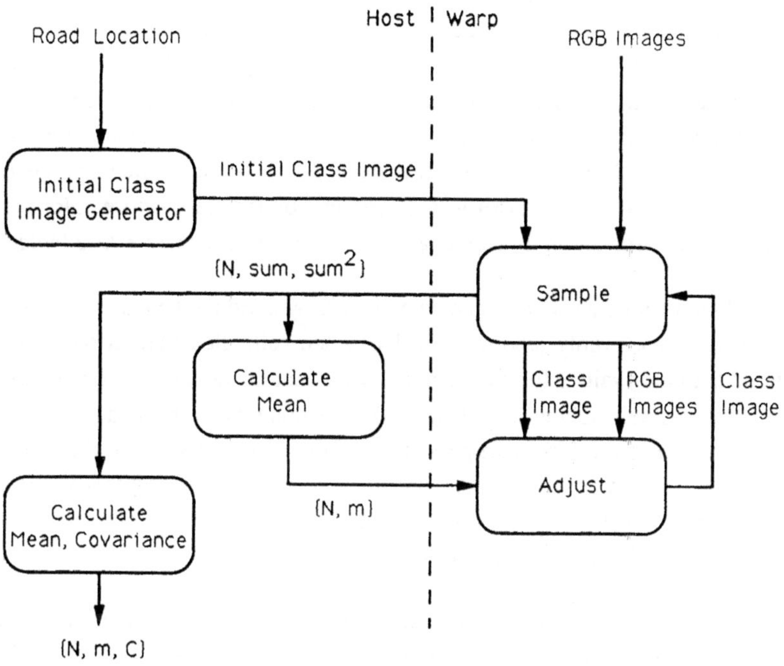

Figure 14.5: Implementation of ISODATA Clustering on the Warp Machine

with a *Class Image* generated on the external host and the input *RGB Images*. This pair was called several times before the new color models were formed. After the last iteration, the *Sample* was executed one last time to generate the sums required for the final color models to be calculated.

The Sample Module computed sums of pixel values from a labeled input training set. The labels for each image pixel were stored in the input *Class Image* which either labeled each pixel as one of the road or off-road classes, or as unknown. It also read the *RGB Images*. From this it accumulated the sums, $\mathbf{sum}_j$, and the squared sums, $\mathbf{sum}_j^2$ of the red, green, and blue pixel value for each class w_j. It also counted the total number of samples per labeled class, N_j. Color values were calculated for each class using samples labeled from the *Class Image*. From this information, the mean color of each road and off-road classes were calculated by an external host routine.

The Adjust Module adjusted the *Class Image* by using the current mean colors road and off-road classes. It read the mean class colors from the external host, the old *Class Image*, and the input *RGB Images*. Any pixel that was labeled as unknown in the old *Class Image* remained labeled as unknown. If the pixel was labeled as one of the road classes, then the pixel color (from the input color images) was compared with each of the road mean colors. The pixel was then re-labeled as the class whose mean color most closely matched the pixel color value. Similarly, if the pixel was labeled as off-road in the *Class Image*, then the new *Class Image* label was determined by the class whose color mean value closest to the original pixel data. This module wrote a new *Class Image*.

The Sample module was implemented in W2 by dividing the *RGB Images* and the *Class Image* evenly among the cells. Each cell calculated its own partial sum of the color values for each class, partial sum of the color values squared, and number pixels having each label. The resulting sums from each cell were accumulated as the values were passed out of the Warp array and to the external host. The host then calculated the statistical color models using this values by the standard statistical equations for mean and covariance. The new mean values were then passed into the Adjust model.

The Adjust module was implemented in W2 by also dividing the input *Class Image* and *RGB Images* evenly column-wise among the cells. The mean values were copied to all of the cells. Therefore each cell produced a column stripe of the output *Class Image*.

14.4.3 Performance of SCARF Implementations

SCARF was picked for implementation on the prototype Warp machine in January of 1987. At that time, the system was already implemented on a Sun 3 and was processing images in about thirty seconds per image. During the time that W2 code was implemented, the C code was optimized, giving a final Sun 3 time of about twenty seconds.

The first implementation of SCARF used the prototype Warp machine and was completed in March of 1987. This implementation used only eight of the ten available cells so that the column data could be distributed evenly among the cells to simplify the implementation. Initially, the modules were implemented as described above. However, the time for downloading microcode to the Warp cells and the time required for passing data to and from the cells prevented any speed up of the implementation on the Warp machine over the Sun 3 implementation. This implementation required about fourteen downloads of Warp microcode and image data.

To improve the processing speed, the microcode for individual modules were linked together forming three microcode blocks. As a result, the microcode was downloaded only three times per image. To improve this rate, we implemented the two Roberts operators in W2 to reduce the size of the microcode required for these modules. At this time, the microcode could be linked into two separate sections and then required only two downloads per image.

We also noticed that we were passing in input image data repeatedly to the Warp array. By locking the Warp machine so that no other users could access the machine, and declaring the repeated input as global data in the same position in each W2 module, the input could be loaded once, and then used by different modules without being reloaded. With this modification we started to see an improvement over the Sun 3 times.

We implemented a version of *Image Pyramid Generator* on the cluster processors. This removed some of the load from the Warp array and had the additional advantage that the data from the frame buffer did not need to be copied to the external host before it was transferred to the cluster memory. Instead, the reduction implementation read the first of its inputs directly from the frame buffer and this data was never copied. In order to be as fast as possible, however, this implementation only approximated the averaging that was done in the Sun implementation.

Using microcode linking, common global storage, and by hand optimizing the W2 code, we were able to get the system running in about ten seconds per image. This speedup was small, but the results were promising for future implementations on the production Warp machine.

The next implementation of SCARF was more computationally expensive than the original version. It used two color camera inputs rather than the original one camera. It also classified 60×64 images rather than the original 30×32 images. The image pyramid generator now had twice as much data to process, and the classifier and the color model update had four times as much data since the vectors were now six dimensional rather than four as before. This implementation ran in about sixty seconds on a Sun 3.

This new SCARF was implemented on the new production machine by Chen and Crisman. On this machine, we had eight times more cell memory for global data and for programs. This machine used DMA for faster I/O from the external host to the Warp cells. The image data was now divided by rows on the cells rather than by columns. This allowed an even division of the 60 rows among ten cells rather than 64 columns divided among eight cells.

The new machine allowed successful use of the reverse data path feature where data could be accumulated on one data path onto the last cell, then the data could be passed back and copied into the other cells. This then allowed us to combine the *Sample* and *Adjust* modules from the Color Model Update into one W2 module. To do this the *Sample* module was modified so that after all of the sums were passed to the last cell, the last cell would calculate the new mean values and pass then back to all of the other cells. Then the *Adjust* module can be run without the intervention of the external host.

The larger memory of this machine also allowed us to store the *RGB Images* in a global memory location and then only input this data once per processing step. The increase in program memory allowed all of the modules to be linked into one microcode which was only downloaded to the Warp machine once before any image processing began. This implementation required about ten seconds per image, which was a speedup of six over the Sun 3 implementation.

The last implementation of SCARF on the Warp machine was completed in September of 1989 on the production machine. This version still used 60×64 images; however, it returned to the original one camera version of the code. The entire SCARF loop was implemented in one W2 function and one function was set up to initialize the whole system. The initialization process read in a 480×512 image and created a 60×64 image in Warp's cell memory. The W2 SCARF loop function started by doing the *Color Model Generator* on the resident image in memory. Then the new color image was read into the Warp cells and reduced. Next the *Classifier* and the *Road Hough* was applied to the input image. Only the resulting road location was passed out of the Warp array. Therefore, the main W2 loop function read only the full size color images, and wrote a couple of floating point numbers representing the road location in the image.

This version of SCARF ran in one second of Warp machine time, and a total of three seconds of time which include some limited displays and sending motion commands to the robot vehicle. This implementation is challenged only by a similar Sun 4/Androx implementation which processes lower resolution 30×32 images in 3.5 seconds.

14.5 ALVINN

ALVINN (Autonomous Land Vehicle in a Neural Network) applied connectionist techniques to the same problem addressed by SCARF, that is, road following using color images [13]. The key difference is that while SCARF was "trained" by hand, adapting standard vision algorithms to the recognition of a road, ALVINN used a neural network learning algorithm to automatically learn what image features were useful to discover the position of the road.

The development of ALVINN on the Warp machine went through two phases. In the first phase, from approximately February through May 1989, a road image generator was implemented and used to generate training images that were fed, together with the correct road position, to a standard back propagation learning algorithm. The back propagation algorithm ran on Warp, off-line; training runs were done overnight in the lab. After training, the learned network was used on Navlab to control the vehicle. (The network was run on a Sun 3, since application of the network, once it was learned, required relatively little computation).

This training technique fully exploited the power of the Warp machine; eight hour runs were used of the Warp machine, with approximately three-quarters of the time during these runs representing actual Warp machine time. Comparable training on a VAX 780 would have taken months.

Training using this method demonstrated the feasibility of using a neural network to control a robot vehicle. But the method suffered from a serious problem. Essentially, the process of adapting computer vision techniques to road recognition was replaced by the process of adapting computer graphics techniques to road image generation. This "forward" generation problem was easier than the "inverse" recognition problem, at least for the simple roads in the park, but it still required human intervention, so that the generated road images accurately represented the range of images that would be presented to Navlab. For successful navigation in more varied environments, the road image generation code would have to become more and more complicated and difficult to program and test.

To overcome this, training "on the fly" was attempted, starting in June 1989. Road images were taken directly from the camera, reduced in size, and presented to the neural network together with the current driver's steering angle (with the vehicle under human control). A clever technique was used to create many example images from one road image and steering angle. With the Warp machine on Navlab, backpropagation was used to modify the neural network weights as the vehicle was driven up the road.

Remarkably, it was found that with a short sequence of a few tens of images, the network could be trained successfully to follow the road. The eight-hour runs on the Warp machine in the lab were replaced by short runs driving Navlab for about ten minutes. Apparently, the intense training of the network in the long runs was unnecessary; in fact, the road following problem was much easier than it had appeared based on the long runs.

14.6 Evaluation of the Warp Machine on Navlab

We now critically evaluate the Warp machine in light of the Navlab experience. We will treat hardware and software separately.

14.6.1 Warp Hardware

The Warp hardware consists of three components: the Sun host, the external host, and the Warp array itself.

The Sun Host

The choice of a Sun as the host of the Warp machine was one of the good early decisions made in the Warp project. At the time, the Sun workstation was one of the most powerful general-purpose workstations available; as it turned out, Sun continued to lead the field both in hardware and in software. The Navlab group decided to use Sun workstations as the basic general-purpose computing element on Navlab. Since the Warp machine had a Sun host, Navlab programs could be run on the Warp host whether or not they used the Warp array. This was important because of the limited space and power on Navlab; having to provide power and space for a workstation that could only be used for controlling Warp programs would have been an extra burden. This is demonstrated, in fact, by the eventual decision to remove the Warp machine from Navlab; this happened largely because the Navlab group moved to Sun 4 workstations. Upgrading the Warp host to a Sun 4 would have required extensive changes to the software. Thus, the Warp host would have been unavailable for running the rest of Navlab software.

The External Host

The Warp machine's external host consists of three MC68020 processors in a VME card cage, together with their memories totalling fourteen megabytes. Two of the processors input and output data to the Warp array, through a special board called the switch; these processors are called the "cluster" processors. The third processor performs auxiliary functions; it is called the "support" processor. The external host communicates with the Sun through a VME bus repeater. The external host card cage also held commercial digitizer boards, which were originally Datacube and later Matrox VIP boards.

A key early decision was to use a commercially available system that was programmable in C. This decision has been validated by the ease of code generation by the W2 compiler for the external host input and output routines (using a commercially supplied C compiler) and by the availability of commercial boards for digitization. We had to do little software development for the basic functionality, and hardware development was limited to the switch board.

Because of the use of industry-standard processors and busses, the external host was the weakest part of the Warp machine. In our early versions of FIDO on the Warp machine, this kept us from realizing full use of the Warp array, because of the constraints in rearranging data on the external host.

The decision to include the support processor was questionable. The support processor was intended to be used for auxiliary functions, such as controlling the digitizer board and possibly controlling the driving functions on Navlab; as it turned out, the digitizer boards were controlled by the Sun, and Navlab driving functions were controlled by a separate processor entirely. In fact, the support processor was never used, because of the difficulty of programming it and the absence of almost any debugging facility in the external host. System cost could have been reduced by almost a third by eliminating the support processor, with no loss of functionality.

Many of the external host capabilities were completely unused. For example, it was possible to use the external host to drive an RS232 connection; this connection, or another similar standard interface, could have been used to control Navlab. This was never done, because the Navlab controlling software and hardware was developed independently, and because of the difficulty of adapting such an interface to a new computer like the Warp machine.

Placing the digitizer boards in the external host was also questionable. The intention was to feed data directly from the digitizer boards to the Warp machine under control of the cluster or support processors. In fact, the normal method was to copy data from the digitizer board into the Sun's memory, and then to subsample the data there and pass it to the cluster processor for use in the Warp machine. Only in the most optimized versions of FIDO and SCARF did we actually use the support processor to take data directly from the digitizer board. In all other systems, the data traveled twice over the VME repeater connecting the external host to the Sun.

The large memories in the external host were rarely used in Navlab. Navlab datasets were generally quite small, and there was no need for more than a few megabytes of memory. However, the large memories were useful in order to maintain compatibility with the Warp machines in the lab, where the large memories were used by other programs. When Navlab was docked and connected to the Ethernet, programs could be run interchangeably on the Navlab or the lab Warp machine.

The Warp Cell Array

The overall structure of the Warp cell array, a short linear array, has been validated by our experience on Navlab. The linear array was quite capable for the low- to mid-level vision algorithms we intended to implement on it at the start of the project; as the range of applications increased to include mid-level processing in SCARF and the neural network back propagation algorithm, the same linear array was usable. The key reason for this was the very high I/O rate within the array; this allowed us to overcome its limited connectivity.

The short linear array also lent itself to dealing with the relatively small datasets (32×32 or 64×64 images) in SCARF. Our early applications studies of the Warp machine were oriented towards dealing with standard 256×256 or 512×512 images. We thought that the increased power of the Warp machine would make it possible for the same processing then being done on small images to be done on large images, which would improve the accuracy and utility of color vision. As it turned out, this was not true. No increased vision performance could be obtained by using more spatially dense images; the road was a fairly large object in most of the scene, and where it was small (near the horizon), recognizing it accurately was useless, because of other uncertainties in the system. So we turned instead to processing small images with the Warp machine. The short linear array was just as suitable for this as it was for processing large images; in fact, with small images the relatively small Warp cell memory could be used to store previous images and other datasets, as in SCARF.

The two-way I/O pathway within the array was used to allow the computation of a result known to all cells entirely within the array; this was used in SCARF and in some implementations of the back propagation algorithm. In fact, for many purposes a circular connection (allowing the last cell to communicate directly with the first) would have been preferable.

The Warp cell included hardware floating point; this facility was one of the main reasons for building the Warp machine in the first place, and was one of the most expensive features in the Warp machine. The Navlab application made good use of Warp's floating point hardware. In SCARF, floating point was used extensively in the calculation of road statistics and their application to color pixel classification; in ALVINN, floating point was necessary for a good implementation of the back propagation algorithm. These applications of floating point came from outside the Warp project; independently, the Navlab group began using statistical methods for color classification, and the neural network group required floating point for their work. Without floating point the Warp machine would have been far less effective as a tool on Navlab.

14.6.2 Warp Software

As with hardware, we divide the software discussion into three parts: Warp host (Sun) software, external host software, and Warp cell software (the W2 and Apply compilers).

Warp host

The Warp array was used as an "attached processor" to the Sun. Datasets were downloaded into the external host, and then the Warp array was called to process them, usually while the Sun waited. (In fact, Sun processing could go on in parallel, and this was done in some of the SCARF systems. But generally this feature was not exploited because there was little for the Sun to do from the time the image was captured to the time the road was recognized).

This model was extremely useful in the development of software for the Warp machine. It was implemented using a mechanism that allowed replacement of a subroutine call in C by a single subroutine call in the Warp software package; the subroutine handled all transfer of data to and from the Warp external host, locking the Warp machine for exclusive use, and downloading and call of the Warp program. This could happen even if the Sun executing the call to the Warp machine was not a Warp host; data would be transferred over the Ethernet to a selected Warp host. It is quite likely that the Warp machine would not have been used much at all in real applications without such a simple method for accessing the machine.

However, the attached processor model implies many overheads. The Sun can become a bottleneck for processing. The startup time for the Warp machine can be quite significant. A serial processor, the Sun, must prepare datasets for a much more powerful parallel processor, the Warp array. Data structures must be moved from the Sun into the external host for processing. All of these overheads seriously affected the performance of the Warp machine on Navlab.

For example, images as captured by the Datacube boards were 480×512 in size. They had to be reduced in size for processing, since spatial resolution was not an important factor in road recognition. This could be done on the Sun, in the external host, or on the Warp machine. Existing libraries of software made image reduction on the Sun trivial – in fact, transparent to the programmer. Programming the external host (the logical place) to do the reduction was difficult, and the Warp cell array could do the reduction only if the images were first transferred from the Datacube boards to the external host memory either by the Sun or the external host. These tradeoffs made image reduction usually happen on the Sun, although in some SCARF systems it was implemented on the external host.

The Warp machine's startup time (time to start up a Warp program with the code already downloaded) was about 25 ms. This time was not a significant fraction of processing time for 256×256 or 512×512 images; in fact, the minimum processing time for 512×512 images was about 60 ms, and usually several times longer. But for small, 32×32 or 64×64 images, this time could be a significant fraction of total time, particularly if several Warp functions were applied to process the image and recognize the road, as in SCARF. As a result, the programmer had to spend a lot of time organizing the Warp functions so they could be executed as the result of a single call, to reduce the overhead. We would have been better served had the startup time been significantly reduced; this could have been done by providing special hardware in the Warp machine's interface unit to allow the Warp machine to initialize itself. As it was, the Sun had to issue special commands to do the various stages of initialization.

Moving data structures back and forth from the external host implied considerable programming difficulty, since many of these structures were embedded in various ways in C programs. This was especially true for FIDO, an old program that had been worked on by a number of programmers. All of the data structures had to be "cleaned up" before the Warp programming could begin; and the process of cleaning up the data structures introduced new overheads.

If we had not used the "attached processor" model, we might have taken an "array-centered" view of the Warp machine. In this model, the Warp array would have been viewed as the central processing resource, and other devices, such as the external host, would have been viewed as supplying data for the Warp machine. We could have attached multiple I/O devices, supplying data from different cameras and perhaps the ERIM laser scanner, and coordinated the processing through the Warp machine. This model would have required much more sophisticated Warp software; we would have needed an operating system on the Warp array to manage all these resources. But such a model would overcome the other problems discussed in this section.

External host

The external host software was designed on the assumption that in Navlab speed was of overriding importance, code would be linked together before runtime (i.e., runtime downloading of code was not necessary), and a library of compiled code could be built up that was not changed frequently.

In fact, these assumptions were largely untrue. The W2 compiler made it possible to write new routines and test research ideas using the Warp machine, which made it much more important to allow rapid testing of new code. (The early development of FIDO, with its programming of a few routines in microcode by several programmers over a period of months, much more closely matches the model we had in mind when the external host software was designed.) The difficult programming environment of the external host, which might have been acceptable if the code was not modified much, instead meant that it was reduced to performing I/O to and from the Warp array, using programs generated by the W2 compiler. And testing of new Warp routines could be best done using an interactive system, which meant that the external host software had to be adapted to allow runtime downloading of code.

One of the reasons for the difficult programming environment on the external host was that control of the development of external host software was transferred to Carnegie Mellon's industrial partners at an early stage of the project, well before it was used. This made it difficult to change the software as our applications experience grew. The industrial partners did a competent job of maintaining and extending the external host software as it was originally designed; but the software would have had to be redesigned extensively to be widely used in Navlab.

The use of the external host in FIDO shows its capabilities when the programming difficulties are overcome. Irregular operations were be mapped onto it as part of pre- and post-processing of data from the Warp machine. Also, it performed memory access-intensive but not compute-intensive computations as well as or better than the Warp array, which could also allow the Warp array to be used for something else in the meantime.

Warp array

In this section we discuss the Warp array software (primarily the W2 compiler) as seen by the user. This includes many design decisions that were essentially forced by the Warp cell hardware, and thus are really hardware issues. Distinguishing between W2 issues forced by the hardware and forced by other concerns is appropriate for a paper on the W2 compiler, not an applications paper.

W2 made the Warp machine much more programmable than we expected. This led to major changes in the importance of some parts of the system and made it possible to overcome deficiencies in one area by using another instead. For example, we could modify our programs to accommodate a regular data pattern from the host, which led to higher I/O rates. This was important even in the later versions of the host, which had faster processors and higher data rates, but which could use DMA, which required a regular address pattern. This flexibility was the main reason we were able to observe the predicted performance of FIDO in actual Warp runs.

W2 made it possible to experiment with different algorithms, in the context of a research system such as FIDO, while getting good use of the powerful Warp array. As we programmed more and more of FIDO on the production Warp machine, programmability was essential, especially as it allowed us to make use of more complex programming models that used the powerful Warp array more and require less intervention by the relatively weak host.

Apply was used far less in the Navlab work than we had hoped. This was partly because of the relatively late introduction of Apply into the project; the first true Apply compiler for the Warp machine was not running until the fall of 1987, one year after the Navlab group began working with the Warp machine. By this time much of the programming difficulties Apply addressed had been overcome by learning on the part of Navlab programmers. Just as important,

Apply code tended to be larger than W2 code for the same problem. In order to process the borders of the image properly, Apply duplicated the inner loop of the image processing function once, leading to a doubling of the code size. This was a serious problem when the user was attempting to keep all code for the entire SCARF application, for example, on the machine at the same time. W2 programs were smaller, though no faster than the Apply programs.

Border processing was a problem for the W2 programs, too, however. The C functions processed borders by duplicating rows and columns near the edges of the image. This was hard for the W2 programs to do. It was simpler just to use a constant value (0) for the border of the image. However, this led to spuriously high values of edge detectors like Roberts near the image border. This sometimes affected the accuracy of the road image processing based on the texture image.

W2 did not support function calls until quite late in the project. Macro calls were used instead. This lead to code size problems for transcendental functions, so that approximations were used instead. This often lead to less accurate results than those used on the Suns. In particular, the classification in the SCARF system was often noisier for the W2 implementation than for the Sun implementations.

There were some peculiarities in converting data between the external host and the Warp array. Because the Warp machine's primary processing power was in floating point, all images were best processed in this way. The interface unit had hardware conversion of 8-bit and 16-bit integers to and from floating point, but could not convert 32-bit integers. As a result, 32-bit integer images in the external host had to be treated differently from 8- or 16-bit integer images.

Primarily as a result of the fixed-size queues, it was impossible to use variable-length for loops in W2 programs until the design change that allowed blocking on writing to a full queue, or reading an empty one, was fully integrated into W2. This meant that image sizes were fixed at compile time. This increased code size (for example, in SCARF there were three different versions of the Roberts operator).

It was impossible within W2 to execute a W2 function repeatedly until some condition was met, for example repeatedly classifying image regions until convergence was achieved. This was a consequence of the distributed nature of control in the Warp machine. The result of this was that the Warp host was involved in repeatedly calling W2 programs and testing for convergence, which significantly increased overheads.

Partly as a result of the small datasets used on Navlab, and partly as a result of limited hardware support, we still needed to use speed tricks to generate Warp code that was significantly faster than the Sun code. For example, we had to avoid using division on the Warp machine, especially when doing integer index calculation for arrays.

Given the research that was going on in the Warp project while the Warp machine was being used in the Navlab project, it is remarkable that things worked as well as they did. Navlab programmers commonly had to deal with new features in the W2 compiler, for example, and it is only due to the good support from the compiler group that they were able to overcome bugs that were due to idiosyncrasies of the machine, and were extremely difficult to identify without experience.

14.7 Conclusions

This paper tells the story of a unique experiment; the installation and use of a parallel supercomputer on a robot vehicle. Let us try to summarize and draw some conclusions from this experiment.

- The Warp machine was useful in the Navlab project. The programmable floating point capability it brought to Navlab was unavailable by other means. Key elements of the architecture, such as its high I/O rate and the short linear array, have been validated by the Navlab experience. The high processing rate of SCARF could not have been obtained without the Warp machine, and it was the presence of the Warp machine on Navlab that led to ALVINN.
- Early applications support is essential. The development of FIDO and other vision applications guided the early design of the Warp machine and provided test programs and early demonstrations. Without this early work, the Warp machine may never have been used on Navlab.
- Continuing software support is essential in a project such as this. It is impossible for hardware and software designers to anticipate all of the issues that will turn up in use of the machine, even if applications designers participate early in the project. This is partly because applications can change, and partly because success in one area can affect others. For example, the Warp machine became much more programmable than we anticipated because of the W2 compiler, which made the design of the external host partly obsolete.

- The "attached processor" model used in the Warp machine is natural and easy to use by the programmer; but it leads only with great difficulty to large speedups in programs. Data structures must be redesigned, careful attention has to be paid to small details of implementation, and so on. If we want to see speedups more than a factor of about ten, we must abandon this model.

Acknowledgments

"We" in this paper refers to a very large group of people, indeed; too many to list here. The Warp project at Carnegie Mellon and General Electric and Warp applications development (including the Parallel Vision and Navlab projects) included over seventy people. We also benefited from the support of the Field Robotics Center and from many discussions with the vision group at Carnegie Mellon.

14.8 References

[1] Annaratone, M., Arnould, E., Gross, T., Kung, H. T., Lam, M., Menzilcioglu, O. and Webb, J. A.
The Warp Computer: Architecture, Implementation and Performance.
IEEE Transactions on Computers C-36(12):1523-1538, December, 1987.

[2] Clune, E., Crisman, J. D., Klinker, G. J., and Webb, J. A.
Implementation and Performance of a Complex Vision System on a Systolic Array Machine.
Technical Report CMU-RI-TR-87-16, Robotics Institute, Carnegie Mellon University, 1987.

[3] Dew, P. and Chang, C.H.
Passive Navigation by a Robot on the CMU Warp Machine.
Aug, 1984.
Internal report, Department of Computer Science, Carnegie-Mellon University, Aug. 1984.

[4] Duda, R. O. and Hart, P. E.
Pattern Classification and Scene Analysis.
Wiley, 1973.

[5] Hamey, L. G. H., Webb, J. A., and Wu, I-C.
An Architecture Independent Programming Language for Low-Level Vision.
Computer Vision, Graphics, and Image Processing 48:246-264, 1989.

[6] Kung, H.T., Ruane, L.M., and Yen, D.W.L.
Two-Level Pipelined Systolic Array for Multidimensional Convolution.
Image and Vision Computing 1(1):30-36, February, 1983.
An improved version appears as a CMU Computer Science Department technical report, November 1982.

[7] Kung, H.T.
Systolic Algorithms for the CMU Warp Processor.
In *Proceedings of the Seventh International Conference on Pattern Recognition*, pages 570-577. International Association for Pattern Recognition, 1984.
A revised revion appears as Chapter 3 in *Systolic Signal Processing Systems*, edited by E. E. Swartzlander, Jr., pp. 73-95, New York, Marcel Dekker, 1987.

[8] Kung, H.T. and Menzilcioglu, O.
Warp: A Programmable Systolic Array Processor.
In *Proceedings of SPIE Symposium, Vol. 495, Real-Time Signal Processing VII*, pages 130-136. Society of Photo-Optical Instrumentation Engineers, August, 1984.

[9] Kung, H.T. and Picard, R.L.
One-Dimensional Systolic Arrays for Multidimensional Convolution and Resampling.
In Fu, King-sun (editor), *VLSI for Pattern Recognition and Image Processing*, pages 9-24. Spring-Verlag, 1984.
A preliminary version, "Hardware Pipelines for Multi-Dimensional Convolution and Resampling," appears in *Proceedings of the 1981 IEEE Computer Society Workshop on Computer Architecture for Pattern Analysis and Image Database Management*, Hot Springs, Virginia, November 1981, pp. 237-278.

[10] Kung, H. T. and Webb, J. A.
Mapping Image Processing Operations onto a Linear Systolic Machine.
Distributed Computing 1(4):246-257, 1986.

[11] L.H. Matthies, C.E. Thorpe.
Experience with visual robot navigation.
In *Proc. IEEE OCEANS'84 Conf.*, pages 594-7. IEEE, September, 1984.

[12] Moravec, H.
Obstacle Avoidance and Navigation in the Real World by a Seeing Robot Rover.
Technical Report CMU-RI-TR-3, Carnegie-Mellon University Robotics Institute, September, 1980.

[13] Pomerleau, D. A.
ALVINN: An Autonomous Land Vehicle In a Neural Network.
In Touretzky, D. S. (editor), *Advances in Neural Information Processing Systems*. Kaufmann, 1989.

[14] Roberts, L. G.
Machine perception of three-dimensional solids.
In J. T. Tippett, et al. (editors), *Symposium on Optical and electro-optical information processing*, pages 159-197. IEEE Boston Chapter, 1964.
Published by MIT Press.

[15] Thorpe, C.E.
FIDO: Vision and Navigation for a Robot Rover.
PhD thesis, Carnegie-Mellon University, December, 1984.

[16] Wallace, R., Stentz, A., Thorpe, C., Whittaker, W. and Kanade, T.
First Results in Robot Road-Following.
In *Proceedings of IJCAI*, pages 1089-1093. 1985.

Chapter 15: Outdoor Visual Navigation for Autonomous Robots

Charles E. Thorpe

Abstract

Designers of mobile robots need to consider the special requirements that mobile robots have as physical problem-solving systems. In particular, robot design needs to consider task-specific models; the explicitness of representations; and architectural support. Several of the systems and subsystems built for the Navlab, and for the CMU Planetary Exploration walking robot, illustrate the different models and processing that enable those robots to perform their various tasks.

15.1 Introduction

Robots are *Physical Problem-Solving Systems*, and thus occupy a unique scientific and design niche and require unique processing and models. Robots are certainly physical systems, and robot designers can certainly borrow technology from physical systems such as signal processing and control theory. Many robots are also problem-solvers, and use ideas from symbolic systems, particularly Artificial Intelligence and high-level Image Understanding. But there remains an intermediate level which is the peculiar domain of robotics research. At this "robotic" level, robots do think symbolically; but the symbols and reasoning are likely to be geometric rather than pure logic. And at the robotic level, robots use control systems; but the input to the control system is often planned heuristically rather than calculated optimally.

This multi-level view of robotics has several implications for robot vision and systems, and for the models robots use:

1. **Task Specific Models:** Even at the robotic level, different robots have different goals and therefore need different processing, support, and models. It is crucial, although often overlooked, that the processing and models be chosen to support the particular task for which the robot is designed.

2. **Explicitness of Models:** A particular dimension along which robot models differ is the explicitness of the representation. In general, more difficult problems require more reasoning and need more explicit models; while higher speed requirements allow less processing and therefore need implicit or compiled models.

3. **Architectural Support:** Robot architectures need to support different kinds of processing at different levels. Approaches that ignore either the physical or the symbolic nature of robot problem solving are doomed to have limited scope and applicability.

While systems do and should differ, there are nonetheless several common themes of successful robot design:

1. Use of data, from lots of sensors, processed as streams rather than as isolated frames.
2. Strong use of geometry and demand for geometric tools in the architecture, including representing maps and vehicle motion.
3. Attention to error modeling and filtering.
4. Simple planning, concerned with satisficing rather than optimizing in planning and control.
5. Simple architectures, that include human interactions.

The main part of this paper describes several of the robot systems and subsystems we have built at Carnegie Mellon, and discusses how each is a (good or bad) example of the principles outlined above. We first discuss the architecture used on our Navlab, and systems built on top of it for road following and cross-country traverse. Our last example is drawn from the AMBLER, our walking Planetary Exploration robot. The final section summarizes the approaches taken by those systems, and argues for rational design of robot systems.

15.2 Example Systems

15.2.1 Navlab Controller and Architecture

The architecture we use for our road following and cross-country systems has evolved from a single loop, to a centralized single level "whiteboard", to a decentralized multi-level system. This change in system structure reflects the change in our approach to building realistic mobile robots.

Our first road following systems used a single sensor (video camera) to do a single task (find road edges), and a single control mode (servo to road center) to accomplish a single goal (follow roads) [16]. Moreover, they were written by a single researcher or a small, closely cooperating group. None of the usual functions of of an architecture (resource allocation, focus of attention, conflict resolution, design decomposition) were needed; the systems had nothing that would be recognized as an architecture. Later systems were built around our CODGER whiteboard. We have learned important lessons, both positive and negative, from CODGER, and are now incorporating them into EDDIE, our new architectural framework.

CODGER and the Navlab. When we began to scale up to systems with multiple sensors and multiple tasks, we needed an architecture. Our most sophisticated systems used color images to find roads, a scanning laser rangefinder to check for obstacles, a map to predict intersections and change the image processing appropriately, and a route planner to decide which fork to take at an intersection [14]. Architectural support was clearly needed for communicating between multiple processes, coordinating planning at different levels, and integrating information from different sensors.

Our architecture was designed in three different facets: the module decomposition which separated functions and described the communications among them; the "driving pipeline" which discussed timing of information flow and data dependencies; and CODGER (COmmunications Database for GEometric Reasoning), the central tool which provided database, communications, synchronization, and geometric transformation facilities [4, 12]. The modules consisted of:

- Color vision, to see the road
- 3-D vision, looking for obstacles
- The Pilot, predicting road locations for color vision and planning vehicle trajectories
- The Helm, which managed trajectory execution
- The Controller, which handled the real-time tasks of vehicle driving
- The Map Navigator, responsible for route planning

The driving pipeline was managed by the Pilot. Pieces of the road about four meters long, called driving units, were first predicted, then detected by color vision, then swept for obstacles by 3-D vision, then used for planning by the Pilot, then traversed by the Helm and Controller. Each module could in theory be working on a different driving unit, keeping the pipeline always full and maximizing throughput [5].

All communication between modules went through CODGER. The various modules were looking at the same objects at different times, from different vehicle positions, and in the different coordinate frames corresponding to different sensor positions. CODGER automatically translated geometry into any requested coordinate frame. By keeping a "history list" of vehicle positions, CODGER could take a location specified relative to the vehicle at a particular time, and return the coordinates of that point, either in the world frame, or relative to the vehicle at a different time. CODGER also had a rudimentary mechanism for including uncertainty in locations and transforms, and was designed to eventually use filtering and multiple observations to adjust object positions and reduce uncertainty.

The communications part of CODGER included features such as anonymous data flow (a module could request or generate data of a particular type with no knowledge of other modules listening or sending that data); message formats specified by the user at run time; and a variety of module synchronization and interrupt generation methods.

CODGER Critique. CODGER and its associated architecture were designed to be a general system, to support whatever missions, sensors, vehicles, processing, and representations we would build. At the time, this was a reasonable decision, since we did not have well-formed ideas of future systems. From a vantage point of four years later, there are several flaws in the design: CODGER was a single-level system, with a single level of communications and of map representations. Moreover, CODGER suffers from architectural overkill, and has little-used features that add to the difficulty of system maintenance.

The entire architecture runs on a single level, the "robotic" level. The support built in to CODGER for geometry and transformations is ideal for that level. But CODGER had inadequate support for lower-level processing, such as high-performance real-time control. It was difficult to add reflex-level behavior, such as stopping the vehicle when an obstacle was detected with short-range sonars. The Pilot expected to manage all vehicle control; and CODGER's internal vehicle history mechanism depended on the Helm knowing what the vehicle was doing. So any lower level of interaction would greatly disturb the system.

Not only was communication on a single level, but map representations were also basically flat. Map objects had several descriptors, from bare geometry to symbolic roads. But each map object was described at all levels, so there was no concept of data abstraction or of differentiation between local and global maps. All the geometric objects ever used were retained in CODGER's database. And all the geometric transforms from a particular object back to the world coordinates had to be calculated as a string of transforms representing estimated vehicle motion since the last landmark sighting which tied vehicle coordinates to the world. This started to affect system performance after several hundred meters of vehicle travel, when transform chains became longer and longer and searches for geometric objects had to look at more and more database entries.

Anonymous messages and run-time specification of message formats turned out to be non-issues. Changing a message format always involved recompiling the sending process and the receiving process, and modifying the message template files that CODGER had to read. Recompiling CODGER would have taken a small additional amount of time, and would have greatly reduced its complexity. The only real use of anonymous communications was graphics programs that could be run optionally, to listen to message traffic and display current vehicle position and map objects. In all other cases, the sender and receiver of the message knew each other's identity, and in fact had to somewhat pervert CODGER's message mechanisms to accomplish some types of message sequencing. Moreover, the design for anonymous communications forced all messages through a central process, which created the potential for a bottleneck and limited the speed of low-level communications.

Finally, CODGER is complicated. The Helm module, for example, has 4000 lines of code; CODGER itself is 14000 lines long; and the CODGER libraries that are loaded with each module add an additional 25000 lines. This complexity, caused in great measure by providing features we do not really need, is one of the major reasons for the new system design.

The EDDIE Architecture. Our new EDDIE system (Efficient Decentralized Database and Interface Experiment) addresses the concerns we have with CODGER. It is much more tightly focused on the real issues of local vehicle navigation, and provides a high-speed lower level. Vehicle positions are maintained by the lowest level controller, which has the closest access to the vehicle and therefore the most accurate information. Communications are greatly simplified, and are now point to point, increasing their efficiency. The map is divided into local and global representations. And by splitting CODGER's functions into separate pieces for local communications, vehicle history, and map handling, the individual modules are much smaller and easier to maintain.

The first part of EDDIE is the new real-time controller. This module subsumes the functions of the old controller and the Helm, and in addition maintains the current vehicle position, which supplants CODGER's history mechanism. Vehicle motion commands arrive at the controller labeled as either "immediate" or "queued". The controller parses incoming commands, handles the queue, and talks to the hardware motion controller at the appropriate times to set new steering wheel positions and vehicle velocities. By querying the vehicle's encoders at frequent intervals, the controller is able to maintain an accurate dead-reckoned position estimate (soon to be supplemented with inertial guidance). In EDDIE, no vehicle position history is kept. The only times when it is necessary to know vehicle position are when new data is acquired, or during trajectory planning. It is easier, and more accurate, to dispense with history mechanisms, and instead to query the controller for the current vehicle position each time an image is digitized, and whenever a planner needs to know the vehicle's location.

Closing all position-estimation loops through the controller has a powerful fringe benefit: transparent path modifications. We have implemented a joystick interface that allows a user to modify commanded trajectories. Joystick input is simply summed with computer input, so the user has the sensation of "nudging" the vehicle away from its planned path. The Navlab will also be equipped with a "soft bumper", a ring of proximity sensors to detect nearby objects before collision. The soft bumper will interact with the controller in the same manner as the joystick, by adding its control input to the input from planning, but will have progressively higher gains as the time to collision decreases. Previous systems would have been destroyed by this subversion of planned paths, since CODGER kept vehicle position history by an open-loop expectation of perfect path tracking. In the EDDIE system, all position queries are handled directly by the controller, and are therefore answered correctly even if the path has been modified.

Communications in EDDIE are unexotic and uninteresting, but fast, with point-to-point connections. We still use TCP/IP over the ethernet, but can now go to shared memory or other protocols for particular connections, as needed.

Map design constraints are now separated into different concerns for local maps and for global maps. The global map will use CODGER ideas, including representing object locations and positions relative to the locations from which they were observed. The local map does not really look like a map at all, but instead consists of various internal representations as needed by individual modules and missions. Typical local models are considered locally rigid, with partially known but constant transforms to global coordinates. One version of the local map is simply the most recently observed road segment. Another version, used by the rough terrain planner, is a composite of terrain descriptions from the past four 3-D terrain scans.

Discussion. The main philosophical differences between CODGER and EDDIE reflect our current thinking on architectures:

- Task Specific Models: EDDIE does not impose particular connectivity or map structure, but instead provides tools to let users build their own. In particular, much of the data that CODGER put into a central database properly belongs within a single module or pair of communicating processes, as encouraged by EDDIE.
- Explicitness: The most important models maintained by an architecture are vehicle positions. Where CODGER implicitly assumed open-loop perfection, EDDIE explicitly queries the lowest-level hardware for current position.
- Architectural Support: EDDIE maintains CODGER's geometric tools for maps (at the robotic level), while adding support for soft bumpers, joysticks, and other physical-level control. In both systems we have left AI-level support to be provided by other modules as needed.

EDDIE provides a firm foundation on which to build mobile robots. The following sections describe perception modules and complete vehicle systems which are designed around EDDIE.

15.2.2 Autonomous Mail Vehicle

The current road following system for the Navlab is the Autonomous Mail Vehicle, or AMV. This system draws its inspiration from postal deliveries in suburban or rural areas, which follow the same route day after day, undeterred by "rain nor snow nor dark of stormy night". The mail carriers drive at relatively slow speeds, often on many different kinds of roads. They do gross navigation through a network of roads and intersections, and fine position servoing to mail boxes.

This type of system is an example of a broader class of applications which focus on map building and reuse, positioning, road following, and object recognition. Our AMV project is investigating those issues, including strategies for using different sensors and different image understanding operators for the perception components.

AMV System Design. Our AMV system finds 3-D objects, using range data; finds roads, using color and using the reflectance data from our laser scanner; and enters them into a map. After the map has been built, the Navlab retraverses the same route more efficiently. Instead of processing range images continuously, it only needs to look when and where objects are predicted. Each time the system sees the road, it generates a two-parameter update to its dead-reckoned position estimate, updating cross-track position and heading relative to the road but not along-track position. Each time the system sees a landmark, it gets a full three-parameter position update. Rather than plan paths anew each image, the system plans its desired path for the entire run before it begins, and only does minor servo modifications when the vehicle position is updated. This system is currently being debugged and tested on our Schenley Park roads, using trees as the landmarks. The system needs to be further tested, and we need to make changes in the map structure (currently a linear list) to enable random access, representations of longer roads and intersections, and travel in either direction.

Additional features of the AMV system include human interactions and vehicle safeguarding, provided by EDDIE's joystick controller and soft bumper. The AMV will include reasoning about unexpected objects in the scene, or about low confidence in perception results, to shut down the autonomous system and request human intervention. The combination of soft bumper and joystick control will allow the vehicle to autonomously and safely stop for an obstacle such as a parked car, ask for and receive help, and continue on its appointed rounds.

Objects and Actions. Objects in the map are tagged with actions to be executed when the objects are observed. These actions can be vehicle control ("stop at this object, it's a stop sign"); perceptual ("look to see if the traffic light is red"); or system strategy ("this intersection is hard to see with color vision, so use range data and dead reckoning until clear"). The first ad hoc prototype of a vehicle control hook is now complete. We are designing and implementing more generic action descriptions and hooks, and interactive methods for adding such information to the map. More importantly, we are adding some of this information automatically. For instance, the "traffic light recognition" perception module could look at every object detected during a mapping run, and annotate the map with which objects were really traffic lights. Then on later runs, the system would automatically know when to invoke that module.

Following Structured Roads. Although the mobile robot research community has several road following systems, none of them handles the variety of roads that will be encountered by the AMV. Most existing road trackers work well for only a particular road, or only under benign illumination. They have impoverished models that do not allow them to reason about failures in their low level feature trackers. Weak models and weak or nonexistent high level make them brittle in the presence of disturbances such as disappearing features or illumination changes.

We have two approaches to road following in the AMV, depending on the amount of structure on the road. Our structured road tracking effort uses model-based vision for tracking roads and highways [8]. Few other groups have used a really strong model. Dickmann's group, for instance, uses a strong model of geometry, but has no explicit description of appearance [3]. The Martin Marietta road follower (VITS) includes implicit assumptions about road shape, but no explicit model of either shape or appearance [15].

Yet it is important to build and to use explicit road models. Highways, freeways, rural roads, even suburban streets have strong constraints. Modeling these explicitly makes reasoning easier and more reliable. When a line tracker fails, for instance, an explicit model of road and shoulder colors adjacent to the line helps in deciding whether the line disappeared, became occluded, turned at an intersection, or entered a shadow. This kind of geometric and photometric reasoning is vital for building reliable and general road trackers.

The key to vision for driving on structured roads is explicit modeling of features and relationships that describe the road. We have individual knowledge sources that know how to model and track specific features:

- road edge markings (white stripes)
- road center lines (yellow stripes)
- guard rails
- shoulders
- type and color of road surface

We also have an explicit geometry model of the road, consisting of:

- location of vehicle on road
- location of stripes
- type of stripes (e.g. broken or solid)
- presence and location of guard rail
- maximum and current road curvature

Plus, there are other effects that should be noted and specifically modeled:

- shadows
- local changes in road surface, e.g. patches
- global illumination changes, such as the sun going behind a cloud
- camera changes (auto-iris, auto-gain)
- vehicle steering and speed
- 3-D effects such as going up and down hills

Explicitly modeling all these different features is the basis for efficiency and reliability. The system is efficient because the geometric constraints can specify subwindows of the image for each feature tracker, and tracker history from frame to frame can predict appearance and shape. Another reason for efficiency is that many simple trackers can be easily implemented on parallel hardware. Reliability comes first because of the strong geometric constraints among trackers, and the ability to detect and ignore anomalous outputs. More importantly, the system is reliable because one tracker, on discovering a shadow edge or road curvature change, can pass that information to other trackers and keep them from being caught by the same phenomenon.

Following Unstructured Roads. Some of our test "roads" are in reality bicycle paths, and do not have lines or stripes or distinct edges. They consist of narrow (3 meter wide) strips of asphalt, winding through trees and over hills, in places deeply shaded and partially covered with leaves, twigs, and dirt. We have two methods for following those roads, both emphasizing fast processing rather than deep reasoning. Since we do not have strong shape or appearance models, and since we must contend with unknown and varying illumination, the best we can do is to model current road location and appearance, and to process rapidly enough that changes between successive images are relatively small.

SCARF, which stands for Supervised Classification Applied to Road Following, tracks roads by adaptive color classification [2]. SCARF typically uses four color classes to describe road appearance, and four for off-road objects. Each pixel is compared to all eight classes to determine its most likely class. Classified pixels vote for (or against) all road locations that would contain them. The road with the most votes is used both for steering, and for recalculating the color classes using a variant of Isodata.

The most interesting aspect of SCARF from a system standpoint is its simple road model. Roads are represented as triangles in the image. The apex is constrained to lie on a particular image row, corresponding to the horizon, and the base of the triangle has a fixed width, dependent on road width and camera calibration. There are two free parameters: the column in which the apex appears, and the skew of the triangle in the image. While this simple 2-parameter model does not represent curves or hills or road width variations, it does approximate the road shape well enough to allow reliable driving. It is especially effective because the voting procedure uses all pixels, not just those on the edges, and is therefore insensitive to even large amounts of misclassifications. A model with more free parameters could represent more potential road shapes, but would often be led astray and find curves or branches where in fact all that exists is noise. Furthermore, the simple model allows for fast voting and functions well with small amounts of data, so SCARF can process highly reduced images (typically 60 by 64 or 30 by 32) at high rates (approximately 1 frame per second). Processing images closely spaced along the road means that small errors in road representations are corrected before the vehicle arrives at the mistaken locations. Processing images closely spaced in time means that even the drastic illumination changes caused by clouds covering the sun appear as gradual shifts in road appearance, and so do not derail SCARF.

ALVINN, for Autonomous Land Vehicle in a Neural Net, uses an even more condensed representation of the road's shape and appearance [11]. The product of the CMU Connectionist group, ALVINN does not even have an implicit model of the road. Instead, it learns associations between visual patterns and steering commands in its training phase. Then, when it runs, it directly outputs steering wheel angles, with no reasoning about road location. ALVINN uses 30 by 32 pixel images, and runs even faster than SCARF. Many of the same design philosophies hold for both systems: simple models, fast processing, reprocessing to correct errors due to simplistic models, small images, design for road following rather than for general scene analysis.

Discussion. Most of our road following systems use as strong models as are available. If the only features of the road are current appearance, then colors are explicitly modeled. If the roads have more structure, we model relative locations and appearances of the different features.

Following structured roads gives us an opportunity for machine learning, to derive better rules for which features to track and how to track them under different circumstances. Unsupervised learning requires the ability to detect errors, and alternative actions to choose from to correct those errors. Our explicit models give us both. Since there are many features being tracked in any image, we can detect the correct road location, and find which trackers succeeded and failed, even in difficult images. We are beginning to work on learning rules for individual features and trackers ("track double yellow lines in shadows with the oriented edge operator"), and for sets of features ("if the white line changes to all asphalt, suspect an intersection, and look for a gap in the double yellow line").

Much of this reasoning is still based on feature appearance. If we had better models of illumination, cameras and digitizing, and object reflectance, we could make explicit what is now packaged in "feature color". As color constancy and related research matures, we intend to fold those results into our road modeling.

15.2.3 Generic Cross-Country Vehicle

The Generic Cross-Country Vehicle (GX-CV) packages 3-D perception and local trajectory planning, plus the EDDIE architecture and controller, into a solid foundation for off-road navigation. The basic GX-CV system enables the Navlab to travel in a general direction, planning and executing vehicle trajectories around obstacles and across rough terrain. Hooks in the GX-CV planner will enable missions such as cross-country traverse with global navigation, or cross-country mapping.

The heart of the GX-CV consists of three parts: the EDDIE architecture and controller, medium-resolution 3-D mapping, and planning with an explicit vehicle model. We have prototypes of each of these modules. We are currently refining the modules, particularly to increase efficiency, and are performing the first system tests.

Medium Resolution Mapping. Vehicle navigation over rough terrain requires terrain models at appropriate scale. We have built three levels of terrain mapping. If the environment is guaranteed flat with a few, large, discrete obstacles (e.g. trees), then a fast low-resolution 3-D scan is adequate for obstacle avoidance. If the vehicle is capable of traveling in very rugged terrain or must reason about small objects, then high-resolution maps are required. For the Navlab on the kinds of terrain it can traverse, planning requires representing the world at intermediate resolutions.

Our 3-D perception system is based on a scanning laser rangefinder. Every half second, the rangefinder produces a digital image of 256 columns and 64 rows, where each pixel encodes the range to the nearest object along a ray in space. Our software converts the range data from spherical sensor-centered to cartesian vehicle-centered coordinates, then fits surfaces to the 3-D points and segments the scene into a mesh of polygonal faces. The accuracy of the fit, and the amount of deviation allowed before subdividing a polygon, are user-controllable. Various versions of the software also calculate derived characteristics, such as surface slope or roughness, for each polygon [6, 7].

Planning With Explicit Vehicle Model. Once perception builds the 3-D polygonal mesh, the planner has to find a trajectory for the vehicle. Typical cross-country planners consider vehicle traversability constraints, such as finding areas that are too steep, have too large a vertical step, or have vertical spikes that would hang up the vehicle's undercarriage. The GX-CV planner, in addition, considers accuracy constraints (for both perception and vehicle motion) and sensor positioning. Two distant objects may appear to be far enough apart to allow safe passage, but the planner may not be able to guarantee that the vehicle can move to and past the objects accurately enough to miss them. In that case, the planner must generate a path that moves closer to the objects, orients the vehicle so that sensors can see the objects and update their relative positions, and replan to go through. The planner also reasons about non-holonomic motion constraints (limited turning radius) and variation of traversability with vehicle orientation (both because of wheel orientation and vehicle shape). So, for instance, a particular ditch may be traversable if the vehicle approaches perpendicularly, but would entrap a wheel if the vehicle's path were nearly parallel to the ditch.

Obstacles are represented in the three dimensions of vehicle x, vehicle y, and vehicle heading. The search starts at the current vehicle configuration and expands, following the constraints of turning radius, to reach the goal. Each path is "fattened" by maximum expected error, which turns a planned "ray" in configuration space into a "cone" of possible trajectories. In order for a path to succeed, all paths within that cone must arrive within the goal configuration envelope without encountering obstacles. If that is impossible, the planner must select an intermediate goal and replan. The search is made efficient by an oct-tree representation of obstacles and free space, and by considering various pruning strategies.

Discussion. The GX-CV illustrates the importance of modules agreeing on task requirements, and the importance of model explicitness. The planner includes detailed, explicit models of the vehicle and its capabilities. This, in turn, drives the design of perception and the selection of parameters governing resolution and coverage.

15.2.4 Planetary Exploration by Walking Robot

We use the same sensor both to provide medium resolution 3-D maps for the Navlab, and in a completely different manner for a completely different robot with completely different requirements. Our AMBLER Planetary Exploration robot is a novel walking machine configured for the rough terrain, soft soils, extreme reliability and autonomy requirements, and low power budget of a mission to Mars [1]. Footfall selection demands high-resolution maps of very rough terrain, and navigation requires large-scale maps of the area around the vehicle. The planning to select which foot to move, and where to place it, needs to be aware of vehicle balance and safeguard, and must consider interactions with other feet and with the surrounding terrain. These requirements lead to new perception and planning systems.

High Resolution, Large Scale Terrain Mapping. The difference between a good foothold and a bad foothold can be measured in inches, on the order of the same scale as the sensor resolution. While fine-scale roughness on relatively flat surfaces will not affect the tires of a wheeled vehicle, the small details may make the difference between a feasible toehold and a slick slope for a legged mountain-climbing robot. At the same time, aerial photographs of Mars are likely to have resolutions measured in meters, or nearly the same scale as an entire scan with a surface-based rangefinder. So locating a robot by matching locally perceived features to global maps will require building large scale composite local maps from many images.

The dual requirements of fine resolution and large scale are addressed in our Mars mapping system by the *locus* algorithm. The common need for both requirements is to retrieve accurate elevation data, and measures of elevation uncertainty, at an arbitrary location. If the system can interpolate precisely, it can provide high-resolution footfall information. And if it can retrieve uncertainties at arbitrary locations, it can find the best transform between adjacent scans or between a new scan and a map, and can build larger-scale maps. The locus method provides the solution by incorporating a model of the sensor, its error characteristics, and its viewpoint, to provide interpolation at arbitrary resolution without making any assumptions about the terrain shape other than the continuity of the surface. The locus method works in the coordinates of the sensor. A query for the elevation at a particular xy point is answered by looking at all sensor pixels whose rays in space intersect a vertical line passing through that xy location. Interpolation is straightforward, and "range shadows" (unseen depressions) are made explicit [9, 10].

Using the locus method, we have developed several terrain evaluation strategies for selecting footfalls. We have also built terrain maps that combine information from over one hundred individual range scans, to produce maps that cover two hundred meters. Work in progress will evaluate the accuracy of those maps, and seeks to find more computationally tractable variants of the algorithms.

Gaitless Footfall Planning: Putting Your Best Foot Forward. Once the terrain has been mapped, and the solidity of footing calculated at each location, there still remains the question of where to put the next foot. Our earliest designs followed the traditional tenets of system design: the system was modular and hierarchical, with multiple cooperating processes. We decided on 4 modules: the Gait Planner (GP), which picks a foot to move; the Footfall Area Selection Planner (FSAP), which finds a general area in which to place the foot; the Footfall Location Optimization Planner (FLOP), charged with consulting the terrain map to find the best footfall; and the Leg Recovery Planner (LRP) that checks to see if that position is reachable, then plans trajectories for the leg. All intermodule communications for the AMBLER go through a central planner and coordinator, called Central.

The problem with this design is that it imposes a hierarchy on a problem which has a more natural flat structure. There is no particular reason that checking if a position is reachable should be the last step; indeed, in cases where other legs or projecting boulders intrude into the working volume, the LRP could have the tightest constraints. The planner could then enter a long loop of suggesting footfalls, finding they are unreachable, suggesting different footfalls, trying different legs, etc. Each of these steps would involve planning (by the FSAP, FLOP, or GP), communications through the TCA, checking for feasibility, and returning failure notifications back through the TCA. It may be possible to reduce the amount of thrashing by returning an explanation for failures. But in order for a higher-level module to understand why a lower-level module failed, one of the two modules would have to know enough about the internal processes of the other to make meaningful suggestions for alternatives.

The new design folds all the footfall selection into a single module, with other modules generating constraints. The LRP generates a binary map of which footfalls are reachable and which are not; the FLOP produces a binary map of infeasible locations (too high or low for the legs to reach) and calculates a real-valued map of footfall desirability (softness, slope, etc.); and GP produces a real-valued map that considers vehicle stability and range of motion for each footfall. Then finding the best footfall is simple a matter of eliminating infeasible footfalls and selecting the location which maximizes a combination of the real-valued evaluations. Constraint evaluation is now parallel, rather than serial; interactions are once per step, rather than once per candidate footfall; and the flow of information is not confounded with flow of control [13].

Discussion. Navigating a walking machine requires different models from navigating a wheeled vehicle. Concerns such as sensor resolution and viewpoint, completely ignored in the Navlab, must be explicitly modeled for the AMBLER.

The planning systems appear quite different, but in fact share the same philosophical underpinnings. Both rely on geometrical representations of terrain and vehicle, and on straightforward search to find adequate actions. Neither planner needs elaborate architectures, nor access to external maps or representations.

15.3 Discussion and Conclusions

Each of these systems and subsystems takes a different position on the design questions of what is modeled, how explicit are those models, and which architectural features are supported or required. This is to be expected, since each of these systems is designed for a different task. And in an environment that requires high performance (in terms of processing speed, difficulty of scenes, and precision of output), specialized perception and architectures are not only inevitable but desirable from a design standpoint.

It is somewhat wearisome to listen to debates on "the nature of general robot intelligence." The problems with which such discussions usually grapple are far too small in scope. First, many of the proponents put forth views with a particular intellectual bias, such as "the AI approach" vs. "connectionist architectures" vs. "control theory as the basis" vs. "subsumption". Each of these approaches is necessarily incomplete, since each embraces tools that are appropriate for at most part of the problem. But in a larger sense, the arguments and discussions are futile, for they aim at a "general robot intelligence" that does not now, and may not ever, exist.

The best way to discover the true nature of perception and intelligence for mobile robots will be by induction. We have far too few mobile robots, especially operating in unconstrained outdoor environments. And the best of our robots to date are still woefully inadequate at most tasks. We need to persist in designing robots for particular domains, such as following certain kinds of roads, or traveling across rough terrain at slow speeds. We need to continue to build perception, architectures, and systems that work, and to justify our robots on the problems they solve and the insights they provide, rather than according to a particular school of thought or bias. And we need to continue to concern ourselves with task specific models, explicitness of models, and architectural support.

15.4 Acknowledgements

The systems described here are the product of many people. Navlab planning and systems have been done by Tony Stentz and Eddie Wyatt. The new controller is the work of Omead Amidi. Martial Hebert is the CMU expert on 3-D perception, including the Navlab's medium resolution mapping. Dave Simon built the first AMV prototype, and Jay Gowdy continues development. Karl Kluge is following structured roads with explicit models, while Jill Crisman and Didier Aubert work on unstructured roads with simple appearance models. Dean Pomerleau, a student of Dave Touretzky, does neural nets on the Navlab.

The Planetary Exploration work is directed by Takeo Kanade, William Whittaker, and Tom Mitchell. The locus method and the large-scale mapping are the work of InSo Kweon, and footfall evaluations have been designed by Claude Caillas, Regis Hoffman, and Bala Kumar. Eric Krotkov directs Planetary Perception. Planning footfalls has been done by Dave Wettergreen, Hans Thomas, and Ben Martin; the LRP is the work of Lonnie Chrisman, and the Central was designed by Reid Simmons.

Thanks also to those who keep the Navlab alive and productive: especially Jim Frazier, Jim Moody, and Eric Hoffman.

15.5 References

[1] J. Bares, M. Hebert, T. Kanade, E. Krotkov, T. Mitchell, R. Simmons, and W. Whittaker.
Ambler: An Autonomous Rover for Planetary Exploration.
IEEE Computer , June, 1989.

[2] J. Crisman and C. Thorpe.
Color Vision for Road Following.
In *SPIE Conference on Mobile Robots*. November, 1988.
A revised version appears in this book.

[3] E. Dickmanns and A. Zapp.
Autonomous high-speed road vehicle guidance by computer vision.
In *Proc. 10th IFAC*. Munich, 1987.

[4] Y. Goto and A. Stentz.
Mobile Robot Navigation: The CMU System.
IEEE Expert , 1987.

[5] Y. Goto, S. Shafer, and A. Stentz.
The Driving Pipeline: A Driving Control Scheme for Mobile Robotics.
Technical Report CMU-RI-TR-88-8, Carnegie Mellon Robotics Institute, June, 1988.
A revised version appears in this book.

[6] M. Hebert and T. Kanade.
3-D Vision for Outdoor Navigation by an Autonomous Vehicle.
In *1988 Proc. of Image Understanding Workshop.* April, 1988.

[7] M. Hebert, T. Kanade, and I. Kweon.
3-D Vision Techniques for Autonomous Mobile Robots.
Technical Report CMU-RI-TR-88-12, Carnegie Mellon University, The Robotics Institute, August, 1988.

[8] K. Kluge and C. Thorpe.
Explicit Models For Road Following.
In *IEEE Conference on Robotics and Automation.* May, 1989.
A revised version appears in this book.

[9] I. Kweon, M. Hebert, and T. Kanade.
Perception for Rough Terrain Navigation.
In *Proc. of SPIE on Mobile Robots.* SPIE, November, 1988.

[10] I. Kweon, M. Hebert, and T. Kanade.
Sensor Fusion of Range and Reflectance Data for Outdoor Scene Analysis.
In *Proc. of SOAR'88 Space Operations Automation and Robotics.* NASA, Wright-State University, Dayton, Ohio, July, 1988.

[11] D. Pomerleau.
ALVINN: An Autonomous Land Vehicle In a Neural Network.
Advances in Neural Information Processing Systems I.
Morgan Kaufmann, 1989.
A revised version appears in this book.

[12] S. Shafer, A. Stentz and C. Thorpe.
An Architecture for Sensor Fusion in a Mobile Robot.
Technical Report CMU-RI-TR-86-9, Carnegie-Mellon University, the Robotics Institute, 1986.

[13] A. Stentz and C. Thorpe.
Against Complex Architectures.
In *6th International Symposium on Unmanned Untethered Submersibles.* June, 1989.

[14] C. Thorpe, M. Hebert, T. Kanade and S. Shafer.
Vision and navigation for the Carnegie-Mellon Navlab.
IEEE PAMI 10(3), 1988.

[15] M. Turk, D. Morgenthaler, K. Gremban and M. Marra.
VITS--A Vision System for Autonomous Land Vehicle Navigation.
IEEE PAMI , May, 1988.

[16] R. Wallace, A. Stentz, C. Thorpe, H. Moravec, W. Whittaker, and T. Kanade.
First Results in Robot Road-Following.
In *Proc. IJCAI-85*. August, 1985.

Index

GPSR Compliance

The European Union's (EU) General Product Safety Regulation (GPSR) is a set of rules that requires consumer products to be safe and our obligations to ensure this.

If you have any concerns about our products, you can contact us on ProductSafety@springernature.com

In case Publisher is established outside the EU, the EU authorized representative is:

Springer Nature Customer Service Center GmbH
Europaplatz 3
69115 Heidelberg, Germany

Batch number: 10755284

Printed by Printforce, the Netherlands